FARM
MACHINE

Collec

BY

CLAUDE CULPIN,

M.A., DIP.AGRIC.(CANTAB), M.I.B.A.E.

AGRICULTURAL AND HORTICULTURAL SERIES

FARM MACHINERY

British Library Cataloguing-in-Publication Data
A catalogue record for this book is available from the
British Library

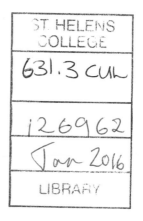

Contents

Agricultural Tools and Machinery

Farming has an incredibly long history. Beginning around 3000 BC, nomadic pastoralism, with societies focused on the care of livestock for subsistence, appeared independently in several areas in Europe and Asia. This form of farming utilised basic implements, but with the rise of arable farming, agricultural tools became more intricate. Between 2500 and 2000 BC, the simplest form of the plough, called the ard, spread throughout Europe, replacing the hoe (simply meaning a 'digging stick'). Whilst this may not seem like a revolutionary change in itself, the implications of such developments were incredibly far reaching. This change in equipment significantly increased cultivation ability, and affected the demand for land, as well as ideas about property, inheritance and family rights.

Tools such as hoes were light and transportable; a substantial benefit for nomadic societies who moved on once the soil's nutrients were depleted. However, as the continuous cultivating of smaller pieces of land became a sustaining practice throughout the world, ploughs were much more efficient than digging sticks. As humanity became more stationary, empires such as the New Kingdom of Egypt and the Ancient Romans arose, dependent upon agriculture to feed their growing populations. As a result of

1

intensified agricultural practice, implements continued to improve, allowing the expansion of available crop varieties, including a wide range of fruits, vegetables, oil crops, spices and other products. China was also an important centre for agricultural technology development during this period. During the Zhou dynasty (1666–221 BC), the first canals were built, and irrigation was used extensively. The later Three Kingdoms and Northern and Southern dynasties (221–581 AD) brought the first biological pest control, extensive writings on agricultural topics and technological innovations such as steel and the wheelbarrow.

By 900 AD in Europe, developments in iron smelting allowed for increased production, leading to improved ploughs, hand tools and horse shoes. The plough was significantly enhanced, developing into the mouldboard plough, capable of turning over the heavy, wet soils of northern Europe. This led to the clearing of forests in that area and a significant increase in agricultural production, which in turn led to an increase in population. At the same time, farmers in Europe moved from a two field crop rotation to a three field crop rotation in which one field of three was left fallow every year. This resulted in increased productivity and nutrition, as the change in rotations led to different crops being planted, including vegetables such as peas, lentils and beans. Inventions such as improved horse harnesses and the whippletree (a mechanism to distribute force evenly through linkages) also changed methods of cultivation. The prime modes of power were animals; horses or oxen, and the elements; watermills had been initially developed by the Romans, but were significantly improved throughout the Middle Ages, alongside windmills – used to grind grains into flour, cut wood and process flax and wool, among other uses.

With the coming of the Industrial Revolution and the development of more complicated machines, farming methods took a great leap forward. Instead of harvesting grain by hand with a sharp blade, wheeled machines cut a continuous swath. And instead of threshing the grain by beating it with sticks, threshing machines separated the seeds from the heads and stalks. Perhaps one of the most important developments of this era was the appearance of the tractor; first used in the late nineteenth century. Power for agricultural machinery could now come from steam, as opposed to animals, and with the invention of steam power came the portable engine, and later the traction engine; a multipurpose, mobile energy source that was the ground-crawling cousin to the steam locomotive. Agricultural steam engines took over the heavy pulling work of horses, and were also equipped with a pulley that could power stationary machines via the use of a long belt. They did operate at an incredibly slow speed however, leading farmers to amusingly comment that tractors had two speeds: 'slow, and damn slow.'

From this point onwards, it has been the methods of powering machines, rather than the agricultural machines themselves, which have been the biggest breakthroughs in farming practice. The internal combustion engine; first the petrol engine and later diesel engines, became the main source of power for the next generation of tractors. These engines also contributed to the development of the self-propelled, combined harvester and thresher, or 'combine harvester' (also shortened to 'combine'). Instead of cutting the grain stalks and transporting them to a stationary threshing machine, these combines cut, threshed, and separated the grain while moving continuously through the field. Combines might have taken the harvesting job away from tractors, but tractors still do the majority of work on a modern farm. They are used to pull various implements – machines that till the ground, plant seed and perform other tasks. Besides the tractor, other vehicles have been adapted for use in farming, including trucks, airplanes and helicopters, for example to transport crops and equipment, aerial spraying and livestock herd management.

The basic technology of agricultural machines has changed little in the last century. Though modern harvesters and planters may do a better job or be slightly tweaked from their predecessors, todays combine harvests still cut, thresh, and separate grain in essentially the same way. However, technology is changing the way that humans operate the machines, as computer monitoring systems, GPS locators, and self-steer programs allow the most advanced tractors and implements to be more precise and less wasteful in the use of fuel, seed, or fertilizer. In the foreseeable future, there may be mass production of driverless tractors, and new advances in nanotechnology and genetic engineering are being used in the same way as machines, to perform agricultural tasks in unusual new ways. Agriculture may be one of the oldest professions, but the development and use of machinery has made the job title of *farmer* a rarity. Instead of every person having to work to provide food for themselves, in America for example, less than two percent of the population works in agriculture. But today, a single farmer can produce cereal to feed over one thousand people. With continuing advances in agricultural machinery, the role of the farmer continues on.

ACKNOWLEDGMENTS

The author wishes to acknowledge valuable assistance kindly given by many individuals and firms in the preparation of the illustrations and in supplying information. Where possible the names of manufacturing firms or suppliers of copyright photographs are indicated beneath the respective illustrations.

CHAPTER ONE

TRACTORS: DEVELOPMENT AND PRINCIPLES OF OPERATION

"There appeared no valid reason why locomotive engines should not be made suitable for moving agricultural machinery, whether threshing, ploughing by means of windlasses, or for other purposes for which the farmer requires motive power; and it was with the view of encouraging the manufacture of such engines that the Society determined this year to offer a prize, not for a mere locomotive, but for 'the best agricultural locomotive engine applicable to the ordinary requirements of farming.'"

"Trials of Traction Engines at Wolverhampton."

Jour. Roy. Agric. Soc. England, 1871.

If early experiments with steam tractors are excluded, the history of the development of the use of tractor power began in 1890, when one of the first tractors powered with an internal combustion engine was built and used in the United States. An early British oil-engined tractor was built by Ruston-Hornsby in 1897, and was awarded a silver medal by the Royal Agricultural Society of England at the Manchester show in that year. This was soon followed (1901) by the "Ivel", a 3-wheeled tractor fitted with a horizontally opposed 2-cylinder 4-stroke water cooled engine. Since that time there has been a more or less continuous development of tractors in this country, but the design of the modern machine owes much to American influence. The first tractor to offer a serious challenge to horses for draught purposes in this country was the Ford, which was introduced from America in 1917. This differed greatly in design from anything that had preceded it, and was the forerunner of the type commonly employed to-day.

After the 1914-18 war there was a temporary loss of interest in tractors in this country, though a large number of "food production" Ford tractors continued to be used until they became derelict. A reason for the decline in interest was the unfortunate experience of many farmers with unreliable and badly serviced machines during the food production campaign. Progress in America, however, continued to be rapid, and in addition to important advances in general design, such new departures as the production of the original International "Farmall" were made. In more recent years, noteworthy developments have included the introduction of low-pressure pneumatic tyres, together with many other improvements in design and manufacturing precision.

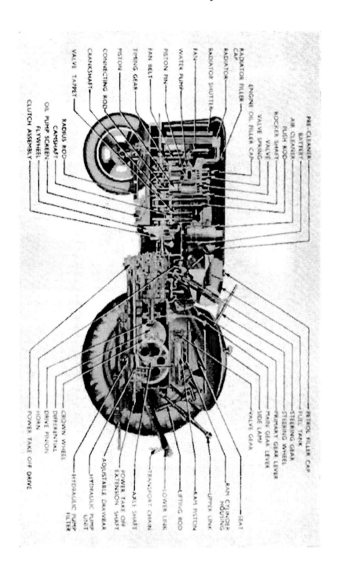

FIG. 1.—SECTION OF MODERN MEDIUM-POW-
ERED ALL-PURPOSE TRACTOR WITH HYDRAULIC
LIFT. (FORDSON.)

*FIG. 2.—LIGHT-MEDIUM ALL.-PURPOSE, TRACTOR
WITH DIRECT-COUPLED PLOUGH CONTROLLED
BY HYDRAULIC MECHANISM. (FERGUSON.)*

The range of tractors has been widened by the production
of high-powered tracklayers at one end of the scale and small
four-wheeled, two-wheeled and tracklaying machines at the
other. Between these extremes, a wide selection of row-crop
and other tractors has been designed for special uses.

The most significant development of recent years has
been the steady evolution of tractors which are specially
designed for use with "tractor-mounted" or "unit-principle"
implements, mounted directly on the tractor itself and raised

and lowered by means of a power lift. The "row-crop" tractor which was specially designed for work between the rows of growing crops is gradually becoming merged into an "all-purpose" outfit which retains the essential row-crop features. The modern tractor has, indeed, become a kind of multiple-purpose machine tool on which all manner of attachments may be mounted.

A contrasting development is the evolution of various types of specialized self-propelled machines such as the self-propelled combine harvester and the self-propelled tool-bar frame. While some of these self-propelled machines are extremely efficient at their specialized work and valuable on certain types of farm, it should be realized that the necessity of providing a transmission system with each machine—even if engines were made interchangeable—renders any great extension of evolution in this direction unlikely at present.

*FIG. 3.—MEDIUM-POWERED ALL-PURPOSE TRAC-
TOR (DAVID BROWN) WITH P.T.O.-DRIVEN DIRECT-
COUPLED ROTARY CULTIVATOR. (ROTARY HOES.)*

Other important advances in tractor design include
the introduction of satisfactory "half-track" equipment
that can be easily fitted in place of the rear wheels; better
carburettors and governors which lead to improvement in
fuel economy; improved transmissions with a wider range
of speeds and better braking; the more general provision of
electric starting and other aids to operating comfort, and a
steady improvement in general tidiness of design.

Most modern machines can be relied upon to give
long and efficient service. They are, moreover, extremely

adaptable power units, in that power may be delivered either to directly mounted tools or at the drawbar, at a belt pulley, or at a "power take-off".

The Use of Tractors in British Farming.

In the period between the first and second world wars the use of tractors in Britain increased steadily, but by 1939 the total number in use in England and Wales was still no more than 55,000 as compared with 549,000 working horses. A survey of tractor use in the Eastern Counties during the middle nineteen-thirties showed that tractors represented 42 per cent, and horses 45 per cent, of the total power available in tractors, horses and stationary engines, but that horses did 70 per cent, of the work and tractors only 26 per cent.

FIG. 4.—LARGE WHEELED TRACTOR WITH FINGINE DRIVEN PICK-UP BALER. (MASSEY-HAR-RIS.)

By the end of the 1939-45 war the position had been transformed. Tractors had become much more versatile and horses of less importance, both in numbers and in use. By 1951, some 300,000 tractors were in use in England and Wales compared with 249,000 working horses, and to-day there are thousands of farms which rely entirely on tractors for draught work. It seems most unlikely that the change from animal to mechanical power in farming can be arrested or reversed. On the contrary, the steady increase in wages, and the need to obtain increased output from a restricted labour force, make it essential for farm workers to utilize mechanical power on an ever-increasing scale. It is generally agreed that a man is capable of developing approximately $\frac{1}{8}$ th h.p. The cost of this power, with wages at 3s. an hour, is approximately 24s. per h.p. hour. No farmer can afford to rely much on the unaided power of human muscles. The worker who is equipped with a pair of horses can provide power at a cost of about 3s. 6d. per h.p. hour, and with a tractor developing only 15 drawbar h.p. the cost of power at the drawbar is rather less than 6d. per drawbar h.p. hour in average conditions. It may safely be claimed that the future of agricultural progress depends largely on the extent to which mechanical power and machinery can be employed to render labour more productive.

16

In present conditions, some farmers find it best to do part of their work with tractors and part with horses, partly because there is a limit to what one tractor can do in the peak periods. For instance, a tractor can be used quite satisfactorily for drilling; but it often happens that the tractor is needed for ploughing or cultivating when the time for drilling comes, and it cannot do both jobs.

There are also many light jobs, where the tractor is necessarily run under-loaded, at which horses can compete successfully with tractors on almost any count. An example of such work is beet hoeing, where the typical small tractor outfit covers only the same width as a one-horse outfit. Thus, while it is possible, with suitable tractor power units, to do practically any job on the farm without horses, there is still on many farms a certain amount of work that is done more economically if horses are employed. In particular, light transport work where a great deal of stopping and starting is involved, and very light cultivations such as harrowing, are often better done by horses. The chief practical questions that arise are just what sorts of tractor's are best for the farm and the work, how many tractors are needed, and how few horses it is necessary to keep to do those jobs that the tractors cannot manage or the horses can do better.

In considering the influence of the use of tractors on farming organization the whole farming business must be studied. Tractors have attained their present important position in British agriculture not merely on account of lower working costs, but also because of the benefits that arise from ability to do cultivations more thoroughly and in better season, with the production of more or better crops. Extra tractor work is frequently justified by a more intensive system of farming and a greater output.

FIG. 5.—MEDIUM-POWERED ALL-PURPOSE TRAC-TOR (NUFFIELD) WITH HYDRAULIC FRONT-MOUNTED MANURE LOADER. (COMPTON.)

The Cost of Tractor Work.

The cost of operating tractors varies from farm to farm and from district to district. Factors influencing costs, apart from the size and type of machine, include the number of days worked annually, the types of work performed, the care the tractor receives, and several other items. One of the most important factors influencing cost per tractor hour is the amount of use. Low hourly costs are achieved on some farms by using tractors nearly every day of the year and doing all kinds of work with them. On light land it is generally possible to work tractors many more days in the year than on heavy. Low hourly costs achieved by regular use are not, however, necessarily an indication of efficient use, since this may be achieved by using tractors for work that horses could do more economically.

The cost of operating a modern medium-powered paraffin tractor, costing about £450 on pneumatic tyres, for 1,200 hours annually, would be approximately as follows:

	£	s.
Petrol, 95 gallons at 4s. 2d	19	16
Vaporizing oil, 1,350 gallons at is.4$\frac{1}{2}$d.	92	16
Lubricating oil, 40 gallons at 8s. 6d.	17	0
Grease	1	0
Cost of fuel and lubricants	130	12

Repairs (at Garage, £25, on farm, £10)	35	0
Licence and insurance	3	0
Depreciation $\left(\text{6-years life } \dfrac{£450}{6}\right)$	75	0
	£243	12

Cost per hour approximately 4s. od.

The method of charging depreciation in the above example is open to argument. If the "written down" method were used, depreciation chargeable in the first year would

be $28\frac{1}{8}$ per cent, on £450, i.e. £126 us. The sum allowed for repairs is also possibly rather low. Farmers can readily calculate their own running costs by substituting their own figures for fuel, oil, repairs, etc., and a more detailed discussion of methods of costing mechanization processes may be found in another book by the present writer.[1]*

It should be noted that the total cost of operating the tractor considered above must be increased by approximately 3s. per hour, since the driver's wages are not included in the calculation. The conclusion is reached that the total operating cost, including driver's wages, for a modern medium-powered tractor, is of the order of 7s. per hour.

*1. Figures refer to list of references at end of Chapter.

The Choice of a Tractor.

The best basis for choosing suitable tractors is experience, and practical trial in the conditions in which the machines are to work. Nevertheless, it is helpful to consider some of the principles involved in choice and to mention factors that need to be studied.

Type Of Tractor. In most instances the tractor will need to be an "all-purpose" machine which can be applied to almost any kind of farm work, including ploughing, cultivations, sowing, row-crop work, harvesting, transport and belt work. This will apply particularly to small farms where one or two tractors are required to do everything. On larger farms there may be sufficient of particular kinds of work to warrant using special types of tractors, e.g., tracklayers for deep cultivations, or light tractors of the self-propelled tool-bar type for drilling and inter-cultivation of root crops. Market gardeners may require tractors specially adpated for work in vegetable crops, and such special needs are briefly considered in Chapter Two, where the principal types of tractors available are described.

Row-Crop And Standard Tractors. Most modern medium-powered tractors are "all-purpose" machines which incorporate such features as make them suitable for work in all common farm row crops. Some manufacturers offer standard type tractors which arc a little cheaper than the row-crop type because fewer adjustments and fittings are provided. Such machines may be quite suitable for use on farms where row-crop features are seldom or never needed. If the standard type has a lower centre of gravity it may be preferable for use on hillsides.

Tractor-Mounted Implements. If a general-purpose tractor is required, an important consideration is whether the tractor should be equipped with a power lift in order to enable it to use tractor-mounted implements and machines. Practically all modern British medium-powered tractors are or can be equipped with a hydraulic lift, and there is a wide range of efficient mounted implements and machines specially designed for use with them. In general, it may be affirmed that modern medium-powered tractors are most efficient and economic when used as "unit-principle" machines with their mounted equipment, and no farmer should ignore this point. Nevertheless, the use of mounted equipment has disadvantages, one of which is mentioned below.

*FIG. 6.—MEDIUM WHEELED TRACTOR (INTER-
NATIONAL) WITH PAIR OF UNIT TYPE HAND-FED
TRANSPLANTERS (ROHOT) ATTACHED TO GENER-
AL-PURPOSE TOOLBAR. (STANHAY.)*

Tractor Sizes, and Standardization. One of the most difficult problems in choosing tractors for general farm work is to decide whether all the tractors should be of one make and size, or whether a range of makes and sizes should be selected to suit the various jobs. This is a problem on any farm where more than one tractor is needed, and is perhaps most difficult on medium-sized farms where three or four tractors are used. The advent of tractor-mounted implements has accentuated the problem, for though a group of manufacturers of large/medium tractors have power-lift linkages which will accommodate equipment designed for any one of them, two other important manufacturers produce tractors and equipment which are not interchangeable with the first group or with one another. The British Standards Institution has now introduced British Standards for the three-link system which will in time help to overcome these difficulties; but the present situation is that if a farmer wishes to operate two different sizes of unit-principle tractors he also has to equip himself with two distinct sets of mounted implements, each of which can only be used with the appropriate tractor. This leads to a high expenditure on equipment, and lack of flexibility in operation. So long as these conditions hold, there is much to be said for sticking to one make of tractor on small and medium-sized farms, and in these circumstances choice will frequently depend as much on the range of mounted equipment that is available to go with the tractor as on the size or other features of the tractor itself.

FIG. 7.—LIGHT ROW-CROP TRACTOR WITH SMALL ENGINE-DRIVEN COMBINE HARVESTER. (ALLIS-CHALMERS.)

On larger farms, there is considerable advantage in having more than one type and size of tractor, since this arrangement allows the matching of the tractor with the job in hand, and permits choosing the best equipment from more than one manufacturer. Operating more than one type of tractor necessitates keeping more spare parts on the farm, and requires the man responsible for maintenance to have a wider knowledge; but these are minor disadvantages compared with the greater operating efficiency where both tractor and equipment really suit the job.

Where it is decided to stick to one make and size of tractor, the decision as to which is most suitable must embrace a study of all the operations that will need to be carried out. Among the most important are ploughing and the basic cultivations, which must be done thoroughly and in good season, and will serve to illustrate the problems. The power needed for ploughing and other cultivations varies greatly according to the nature and condition of the land. The "ploughing resistance" of various types of soil may be indicated by a figure in lb. per sq. in. which, when multiplied by the total sectional area (sq. in.) of the furrow slices, gives the drawbar pull needed. Ploughing resistances range from about 5 lb. per sq. in. for very light blowing sand to over 15 lb. per sq. in. for heavy clay, an average figure for medium loam being about 10 lb. per sq. in. of furrow section. Thus, in average conditions, the drawbar pull needed to operate a 3-furrow plough with furrows 10 in. wide and 7 in. deep would be 10×7×3×10=2, 100 lb. The drawbar horse-power required may be calculated by using the following formula:

$$\text{Drawbar H.P.} = \frac{\textbf{speed (m.p.h.)} \times \textbf{drawbar pull (lb.)}}{375}$$

At a ploughing speed of 3 m.p.h. the drawbar horse-power required for the above conditions would therefore be

$$\frac{\textbf{3 (m.p.h.)} \times \textbf{2,100 (lb.)}}{375} = \text{approximately 17 H.P.}$$

Such a performance is just within the capabilities of a medium-powered tractor of about 25 b.h.p., provided that conditions for traction are satisfactory. For heavier land, however, especially if the tractor is to operate on pneumatic tyres, it would be necessary to consider either a 2-furrow plough or a more powerful tractor.

Within limits, securing a satisfactory performance from a tractor is more a question of selecting suitable implements and using suitable gears than one of buying tractors of different sizes. Few of the jobs that modern all-purpose tractors are required to undertake need the full power of even the small/medium sizes, and two small/medium tractors are often more useful than one large one. On the other hand, where a large tractor can be given a full load there may be a considerable saving of labour cost by using it—a fact which may often be taken advantage of on large farms and by contractors.

Type of Engine. Many tractors can be obtained with petrol, paraffin or diesel engine. As we have seen earlier, the total cost of operating a medium-powered all-purpose paraffin tractor for about 1,250 hours annually is made up approximately as follows: fuel and oil, just under 30 per cent.; depreciation and repairs, just under 30 per cent.; labour (driver's wages), 45 per cent. Owing to the high rate of tax on petrol, using petrol instead of vaporizing oil increases the fuel cost enormously, and compared with this increased cost, any savings in depreciation, lubricating oil and repairs are quite insignificant. While, therefore, the use of petrol may be necessary for tiny engines whose fuel consumptions are low, it can only be justified in very exceptional circumstances for medium-powered tractors. The choice between a vaporizing oil engine and a diesel engine is a difficult one. The first cost and depreciation on the diesel engine are considerably higher, but this is more than balanced by the saving on fuel cost. (The higher efficiency of the diesel (see Appendix Six) results in appreciably lower fuel consumption). The diesel engine usually costs more to repair when an overhaul is needed, and maintenance of the fuel pumps and injection nozzles is still not as easily carried out as that of the carburettors and electrical ignition equipment of the paraffin engine. The diesel engine usually costs more for electric starter batteries.

On balance, there is very little to choose between a good diesel engine and a good paraffin engine for a medium-powered tractor, and the increasing popularity of the diesel is at least in part due to the fact that many of the paraffin engines of the past have not been as up-to-date in design as the diesels with which they are compared. A good paraffin engine, with a bi-fuel carburettor that permits an immediate switch over from petrol to paraffin or vice-versa, and a manifold which will deal with vaporizing oil really efficiently, can be quite free from the starting and operating troubles which were so common in earlier models; and the choice between paraffin and diesel is now largely a matter of personal preference.

Wheel and Track Equipment. For general farm work, where the medium-powered tractor is to be regularly used for transport as well as for all kinds of field work, pneumatic-tyred wheels are now almost an automatic choice. The tyres are so expensive, however, that some thought would be given to the possibility of changing to steel wheels when long spells of heavy ploughing or other heavy cultivations lie ahead. Modern tractor wheels arc easy to change, and many farmers overestimate the time taken for this simple job. The essential requirement is a flat, firm floor and a good hydraulic jack. Some tractors employ the tractor's built-in hydraulic lift mechanism and a simple linkage for jacking up the rear wheels. Where the tractor is regularly required to run on hard roads the advantages of fitting retractable strakes or wheel girdles for heavy work should be considered.

FIG. 8.—MEDIUM-POWERED DIESEL-ENGINED
TRACKLAYER WITH MULTI-FURROW PLOUGH.
(FORDSON COUNTY.)

For very heavy tillage operations, and for operations on heavy land where damage to soil structure may be caused by tractor wheels, tracklaying mechanisms are most efficient. In really difficult conditions a tracklayer may operate reasonably efficiently where any type of wheel fails completely. A tracklayer enables field work to continue late in the autumn and to start very early in spring, while at other times of the year tracklaying devices enable the tractor to translate a high proportion of the engine power into effective work at the drawbar. Full-track machines are, however, expensive in first cost and maintenance, and non-versatile; and half-track mechanisms which are easily interchangeable with pneumatic-tyred wheels offer attractive possibilities for medium-powered tractors. In general, it is a sound rule to stick to the cheapest type of tractor that will do the job satisfactorily, and only to buy expensive special-purpose tractors when the need has been clearly established. In this connection, the possibility of getting a certain amount of work done by contract should always be kept in mind.

*FIG. 9.—MEDIUM-POWERED ALL-PURPOSE TRAC-
TOR (FORDSON) WITH HALF-TRACKS (ROADLESS)
PULLING HEAVY DISC PLOUGH.*

With regard to price, the cheapest tractor is, naturally, not always the best. Some of the higher priced outfits offer value for their extra purchase price in the form of fittings which on other tractors are extras. A cheap tractor is one that is cheap in operation; and in order to achieve economical operation the tractor must be suited to the needs of the farm. Other factors to be considered are low fuel consumption, low maintenance cost, ease of adjustment and repair, dependability, adaptability, ease of operation and service facilities. There is now a good tractor to suit the special requirements of almost every farmer, and the prospective purchaser in making his choice should carefully weigh the advantages and disadvantages of each type of machine for his own farm, bearing in mind points of construction and operation dealt with in this and the following chapter.

Tractor Power or Capacity.

The measure of a tractor's power or capacity for field work is its drawbar horse power. (See Appendix Four.) Tractors are, however, sometimes described, especially by American manufacturers, according to the number of plough furrows they can pull in "average" land at a speed of about 3 m.p.h. In such conditions approximately 5 drawbar h.p. is required to pull a single furrow of normal depth and width, and a tractor of 15 drawbar h.p. may be described as a "three-plough" tractor. Since ploughing conditions vary so greatly it is probably more satisfactory for those who understand the meaning of drawbar horse power to use the horse power figure to indicate capacity.

When it is known how many furrows a given tractor will normally pull on any particular soil, the width of other standard implements that the tractor may be expected to handle can be approximately estimated. A tractor that is capable of pulling three 10-inch furrows can generally handle three times the width of cultivator or disc harrow (7 ft. 6 in.), six times the width of heavy harrow (15 ft.) and five times the width of drill or binder (a 12-ft. drill or two 6-ft. binders).

If it is desired to calculate the rate of working of implements other than ploughs, the following formula, which allows 20 per cent, for time lost in turning at the headland, etc. may be used:

$$\text{Rate of working (acres per hour)} = \frac{\text{Working width of Implement (ft.)} \times \text{speed (m.p.h.)}}{10}$$

This formula is not applicable to ploughing or to other implements with which long stops are necessary for setting, marking out, filling up, etc.

Tractor Testing. In 1920 a scheme for the compulsory testing of all types of tractors marketed in the State of Nebraska, U.S.A., was inaugurated, and these tests, which are carried out at the University of Nebraska College of Agriculture, have come to be recognized throughout the world as a reliable guide to the capabilities of the machines tested. Among the important figures obtained from the tests are the rated belt horse powers and the rated drawbar horse powers of the tractors. The rated powers are determined according to a Standard Farm Tractor Rating Code, drawn up by the American Society of Agricultural Engineers (A.S.A.E.) and the Society of Automotive Engineers (S.A.E.). Without entering into the technicalities of how the maximum powers are measured, it may be briefly stated that the *rated belt horse power* must not exceed 85 per cent, of the maximum belt horse power developed continuously on test, and that the *drawbar rating* must not exceed 75 per cent, of the maximum drawbar horse power.2

There are many good reasons for rating tractors below the maximum powers developed on test. For example, the tractors are tested for drawbar power on tracks which are level, straight, and have good surfaces for adhesion; and the maximum load is gradually applied by electrical and other special devices. Working conditions in the field are always much less favourable, and tractors cannot generally develop their maximum test horse powers continuously when in practical use.

In Britain, tractor testing is not compulsory, but all the leading tractor manufactures make use of the testing facilities provided by the National Institute of Agricultural Engineering. The N.I.A.E. carries out two main types of tractor test in addition to any special test which may be undertaken to assist manufacturers in their development work. The first—the N.I.A.E. test—is a comprehensive one. In addition to the usual belt test and drawbar tests in a range of soil conditions there are also tests on a hillside, and a ploughing test. The drawbar tests are carried out on firm grassland on heavy clay, on stubble on light land, and on loose, freshly cultivated soil; and the results are shown partly in the form of graphs. These N.I.A.E. tests give a great deal more information than the Nebraska tests, or the British Standard Test referred to below, concerning the pulls and powers that can be exerted by the tractor on farm land. A detailed test report which includes a specification is prepared, and provided the manufacturer agrees, the test report is subsequently published by the N.I.A.E. in the form of a small booklet.3

The second type of test—the British Standard Tractor Test—is also carried out by the N.I.A.E. but has a more restricted scope. The test conditions are laid down in British Standard No. 1744: 1951. The drawbar tests of pneumatic-tyred tractors are carried out only on dry, level tarmacadam, while tracklayers and steel-wheeled tractors are tested on dry, level, mown or grazed grassland on heavy clay. The results of the drawbar tests carried out under the British Standard Test regulations are roughly comparable with those produced by the Nebraska testing methods, but the actual performances recorded are obtained from a series of performance curves, and not from "spot" readings.

Interpretation of Tractor Test Results. It is not possible here to give detailed guidance on how to interpret tractor test reports, but some of the factors that need attention should be mentioned. The reader must first learn to appreciate the significance of drawbar pull, tractor speed and drawbar horsepower—terms which are explained in the Appendix. Maximum drawbar pulls on a test track are usually limited not by the power of the engine, but by wheel slip, and they can be greatly increased at low speeds merely by adding weight to the tractor. With pneumatic tyres on firm ground the addition of 1,000 lb. to the rear wheels will usually increase the maximum drawbar pull by about 500 lb. It is, therefore, important to know how much weight has been added, and whether the addition of such weight would be practicable in farm conditions. It must be realized that it would be possible for a tractor fitted with wheels quite unsuitable for normal farm work to put up an excellent drawbar performance in the Nebraska test or the British Standard Test. It is, therefore, important to study the results of ploughing and other agricultural tests when these are available.

Increasing the maximum drawbar pull by adding extra weight does not necessarily increase the maximum drawbar horsepower developed by a tractor. In the case of pneumatic-tyred tractors, as is shown later, the maximum drawbar horse-power is attained at fairly high speeds and low drawbar pulls, and the limiting factor at these higher speeds is engine power and not wheel slip. The addition of extra weight at these higher speeds will usually, in fact, slightly reduce the maximum drawbar horse-power. The reader of a test report will, therefore, need to study drawbar pulls and drawbar horse-power over the range of working speeds. Fig. 10 shows characteristic test results for a medium-powered tractor in first and second gears when fitted with pneumatic tyres and steel wheels. It illustrates many points referred to elsewhere, e.g. the higher drawbar pull with steel wheels, and the higher drawbar H.P. obtainable in second gear on pneumatic tyres. Other graphs included in test reports may show the effect of added weight on these factors, and the effects of various combinations of conditions on specific fuel consumptions and on wheel slip.

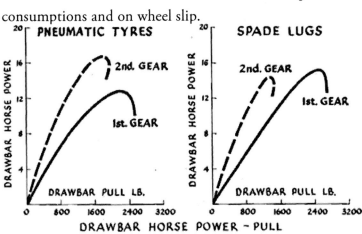

FIG. 10.—*GRAPH SHOWING INFLUENCE OF WHEEL EQUIPMENT AND WORKING GEAR ON DRAWBAR PULL AND DRAWBAR H.P. OF A MEDIUM-POWERED TRACTOR.*

Economic Operation: Loading.

So long as a tractor is in use, even though the work is very light, labour has to be employed to drive it and expense is incurred on depreciation, repairs, lubricating oil and fuel. Apart from fuel cost, none of these items is greatly influenced by the load. Thus, labour is a constant charge per hour and it has also been shown that provided there is no overloading of the tractor, depreciation and repairs are very little higher in an engine which works at full load than in one which works at half load, assuming that each works the same number of hours per year. Indeed, the depreciation of a paraffin engine that is not given a full load may be very much greater than that of one working at full load, owing to the serious dilution of the lubricating oil that occurs if the engine is allowed to run too cold. The labour and overhead charges therefore represent an almost constant charge per hour, regardless of how much power is developed; and on an average, these constant charges represent above two-thirds of the total cost of running the tractor at full load. If the tractor is worked at half load or less, the labour and overhead charges account for much more than two-thirds of the total cost, and the only saving of any importance is the reduction in the amount of fuel used. Thus, suppose that the total cost of running a medium-powered tractor at full capacity is 7s. 6d. per hour; about 70 per cent, of this amount (5s. 3d.)

represents a constant charge, incurred whether the load be great or small. For example, it would cost about 6s. per hour (5s. 3d. constant charge, plus gd. fuel) to run the tractor without load, and about 6s. 9d. per hour to run it at half load. It is, therefore, important for economical working to provide the tractor with a full load whenever practicable.

The tractor engine is normally fitted with a governor to ensure that the engine maintains a constant speed. A gear-box provides various definite forward speed ratios, and assuming that the governor keeps the engine speed constant and that there are no complications due to slippage, the tractor has definite forward speeds.

For each gear the tractor has a fairly definite optimum drawbar pull, and this optimum is generally about 80 per cent, of the maximum drawbar pull at that speed. The object should therefore be to provide implements which require drawbar pulls approximating to this optimum. For instance, if a medium-powered tractor pulling a 2-furrow plough at $2\frac{1}{2}$ m.p.h. exerts a drawbar pull of 1,500 lb., it is developing only 10 h.p. on the drawbar. In this case a 3-furrow plough could probably be pulled, and the efficiency of working would be much higher for the reasons explained above. If the implement is small, the load may often be adjusted by putting the tractor in a higher gear, or by adding another implement "in tandem".

An adequate load is also of importance from the standpoint of optimum fuel economy with all types of tractors. Typical specific fuel consumptions of tractors recently tested at N.I.A.E. are as follows:—

Type of Tractor Engine		Specific Fuel Consumption (lb. per b.h.p. hour)		
		At full load	At ½-load	At ¼-load
Medium sized petrol	..	0·65	0·90	1·40
Medium sized paraffin	..	0·72	1·05	1·60
Medium sized Diesel	..	0·47	0·58	0·86

Thus, the fuel consumed per b.h.p. hour at half load is approximately 30 per cent, higher than at full load; while at quarter load it is a little more than double that at full load. The superior fuel economy of diesel engines is more marked at light and moderate loads than at full load.

There are times when it is impossible to load a tractor adequately, but at other times it is possible to increase efficiency greatly by either adding to the load or working in a higher gear. A practical method commonly employed to provide a full load and speed up the work is to hitch implements in tandem. There are often savings in both time and money if operations such as discing and harrowing, drilling and harrowing, rolling and harrowing, etc., are carried out in one journey across the field. There are, of course, operations where it is impracticable to increase either the width of work or the speed, or to hitch implements in tandem. In such circumstances the best solution will be to run the tractor in a higher gear, with the engine throttled down. With a trailer mower, for example, the most economical method of operation may be to run in high gear with the engine throttled down to about half speed.

Wheel Slip.

Slipping of a tractor's drive wheels always wastes power and fuel, and with pneumatic tyres this wastage may be serious, even if the wheels do not spin. A simple method of determining wheel slip when a tractor is working is to make a mark on the tractor wheel and then measure the distance the tractor moves forward in, say, 10 revolutions of the wheel, first under load, and then on the same surface with no load. The percentage slip will be:

$$\frac{(\text{Distance travelled without load}) - (\text{Distance travelled when working})}{\text{Distance travelled without load}} \times 100$$

The amount of slip revealed by such a test is often surprising. Fifteen per cent, slip on pneumatic tyres is hardly noticeable, yet it represents an important waste of time and tractor fuel. Slip can never be entirely eliminated, but it can sometimes be minimized by lightening the load and working in a higher gear, while at other times it may be remedied by adding weight, fitting strakes, or fitting alternative types of wheel or track equipment.

Operating Speeds.

A three-speed gearbox, giving forward speeds of about 2, 3 and $4\frac{1}{2}$ m.p.h. at the standard governed engine speed, was adequate so long as the tractor was normally fitted with steel wheels; but to-day, with almost all new medium-powered tractors going out on pneumatic tyres, the four-, five- or six-speed gearbox has many advantages, especially where the tractor can be easily converted from pneumatic tyres to steel wheels or to half-track equipment.

On steel wheels, the maximum drawbar horse power that can be developed generally falls off rapidly at speeds above 3-4 m.p.h., owing to increased rolling resistance.

Practical tests with steel-wheeled tractors have shown that where 3 furrows can be pulled without undue wheel slip in bottom gear, ploughing in this gear gives a saving of both time and fuel over ploughing with 2 furrows at a higher speed. With pneumatic-tyred wheels, however, the situation is quite different. A study of published tractor tests shows that because of the limitations imposed by wheel slip at high drawbar pulls, modern pneumatic-tyred tractors can only exert their maximum drawbar powers at moderately high speeds. With such tractors, practical experience shows that pulling 2 furrows at a speed of 3-4 m.p.h. is on most soils a better proposition than attempting to pull 3 in bottom gear—an attempt which may prove unsatisfactory even after wheel weights or wheel strakes are employed.

On the other hand, it should be realized that arguments against the general use of high speeds for tillage are provided by tests which show that the draught of implements increases with increase in speed. For example, increase of the speed of ploughing from $2\frac{1}{2}$ m.p.h. to 4 m.p.h. may increase the draught by 10 per cent. Unless the increased draught is justified by better work (a rare occurrence), it represents sheer waste of power. Moreover, high speeds cause increased wear and tear of the implements owing to the greater shocks and increased rubbing friction. It should be added that there are certain jobs such as spraying, hoeing, combining, etc., which cannot be performed at above a certain critical speed without causing very poor work.

For ploughing, special high-speed bodies are required for work at above about 5 m.p.h. Normal plough bodies simply will not stay in the ground at high speeds.

A practical objection to the use of very low speeds is that they necessitate high drawbar pulls to provide a full load. It may be inconvenient to provide large enough implements, and in any case, under ordinary field conditions the tractor may be unable to exert very high drawbar pulls owing to adhesion difficulties. The best practical solution is to select implements which,' in average conditions, provide a full load for the tractor at about 3-4 m.p.h. so that more difficult conditions may be met by the use of a lower gear.

Light, pneumatic-tyred tractors frequently need to be "ballasted" in order to secure a good drawbar performance at low speeds. Methods of ballasting include the use of liquid filling for the tyres, and addition of cast-iron wheel weights. The effect of added weight on maximum drawbar pulls and drawbar h.p. is illustrated by figures from an N.I.A.E. test report. The rear axle weight (2,856 lb.) of a Fordson Major tractor which was already water ballasted 75 per cent, was further increased by 1,600 lb. This increase of rear axle weight produced increased drawbar pulls in first gear ranging from 800 lb. to 1,600 lb. according to conditions. On light land in a condition favouring the performance of a heavy tractor, addition of 1,600 lb. to the rear axle weight increased the drawbar h.p. in low gear by 91 per cent. It should be added that a heavy rear axle weight on a rubber-tyred tractor is not always desirable. It is especially undesirable when the tractor is used on damp seedbeds or on growing crops.

The value of a high road speed on an all-purpose medium-powered tractor is now generally agreed. A road speed of 15 m.p.h. or so is invaluable for rapidly moving to scattered fields, and also for some forms of road transport.

Engine Speed.

The modern tendency is to increase the speed range provided by the gearbox by means of a good, flexible governor control. The governor is made capable of adjustment to give about 50 per cent, more or less than the standard governed engine speed, so that a speed range of say 2 m.p.h. to 10 m.p.h. is extended by means of the governor control to a range of 1 m.p.h. to 15 m.p.h.

Intelligent use of the gears and governor control can result in appreciable fuel economy, as well as in a saving of time.

A modern tractor equipped with an "all-speed" governor and 5-6 gears can work at a normal speed of 3-4 m.p.h. either by use of a low gear and high engine speed, or in a higher gear at lower engine speed. Occasionally there is a choice of more than two gears and engine speeds. N.I.A.E. tests confirm what observant farmers have known for some time—that for fuel economy it always pays to choose a high gear and low engine speed whenever this is practicable. It is, of course, only at light or moderate drawbar pulls that there is any choice, but so much of a tractor's working time is spent on light work that this is a point of some importance.

Maintenance and the Tractor Driver.

Farmers do not always appreciate the prime importance of systematic attention to care and maintenance of the tractor. All manufacturers provide a handbook giving detailed instructions on such matters as adjustments, greasing, changing lubricating oil, correct grade of oil for various parts of the tractor, and so on. This handbook is probably the most important part of the tractor's tool-kit, and the instructions contained in it should be carried out as thoroughly as possible.

The driver is assisted and encouraged to attend regularly to maintenance if he is provided with a log-book in which changes of lubricating oil and dates of the various other types of servicing are regularly recorded. Such a log-book, if properly kept, permits a check on fuel and oil consumption, and makes it easy to see at any time whether the tractor is due to be serviced or not. Farmers who have a number of tractors have frequently found that a properly kept log-book enables them to detect faults in operation and to improve tractor efficiency.

A good driver will take care to use clean water in the radiator and drain it out at night in winter. He will avoid getting dirt or water in the fuel, and will keep the working parts of the tractor reasonably clean. He should also be able to keep the engine tuned up by the adjustment of tappets, sparking plug points, etc., and should be capable of detecting trouble in its early stages and carrying out minor repairs and replacements on the farm. He should be able to carry out an annual overhaul and should generally take interest and care in the running and use of the tractor. In addition, he should be a competent ploughman and should be familiar with the working of such machines as the mower and binder.

It is clear, therefore, that high qualifications are called for; and it may be thought that few such men are available. This is, indeed, true; but it is possible to produce one by choosing a young and keen man and giving him a chance to learn by letting him assist the service agent to carry out an annual overhaul and all repairs. If he possesses the necessary aptitude and intelligence he should be able to do these things himself in a short time. It is sound economy to have at least one such man on the farm and to pay him a good wage. In addition to work on the tractor, there is work on fixed engines and all kinds of farm machinery that calls for considerable skill and is expensive if performed by outside labour.

Tractor and implement maintenance are greatly facilitated if a well equipped fuel and spares trailer is provided. Such a trailer should carry sufficient fuel to last several days, lubricating oil and grease; a semi-rotary pump for filling the tractor or a good funnel and can; a water-can and water-drum; a tow-chain and spare shackles; a good set of tools and spare plough shares, sparking plugs, nuts and bolts, etc. The cost of providing such a trailer is quickly repaid in time saved on servicing and minor troubles. Moreover, the driver is much more likely to attend to routine servicing if he is adequately equipped for the job.

Safety Precautions.

Farmers and tractor drivers should never forget that limbs and lives may be endangered unless proper precautions are taken in handling power-driven machinery. With tractors the chief causes of accidents are carelessness in hitching to implements, and failure to fit proper shields over power drive shafts.

Hitching a tractor to an implement is often a difficult job for one man, especially where the implement drawbar is heavy, as with a disc harrow or a roll. Unless a special hand clutch is fitted, the tractor driver should always remain squarely on the tractor so long as it is in gear. The driver who attempts to operate his clutch and also to guide the implement drawbar over the tractor drawplate takes a big risk. His foot may slip; and if this happens he will be fortunate if he escapes being crushed. Many lives have been lost in this way. Farmers should either see that the driver carries a handy block of wood or a jack to support the implement drawbar, or should send a second person to help with the hitching.

Power drives are seldom treated with the respect they deserve. The square shafts and universal joints, if not properly covered, readily catch up any loose clothing; and if this happens, the operator is lucky indeed if it is only his clothing that suffers. Tractor drivers and all other workers who have to go near machinery in motion should avoid wearing loose clothing.

In the past, accidents have been caused by tractors which reared and overturned backwards if the clutch was too rapidly engaged. This now only occurs if the tractor is prevented from going forward by having its rear wheels in a grip, and is easily avoided by disengaging the clutch. Sideways overturning, however, still causes a number of fatal accidents, the most frequent causes being careless operation on steep hillsides and running too near' the edge of a ditch. Where a tractor is required to work on steep hillsides the wheels should be set out as wide as possible, a sharp lookout should be kept for local hollows in the ground surface, and turns should be made slowly.

Save in very exceptional circumstances no attempt should be made to adjust or lubricate a machine that is in motion. The temptation to save time may be great; but on the farm, as in the factory, the motto should be "Safety

First". It simply does not pay to take unnecessary risks with any power-driven machinery.

REFERENCES

(1) Culpin, C. *Farm. Mechanization: Costs and Methods.* Crosby Lockwood & Son, London, 1951.

(2) "Nebraska Tractor Tests." The University of Nebraska, Lincoln, Nebraska, U.S.A., Bulletin 397. January 1950 and supplementary data sheets on current models.

(3) N.I.A.E. Tractor Tests. Reports of Tests on individual tractors and other farm machines are published periodically by National Institute of Agricultural Engineering, Wrest. Park, Silsoe, Beds.

(4) Influence of Engine Loading on Tractor Field Fuel Consumption. N.I.A.E., Silsoe, Beds. (1951).

(5) Hine, H. J. "Tractors on the Farm." *Farmer & Stockbreeder*, London, 1950.

(6) Jones, F. R. *Farm Gas Engines and Tractors.* (An American text-book.) McGraw-Hill, 1950.

The following deal with all types of farm equipment:

(7) N.I.A.E., Wrest Park, Silsoe, Beds., publishes reports of research work, and agricultural engineering abstracts.

(8) Wright, S. J. "Farm Implements and Machinery." *four. R.A.S.E.* (An annual review of publications on farm power, implements and machinery.)

(9) *Farm Implement and Machinery Review* (London). A monthly journal containing news of new machines and methods.

(10) *Farm Mechanization* (London). A monthly agricultural engineering journal.

(11) *Power Farmer* (London). A monthly agricultural engineering journal.

(12) *Agricultural Machinery Journal.* (London.) A monthly agricultural engineering journal.

CHAPTER TWO

TRACTOR TYPES: CONSTRUCTIONAL FEATURES

"Our business was to award the prize in Glass XVII, which was to be given 'for the best agricultural locomotive engine applicable to the ordinary requirements of farming'. We had, therefore, to judge of the merits of the engines when used to replace portable engines, as a mere implement for driving farmyard machinery, of their merits when used as locomotives upon the high road, and of their merits when used as locomotives upon farm roads, or upon the surface of fields where there were no roads."

"Trials of Traction Engines at Wolverhampton."

Jour. Roy. Agric. Soc. England, 1871.

The Principal Types of Tractors. While a high proportion of the vast numbers of tractors now used on British farms have four wheels and a petrol-paraffin engine of some 20-30 b.h.p., account must also be taken of the other important types and sizes. There is an almost infinite variety of types and sizes between the tiny motor hoes and the giant tracklayers, and to add to the difficulties of classification, some tractors can appear as wheeled, half-track or full-track machines, and may be fitted with petrol, paraffin or diesel engines. The classification shown in Table I is, therefore, an arbitrary one, its aim being merely to convey a general picture of the wide range of tractors available. The "row-crop" or "all-purpose" tractors, being now the most important type, are dealt with first.

Row-crop or All-purpose Tractors.

This group includes both 3-wheeled and 4-wheeled machines specially designed for row-crop work. The medium-powered size (20-30 b.h.p.) is deservedly popular because it is powerful enough and adaptable enough to tackle satisfactorily almost any kind of farm work, and because its manufacture in large numbers in Britain has resulted in an excellent product at a moderate selling price. The modern row-crop tractor possesses the following distinctive features which enable it to work between crops drilled in rows: (1) high ground clearance; (2) wheels with narrow rims; (3) wheels adjustable for various widths of rows; (4) a small turning radius; and (5) special fittings for the attachment of various tools. Other features found on modern row-crop tractors are power lifts for the tools and rear-wheel steering brakes.

TABLE I.—RANGE OF TRACTOR TYPES.

Type of Tractor	Power Group and Approximate Power Range (B.H.P.)	
		H.P.
Market Garden ..	Motor Hoes	$\frac{1}{2}$ - 3
	2-wheeled Ploughing Tractors ..	3 - 8
	Baby Tracklayers	6
	Self-propelled Toolbars	5 - 17
" Row-crop " or	Small	12 - 20
" All-purpose " ..	Medium	20 - 35
	Large	Over 35
Standard wheeled ..	Small	12 - 20
	Medium	20 - 35
	Large	Over 35
Tracklayers ..	Small	Under 25
	Medium	25 - 40
	Large	Over 40

The tricycle type row-crop tractor has advantages for working some crops, not only owing to its smaller turning circle, but also because forward or middle tool-bars can be more easily fitted. Moreover, the single front wheel needs no adjustment for working rows of various widths. The proportion of tricycle type tractors is high in the Fens, where they are found particularly suitable for work in potatoes and sugar beet. On the other hand, most British farmers outside such areas prefer four wheels. A few tractors are fairly easily convertible from four wheels to three and vice versa.

Ease of adjustment of the wheel track widths on a row-crop tractor is an important feature; for it is easy with some tractors to waste half a day on this job. In general, where only open-field work has to be considered, the use of a sliding wheel hub which is fixed to the axle shaft by means of an easily loosened clamp is easiest to adjust. The type which necessitates re-arranging a dished wheel centre on an offset rim has, however, the merit of relative cheapness, and is rather more foolproof mechanically. There is everything to be said for planning the row-widths of crops in such a way that the changing of tractor wheel widths is reduced to a minimum.

FIG. 11.—THREE-WHEELED ROW-CROP TRACTOR WITH TOOL-BAR WHICH CAN BE FITTED EITHER

AS SHOWN, AT THE FRONT, OR AT THE REAR.
(JOHN DEERE AND STANHAY.)

Small row-crop tractors (i.e. those of 12 to 20 b.h.p.) are indistinguishable in their general features from machines of the medium-powered group. As a rule, however, it is only on market gardens that a tractor of the small size is considered large enough for general work. On most farms where such machines are used they are employed for particular specialized jobs such as inter-cultivations, and heavy work is left to larger tractors. Machines of just over 20 b.h.p., on the other hand, are used for almost all kinds of work on all sorts of farms.

As previously stated, the row-crop tractor is gradually becoming merged with the unit-principle tractor into an all-purpose machine which is designed to carry implements of all kinds, e.g. ploughs, hoes, potato diggers, etc., directly coupled to it. With most outfits the implements so attached have ground-engaging depth-control wheels of their own; but with others, notably the Ferguson, the whole range of adjustment is effected by means of a hydraulic control on the tractor, and adjustments in the linkage between tractor and implement.

FIG. 12.—SECTIONAL DRAWING OF FOUR-WHEEL
ALL-PURPOSE TRACTOR WITH POWER LIFT

SHOWN DETACHED. (DAVID BROWN.)

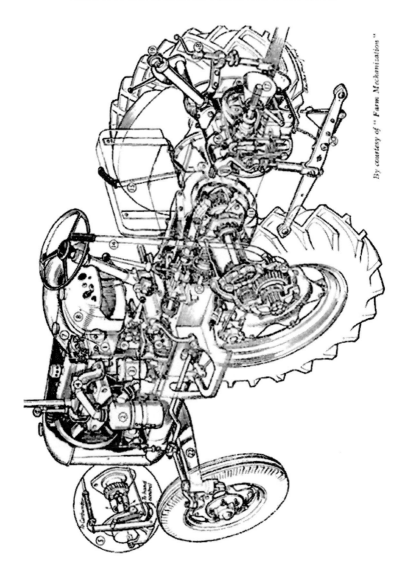

By courtesy of " Farm Mechanization "

KEY TO NUMBERS ON DIAGRAM

1. Overhead valve engine.

2. Oil-bath type air cleaner.

3. Oil filter.

4. Pre-heater.

5. AM-speed governor.

6. Single-plate dry clutch.

7. T.V.O. fuel tank

8. Petrol tank.

9. Fuel change-over cock.

10. Gearbox—4 forward speeds and reverse.

11. Differential unit.

12. Final drive reduction gears.

13. Internal expanding brakes

14. Hand clutch lever.

15. Right-hand brake lever.

16. Hydraulic power lift unit.

17. Power take-off shaft.

18. Levelling lever.

19. Adjustable top link.

20. Left-hand lower link incorporating lateral control.

21. Worm and nut steering.

22. Fabricated front axle mounted on trunnion.

The unit principle has many advantages. In the first place there is a saving of material on implement wheels and control gear generally, and this makes a complete outfit of unit-principle implements cheaper to manufacture than the trailed kind. A further important point in favour of mounted tools is the ease of handling and manoeuvrability. For example, when a wheeled tractor with a trailed plough enters a boggy patch of land it may take half an hour to extricate the outfit. In similar circumstances a tractor with a mounted plough can get through without real difficulty if intelligent use is made of the hydraulic lift control. There are, on the other hand, occasional instances of exceptionally hard soil conditions where a heavy trailed plough will work satisfactorily and a light type of directly coupled plough will fail to penetrate. In soft soil, tractor wheel adhesion may be increased by some mounted ploughs because both the plough weight and downward soil forces pull downwards on the tractor wheels, but in hard ground wheel adhesion may be seriously reduced. In general, the advantage is with the tractor-mounted implement, and there now seems no doubt that there will be a gradual switch over to this type of outfit.

Standard Type Wheeled Tractors.

"Standard" tractors are 4-wheeled machines which have no special provision for doing row-crop work. Little or no adjustment of wheel track width is provided, and the aim is usually to have a low centre of gravity and a cheap, short and sturdy tractor, rather than to secure a high ground clearance, with plenty of room for fitting hoes, etc., between front and rear wheels. Many of the tractors sold in the early nineteen-forties were of this general type, and there are still many farmers, e.g. on hill land, who are not concerned with row-crop work and prefer the general shape of the standard type. Nevertheless, this type is almost obsolete.

As with row-crop tractors, the most popular size is 20-30 b.h.p. High-powered tractors (with engines of over 35 b.h.p.) are much more widely used in some other countries, e.g. U.S.A. and Canada, than in Britain. In Britain they are used mainly on large farms where wide implements can be employed, and tracklayers are not needed. They are also frequently equipped with large pneumatic tyres and a winch, and used for driving threshing machines on contract work in hilly country.

The Transmission.

The ordinary 4-wheeled tractor has the engine, clutch, gear-box, propeller shaft and drive axles all mounted in a rigid iron or steel frame. The two rear wheels are the traction members and the front wheels provide directional control. The rigid frame gives freedom from mis-alignment of the transmission, and may at the same time provide a casing that is dust-proof and water-proof. In well-designed machines, the engine, clutch, etc., are all easily accessible. The power is usually transmitted through an easily-adjustable plate clutch to a simple clash-type gearbox with 3-6 forward speeds and 1 or 2 reverse.

The usual type of final drive to-day is by spiral bevel gears to a high-speed differential shaft on which the steering brakes are mounted, and thence by spur gears to the axle shafts (Fig. 1). The whole of this gearing is usually enclosed in the main gearbox casing, but in a few tractors the final reduction gears are situated at the ends of the axle housings (Fig.12).

Exposed chains were used in some of the early tractors but proved unsatisfactory, mainly because it was impossible to keep them clean or to lubricate them effectively. The roller chains (Fig. 311) on the Case tractor, however, are entirely enclosed, and run in an oil bath. There is much to be said in favour of this type of drive, but it would be surprising if there were any great extension of the use of chains.

Tractors with horizontal engines, such as the John Deere and the Marshall, have a straight-through drive by spur gearing. The belt pulley is carried on the end of the crank-shaft, and the only bevel drive used is that to the power take-off. This type of transmission is very simple and efficient. A conical clutch is used on some tractors.

Whatever type of final drive is employed there is, of course, a differential gear (see Appendix Two) to facilitate turning without any skidding or undue strain. The differential is sometimes a nuisance in field work, because of the fact that the major portion of the tractor power may be delivered to the wheel meeting the least resistance. Thus, when one tractor wheel enters a greasy patch of land which gives poor adhesion it may skid and the tractor becomes bogged because most of the power is delivered to the skidding wheel. Use of a "differential lock" to overcome this difficulty has often been advocated and occasionally tried, but no entirely fool-proof system of providing a solid axle for straight work and a differential for turns has yet been put into production for farm tractors.

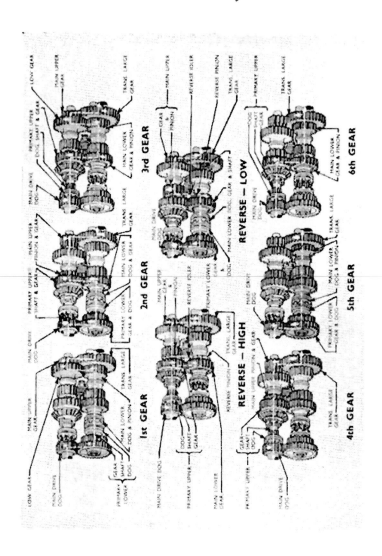

FIG. 13.—GEAR TRAIN DIAGRAM OF CONSTANT MESH GEAR-BOX WITH 6 FORWARD SPEEDS AND 2 REVERSE, CONTROLLED BY TWO LEVERS. (FORDSON.)

The Belt Pulley and Power Take-off.

Modern tractors are equipped with a belt pulley for driving stationary machinery such as threshing machines, grinding mills, etc. The pulley shaft is normally driven through bevel gearing and a separate clutch, the pulley usually being situated at the off side, just in front of the rear wheel. With tracklayers the belt pulley is generally situated at the rear, the drive being taken by bevel gearing from an extension of the propeller shaft. The standard speed for belt pulleys adopted by the British Standards Institution is 3,100 ft. per minute, plus or minus 100 ft. per minute.

The power take-off consists of a shaft passing backwards from the gear-box, which can be fitted with universal joints and telescopic sections so as to drive the mechanism of machines hauled by the tractor. It is an efficient means of applying power to such machines as a binder, for up to 95 per cent, of the power delivered to the shaft may be transmitted, where the shaft is fairly straight, so that little energy is lost in the universal joints. The take-off usually rotates at about half engine speed, and the British Standards Institution has adopted a standard speed of 536 r.p.m., making it possible to design implements so that they can be driven by any make of tractor conforming to this standard.

Power drives should always be adjusted to run as straight as possible. The maximum working angle of drive for a double universal joint should be limited to 22 degrees, but it is sometimes necessary to run power driven mounted machines at greater angles for short periods when the machine is being lifted out of work.

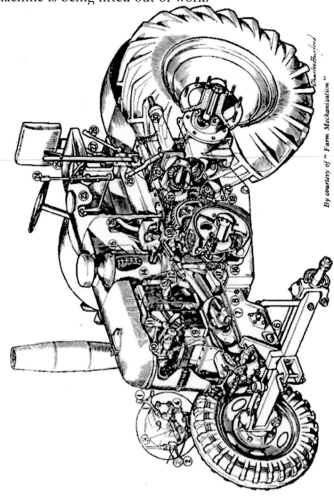

FIG. 14.—SECTIONAL DRAWING OF TRACTOR
WITH SINGLE-CYLINDER HORIZONTAL DIESEL
ENGINE. (MARSHALL.)

KEY TO NUMBERS OH DIAGRAM

1. Ignition paper holder.

2. Handstart valve.

3. Compression release valve.

4. Fuel injector.

5. Cartridge starter.

6. Front axle pin.

7. Transfer port.

8. Piston pad.

9. Radiator.

10. Fan drive.

11. Fan.

12. Flywheel.

13. Oil filter.

14. Air cleaner.

15. Fuel filter.

16. Clutch operating fork.

17. Crankshaft pinion.

18. Gear selectors.

19. First motion shaft.

20. Second motion shaft.

21. Power take-off drive.

22. Differential and final gear.

23. Fuel tank.

24. Fuel control.

25. Brake.

26. Clutch foot control.

27. Clutch hand control.

28. Change gear lever.

29. P.T.O. control.

30. P.T.O. unit.

The Drawbar and Hitch.

For many purposes, a tractor delivers the power of its motor through a fixed drawbar. When the engine is exerting its power in driving the rear wheels, the front of the tractor tends to rise owing to the "torque reaction", i.e., the tendency of the drive wheels to remain stationary while the engine winds the tractor round the back axle. This transfers weight to the rear wheels, the front tending to rear off the ground. The drawbar must be placed below the axle, so that the reaction of the drawbar pull has an *opposite moment (vide*

Appendix One), tending to bring the front wheels back on to the ground. Distribution of tractor weight and location of the drawbar must be such that there is always sufficient weight on the front wheels to make steering possible. The British Standards Institution has adopted various standards for drawbar hitch locations, one of which is that the vertical distance between the ground line and the top of the drawbar shall be 12 to 18 in. Most manufacturers provide a range of both vertical and horizontal adjustments to facilitate hitching to all kinds of implements.

The Wheels.

The 4-wheeled machine normally has 3-point suspension, the front of the frame being centrally attached to the axle by means of a horizontal trunnion. This enables the wheels to follow the contours of the ground quite freely. The design of the traction wheels should aim at securing the best possible adhesion with a minimum absorption of power. The tractor often works on loose surfaces, and the power expended by it in propelling itself over the land is considerable. In general, the less the disturbance of the soil, the lower is the rolling resistance. Other things being equal, the greater the diameter of the traction wheels the less the wheels sink into the soil and the better the grip. But very large or wide traction wheels have many disadvantages, and the normal type is about 4-5 ft. in diameter and 8-12 in. wide.

Adhesion is increased by increasing the weight, but Table II (col. 3) shows how the weight per horse power has been reduced since 1920 while, as column 4 shows, the ratio of maximum pull to test weight steadily increased until the influence of fitting pneumatic tyres caused it to level out at about 66 per cent.

There are many reasons why weight should be kept as low as possible, and much attention has been given to devices such as strakes and spade lugs for increasing adhesion.

TABLE II.—PROGRESS IN THE REDUCTION OF TRACTOR WEIGHT AND FUEL CONSUMPTION

Period in which tested.*			Fuel consumption on drawbar rated load. (lb. per b.p. hour)	Test weight per maximum drawbar horse power. (lb.)	Ratio of maximum pull to test weight
1920	..	..	1·63	349	0·44
1923-6	..	..	1·15	275	0·62
1927-30	..	..	1·01	213	0·69
1935	..	..	1·05	215	0·70
1946	..	..	0·73	213	0·66

* Figures based on all wheeled tractors tested at Nebraska in the periods stated.

FIG. 15.—SKELETON TYPE TRACTOR DRIVE WHEEL (ALLMAN.)

Spade lugs are normally arranged in two staggered rows on a steel rim about 8 in. wide, the lugs being set out so that their tips project beyond the edges of the rim. The efficiency of such wheels depends on the points of the lugs penetrating into a firm layer of soil. They may fail to secure good adhesion if the soil is very loose, so that the lugs cannot reach a firm layer without sinking in. A more usual cause of trouble is sticky soil which builds up between the lugs and eventually completely covers their tips. The first condition is best met by using pneumatic tyres, but the latter are no remedy for sticky soil. To meet this difficulty, many "open", "tip-toe" or "skeleton" types of wheel, which have no broad rim, have been developed. The absence of a broad rim permits the soil to pass between adjacent lugs and so enables the wheel to keep reasonably clean. A further advantage of this type of wheel on heavy land is that it leaves a level seed-bed and avoids producing a wide strip of puddled, impervious soil in the wheel-marks. Such wheels are therefore popular for all kinds of cultivations on many clay farms, and are particularly valuable for seed-bed preparation, drilling and inter-cultivations. On land which is very soft to a great depth the absence of a broad rim may result in skeleton type wheels allowing the tractor to sink up to its axles. They are therefore unsuitable for such conditions.

Pneumatic Tyres for Tractors.

The use of pneumatic tyres on tractors began about 1930, and they have become so efficient and popular that to-day, they are the standard fitting on almost all small and medium-sized tractors.

The tyres used for tractor drive wheels are generally low-pressure types, pressures of 10-15 lb. per square inch being employed. Several types of tractor tyres have been tried, some having a very deep tread designed to secure good adhesion in boggy conditions. The present tendency is towards the "wide-base" type with a broad steel rim, walls rather short in proportion to the width, and a deep tread.

The walls are made fairly light, and as the load increases the tyres flatten and cover more ground. Compared with steel wheels, they have a lower rolling resistance under most conditions. Frequently, however, they do not provide as good adhesion as steel wheels equipped with suitable lugs, and the drawbar pull for a given tractor may sometimes be reduced by the substitution of rubber tyres for steel. The adhesion between the rubber tyre and the soil is chiefly frictional, and therein differs from that of a steel wheel in which the lugs behave rather like the teeth of a gear. In some conditions, such as loose sand, rubber tyres may permit higher drawbar pulls than steel wheels; but generally, the drawbar pull which can be transmitted is limited to a fairly well defined maximum.

Most drive-wheel tyres are marked with an arrow to indicate the direction in which they should travel. When correctly fitted, the flexing of the tread bars which occurs when the tractor is at work, assists in freeing soil from the spaces between the bars, and so produces a self-cleaning effect. If the tyres are fitted the wrong way round, and the conditions arc at all sticky, the spaces between the tread bars rapidly become filled with soil, and wheel spin occurs.

Where conditions for adhesion are difficult, tyre pressures should not be too high. The lower the pressure, the greater the area of tyre in contact with the soil and the better the grip. Pressures must not, however, be reduced below a certain limit—generally 10-12 lb. per sq. in., since further reduction results in serious tyre "wrinkling" and rapid disintegration of the tyre walls.

The adhesion may be increased by the addition of weights to the wheels. Another method of weighting the tractor is to use water as ballast in the drive wheels. The tube, fitted with special valve equipment, may be half, three-quarters or almost completely filled with water, the air being kept at the normal pressure. In frosty weather it is necessary to ensure that the water does not freeze, and a concentrated solution of calcium chloride may be used. Instructions for the use of such liquid fillings are provided by all tractor tyre manufacturers.

American studies of 100 per cent, liquid filling show that diffusion through the inner tube walls is so small that the rate of loss of pressure is greatly reduced compared with a 75 per cent, liquid filling or 100 per cent. air. On the other hand, wide changes of temperature have a big effect on the pressure of tyres that are completely liquid-filled, and adjustment of pressures may be necessary where temperature changes are extreme. A disadvantage of complete liquid filling is that resistance to bruising when a tyre strikes an obstruction is decreased. The effects of 75 per cent, filling and 100 per cent, filling on riding qualities are complicated. In general, 75 per cent, filling gives a cushioning effect, and riding quality on rough ground is good. With 100 per cent, liquid filling, riding quality is definitely lowered, but the effects are not serious except on rough roads at high speed.

One hundred per cent, filling is advantageous where heavy tractor-mounted implements are employed. When such an implement is lifted, an extra load is placed on the tyres, any air that is contained in them is compressed, and the deflection of the tyre where it is in contact with the ground may be so excessive as to cause the tyre walls to break down. Where 100 per cent, filling is used, the additional deflection caused by raising the implement is much smaller, since water is practically incompressible; so the danger of overloading the tyres is lessened.

Unfortunately, 100 per cent, filling is not easily effected. It requires special motor-driven apparatus which exhausts the air from the wheel before the liquid is pumped in. The filling is therefore usually a job that must be done at a garage, and even when the utmost care is taken the final result falls short of 100 per cent., since it is impossible to exhaust all the air, or to ensure that the liquid introduced does not contain some air.

FIG. 16.—RETRACTABLE STRAKES FOR WHEELS FIT-TED WITH PNEUMATIC TYRES. (STANHAY.)

There is, therefore, much to be said in favour of a simple piece of equipment which permits filling tyres to about 95 per cent. The only apparatus required, apart from a liquid pump or gravity tank, is a special air-water valve equipped with a flexible extension-piece which allows air to continue to escape from the top of the tube until the level of the water is well above the valve, and the tube is almost completely filled with liquid. When filling is complete the adaptor is removed and rapidly replaced by the air valve. Whenever a substantial amount of extra weight is added to wheels which are wholly or partly air-filled, it is necessary to increase the pressure in the tube in order to limit tyre deflection. Adding a 95 per cent, filling of liquid ballast to an 11 in. ×36 in. wheel increases the weight on each wheel by about 485 lb. and requires pressure to be increased by about 2 lb. per sq. in.

Generally speaking, rubber tyres give a higher drawbar efficiency than steel wheels and lugs at high speeds, while the converse holds at low speeds. For a given tractor with steel wheels, the maximum drawbar horse power generally falls rapidly at speeds above 3-4 m.p.h., whereas with pneumatics the maximum drawbar horse power may progressively increase for speeds up to 5-6 m.p.h. So long as rubber tyres are used *within the limits of their tractive effort,* they are usually more efficient than steel wheels in transmitting power, and this greater efficiency may in specially favourable circumstances lead to economy in fuel consumption of up to 20 per cent. But in considering the relative merits of the two types of wheels, appreciably increased fuel economy cannot be relied upon with certainty.

FIG. 17.—SPIRAL STRAKES FOR PNEUMATIC-TYRED

DRIVE WHEELS.(KENNEDY & KEMPE.)

On wet surfaces which are at all greasy, pneumatic tyres may fail completely. They are seen at their worst on ground which is slippery on the surface and hard beneath, so that the "tread" of the tyre cannot penetrate into the soil to obtain a grip.

In soft ground, on the other hand, pneumatic tyres with a deep tread may perform very much better than steel wheels. In those circumstances where rubber tyres fail to obtain a grip, they may be fitted with various types of lug chains or special strakes to give them adhesion; but in such conditions they lose many of their special advantages.

The great advantage of pneumatic tyres is that a tractor equipped with them is a maid of all work, instantly adaptable to transport, haymaking, harvesting or other of the varied jobs the modern all-purpose machine is required to do. For general farm work the increase in cost and depreciation compared with steel wheels is justified by better performance. Moreover, increased depreciation on the tyres themselves may be more than offset by reduced wear and tear on the transmission system.

For the present, many farmers find that the best plan is to have two sets of drive wheels, one rubber and one steel, the steel wheels being fitted only when prolonged spells of heavy work lie ahead. For the future, it may be expected that the alternative equipment will be on the one hand pneumatic tyres, and on the other, easily fitted half-tracks.

Tractor Engines.

The most common type of engine used in tractors is the 4-cylinder 4-stroke petrol-paraffin type, but multi-cylinder 4-stroke diesel engines arc becoming increasingly popular for medium-powered tractors, as well as for large. Reference should be made to Appendix Six for an account of the working principles and general constructional features of such engines.

Crankcase Oil Dilution.

One of the primary causes of wear in paraffin engines is bad vaporization and incomplete combustion of the fuel, which may lead to heavy dilution of the crankcase oil with unburnt paraffin. It is unfortunately impracticable to adjust vaporizer temperature to all variations of load, so a certain amount of oil dilution is inevitable. With careful operation, dilution can be kept down to about 12 per cent., and the viscosity of the engine lubricant recommended by the manufacturer will be such that with this normal dilution, full protection is given to all bearing surfaces.

With careless operation, however, dilution can exceed 25 per cent, and serious consequences may ensue. In order to minimize the harmful effects of oil dilution, the following recommendations should be followed:

Where the vaporizer has a temperature control lever, set this to suit the tractor load and the weather conditions.

Change crankcase oil regularly, as recommended by the manufacturer.

When starting up, get the engine hot before switching over from petrol to paraffin. Get the tractor to work quickly, and do not leave the engine idling needlessly.

Use the radiator blind or shutter control to keep the cooling water as near boiling point as practicable.

For long spells of light work, close down the adjustable jet on the carburettor (where fitted).

In cold weather it is advisable to run the tractor extra hot towards the end of the day, so as to evaporate fuel from the lubricating oil. It is always advisable to run on petrol for the last two minutes before stopping, so that all vaporizing oil is cleared out of the carburettor and vaporizer ready for re-starting.

Carburettors.*

With most paraffin tractors, the petrol and paraffin are fed to the carburettor by means of a 2-way tap. Such a system is not at all well suited to frequent stopping and starting of the tractor; for all paraffin must be drained from the carburettor float chamber before attempting to re-start, unless the engine is really hot. With a steady increase in the use of electric starters and the utilization of tractors for all kinds of farm work, more convenient carburation systems are of interest.

Some modern tractors employ a "bi-fuel" carburettor which consists essentially of a simple petrol carburettor added to the paraffin carburettor, each having its own independent fuel line and being put into or out of operation by a simple push-pull control from the driver's seat. With the control lever pulled right out, only the petrol starting carburettor is used, and when the control is pushed in the paraffin carburettor comes into operation. Such an arrangement, in conjunction with electric starting, makes stopping and starting almost as straightforward as it is with a petrol engine.

The engine is started up on the petrol carburettor, and when it has warmed up sufficiently the control rod is pushed in to change to paraffin. If, for any reason, the engine fails to run properly on paraffin, it can immediately be switched back on to petrol. This arrangement operates well in practice, and is worth the slight additional expense. It is to be hoped that something of the kind will become generally available for paraffin tractors.

* See also Appendix Six.

Carburettor Adjustments. For efficient operation the carburettor jet must be correctly set, and this setting must be done while the tractor is at work, since the optimum setting varies according to the load. With the carburettor illustrated (Fig. 18) the method of adjusting is to get the engine thoroughly hot and then gradually cut down the amount of fuel drawn through the main jet by screwing in the needle valve until the engine begins to falter. The needle should then be unscrewed slightly, about one-eighth to a quarter turn, until the engine runs normally again. Fuel economy can often be considerably improved by attention to carburettor adjustment.

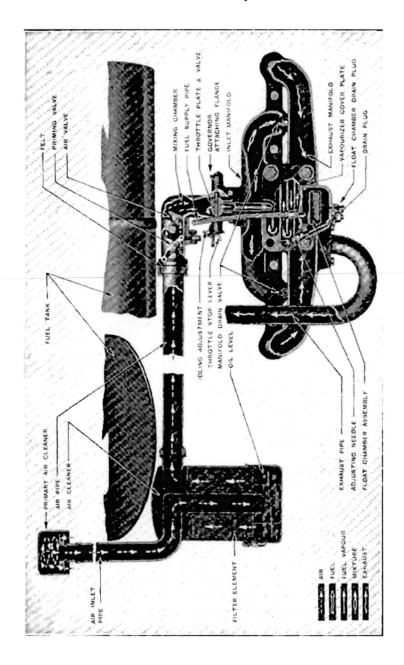

FIG. L8.—AIR CLEANER AND FUEL SYSTEM OF A PARAFFIN TRACTOR ENGINE. (FORDSON BEFORE 1952.}

The fuel consumptions of tractors may vary considerably owing to differences in adjustment and maintenance; and it has been found that an increase of efficiency can be expected by adjusting the carburettor and governor, and servicing air cleaners, grinding in valves, and so on.

Diesel and High-Compression Petrol Engines.

While the petrol-paraffin engine is the type most commonly employed on modern tractors, many manufacturers arc not satisfied that it is the ideal, and engine design during the last few years has been proceeding along divergent lines. Some engines are of the full Diesel type, and these are popular in the large machines in this country owing to the high efficiency with which they burn a cheap fuel. In Germany the most important make of tractor has a heavy single-cylinder low-speed 2-stroke semi-Diesel or "hot-bulb" engine. But it is in the United States that the controversy concerning the best type of engine rages most fiercely, for in most States petrol which is used for agricultural purposes is free from tax, and high-compression engines are being used in some tractors. These engines burn "high-octane" petrol, and apart altogether from the fact that lubrication difficulties due to crank-case dilution are almost eliminated, some of the engines have established such high performance figures as to compel consideration. While the developments with respect to such engines can have little direct effect on tractors used in Britain owing to the high price of petrol compared with other fuels, American tractors have in the past occupied an important place in British farming, and the movement may have a considerable indirect influence in this country.

An interesting type of tractor engine that may have a big future is the "blown" 2-stroke Diesel. As it nears the bottom of the power stroke the piston uncovers intake ports in the cylinder wall, and large exhaust valves of the usual overhead poppet type in the cylinder head are opened. A powerful blower sends a blast of cool air in through the intake ports and out through the exhaust valves, with the result that the exhaust gases are swept out very effectively. The piston on the upward stroke then covers the intake ports, the exhaust valves close, and fresh air is compressed ready for another power stroke. The operation of this type of engine may be easily understood by reference to the sections of Appendix Six dealing with the 2-stroke cycle and Diesel engines.

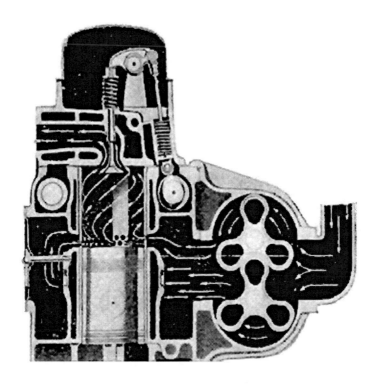

FIG. 19.—SECTIONAL VIEW OF 2-STROKE DIESEL ENGINE INCORPORATING ROTARY BLOWER. (ALLIS CHALMERS.)

The Air Cleaner.

An efficient air cleaner is one of the most important desiderata in the tractor engine. Before air enters the carburettor it must be freed from grit, for in dusty conditions the life of an engine is short if grit is allowed to pass unimpeded into the cylinders. The air cleaner must be completely effective and easily cleaned. Grit may be removed from the air either by centrifugal action or filtering, or both. Filtering may take place through a dry material, but the more usual method is to filter through oil. A satisfactory type of cleaner in general use is the oil-saturated wire wool type. The inlet pipe is sometimes placed high above the tractor to prevent dust created by the outfit from being drawn into the engine.

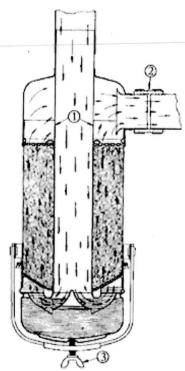

FIG. 20.—INTERIOR VIEW OF AIR CLEANER.

1, Air inlet; **2,** Air outlet; **3,** Sediment bowl. (Case).

The air first passes through a primary cleaner, so constructed that the induced air is given a swirl, thus flinging the heavier particles of grit through louvres around the edges, or into a glass jar at the side, which may be regularly emptied. The air then passes down into oil, and carries some of the latter up into the filtering element. Dust particles are removed by passage through the oil and the oil-saturated filtering element. The latter is kept in place by means of gratings, and washing out and refilling with a light engine oil or used crank-case oil is a simple operation. Washing out and cleaning should be carried out regularly according to the makers' instructions. Frequent changing of the oil is especially important in dusty conditions. The gratings and filtering element may become blocked in extremely bad conditions, and if this occurs they should be removed and washed in paraffin.

It is essential that the correct grade of oil be used in air cleaners. Use of too heavy an oil may result in bad running and excessive fuel consumption.

The Oil Filter.

Grit and other foreign materials should be removed from the lubricating oil by means of an efficient oil filter. The oil pump situated in the crank-case forces oil through the filter before it is delivered again to the main bearings. Foreign matter remains on the filter, which should be so designed that cleaning is a simple operation. The filtering element sometimes consists of a cylinder of felt surrounding a central tube. Other types of filter are filled with fine brass wire which can be washed, or cotton waste, which is thrown away when dirty. Most "full-flow" filters have the oil pass from the inside of the element to the outside, but partial-flow filters often work in the reverse direction. Pressure-regulating and safety devices must be included in order that lubrication may not fail if the filtering element becomes clogged. A simple safety device consists of a ball valve which is normally held on to its seating by the force of a light spring, but rises from its seating and allows the oil to be short-circuited direct from the inlet to the outlet pipe when the pressure in the inlet pipe becomes much greater than that in the delivery pipe. With the full-flow type the filtering element itself is usually spring-mounted, and is automatically by-passed if excessive pressure develops.

The Governor

is a device which automatically ensures a fairly constant engine speed under varying loads. The mechanism is so arranged that when the load is removed and the engine begins to go faster, the increased centrifugal force on the spring-loaded governor weights causes them to move farther from the axis of the governor shaft about which they rotate. This movement causes a collar to slide along the governor shaft and operate a linkage mechanism which closes the throttle and causes the engine to run more slowly. Conversely, if a heavy load slows down the engine, the movement of the governor mechanism automatically opens the throttle and tends to restore the engine speed to normal. Closing of the hand throttle lever, however, puts the governor out of action and allows the engine to "idle".

Petrol-paraffin tractor engines usually run at 1,000-1,500 r.p.m. On many modern tractors it is possible to vary the engine speed between about 600 and 2,000 r.p.m. by adjustment of the governor. This adjustment is useful in operations such as threshing, when it is important to run the machine at the correct speed.

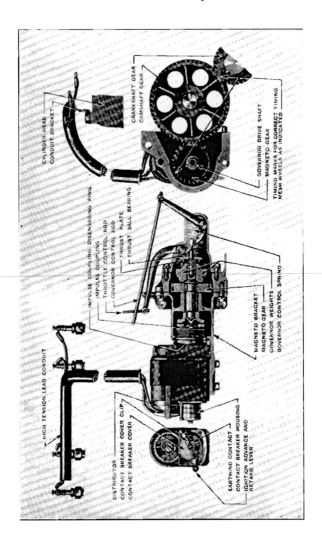

FIG. 21.—DIAGRAM SHOWING SECTION OF GOV-
ERNOR, AND RELATIVE POSITIONS OF GOVERNOR,
IMPULSE STARTER AND MAGNETO ON A TRACTOR.
(FORDSON BEFORE 1952.)

FIG. 22.—PNEUMATIC; GOVERNOR USED TO CONTROL FUEL INJECTION PUMP ON DIESEL ENGINE. (FORDSON.)

Pneumatic governors are frequently fitted to diesel engines, and their mode of operation may be understood by reference to Fig. 22. Two pipes lead from the induction manifold to the governor casing, which has a diaphragm at one end. When the engine is idling, the throttle is almost closed, and the suction in the manifold is at a maximum. This negative pressure, transmitted by one of the pipes direct to the governor casing, is sufficient to draw the diaphragm to the right, against the force of the compression spring. This action pulls the control rod to the minimum fuel position. As the throttle is opened to increase engine speed, the manifold depression decreases, the diaphragm moves to the left, and the delivery of fuel is progressively increased by corresponding movement of the control rod.

When the engine is at work and the throttle is held in a fixed position, engine speed does not remain exactly constant under varying load, but tends to increase if the load is lessened, and to fall if the load is increased. As soon as the engine starts to go appreciably faster, the manifold depression is increased, and the consequent movement of the governor control rod diminishes the fuel supply. Conversely, a fall in speed, resulting in reduced manifold depression for a given throttle setting, leads to an increased supply of fuel and substantial restoration of the speed to the governed setting.

The second suction pipe is connected to the inlet manifold at a point just outside the throttle; so when the throttle is nearly closed, the air in this pipe is approximately at atmospheric pressure. At the governor end, the pipe leads to a damping valve which is connected to the diaphragm. Extreme movement of the diaphragm in response to a sudden depression of manifold pressure allows air to pass *via* the damping valve into the governor casing, thus reducing the suction in the chamber and preventing excessive fluctuation of the control rod, i.e. "hunting".

Power Lifts.

A power lift for the tools is now becoming a regular fitting, and this equipment must be considered one of the most important features of a modern all-purpose machine. The most common type of lift is a hydraulic device operated by the power drive shaft. Other types of lift include mechanical and pneumatic devices.

The hydraulic lift consists essentially of (i) a pump to provide oil under pressure; (2) a control mechanism which regulates delivery of the oil; and (3) a ram on which the oil pressure acts to lift the implement. The fundamental principles of the action of hydraulic lifts are briefly indicated in Appendix One.

The ram is frequently built in at the rear of the tractor, and operates a horizontal cross-shaft to which are attached the lift arms (Fig. 24). Alternatively or additionally, oil may be fed by a flexible tube to an external lifting cylinder, which may be used for the operation of a middle or forward tool-bar, or a trailed implement (Fig. 27). In some circumstances, as when forward and rear tool-bars are being used simultaneously, it may be necessary to raise and lower them consecutively or independently, and some types of control mechanism are designed to permit this.

The main types of power lift pump are (1) the multi-cylinder reciprocating type, the plungers being driven by cams or eccentrics (e.g. the Ferguson); (2) a simple gear pump similar to that commonly used for engine lubrication, but much more accurately constructed (e.g. Fordson, David Brown, Nuffield, John Deere and International); and (3) various special types of high-efficiency rotary pumps. The pump must be capable of operating at a fairly high pressure, and must deliver a sufficient volume of oil to complete the lift in not more than 2 seconds.

The method of control naturally depends on the application required. If the only operation to be reckoned with is the raising and lowering of lift-arms by means of a built-in ram, a fairly simple form of control will suffice. This consists of a valve which in one position allows the oil to pass from the pump to the ram cylinder, and in the other position releases the oil from the cylinder and allows that coming from the pump to pass straight back to the reservoir.

Fig. 23 illustrates a control system of a type now commonly employed, which permits the implement to be fully raised, fully lowered, or held in any intermediate position. A single control lever can be put into any one of three positions—"up", "down" and "neutral", and there are two valves—one a by-pass and one a non-return valve. The by-pass valve, D, can pass oil from the pump directly back to the sump at all times except when it is held shut by plunger H, when the control lever is in the "up" position. The non-return valve, B, is operated by the oil pressure in the system at all times except when it is held open by plunger C, when the lever is in the "down" position.

The diagram shows the "up" position, valve D being held closed and oil passing through B to the ram. If the lever is moved to neutral, valve D is allowed to open, the pressure of oil in the ram cylinder closes valve B, and the implement is maintained in position. To lower the implement, valve B is held open by plunger C, while oil from the pump is allowed to by-pass through valve D.

The principle of operation of the Ferguson system of implement control may be briefly explained by reference to Figs. 24 to 26. A feature of this system is maintenance of the draught of the implement at a constant value which may be varied and pre-determined by the setting of the control lever in its quadrant.

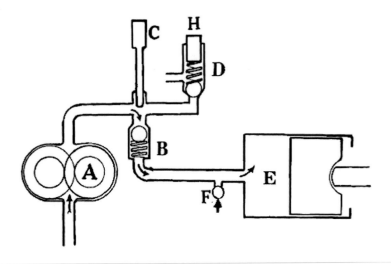

FIG. 23.—HYDRAULIC LIFT CONTROL SYSTEM.

A, pump; **B,** non-return valve; **C,** lowering plunger; **D,** by-pass valve; **E,** ram cylinder; **F,** safety relief valve; **H,** raising plunger.

An oil pump supplies oil to a ram cylinder which operates a transverse lifting shaft, in the way now usual with the 3-link system.

The hydraulic control linkage functions in three distinct ways: (a) hydraulic lift; *(b)* hydraulic depth control; *(c)* safety device to protect tractor and implement from hidden obstructions.

Hydraulic Lift. Fig. 25 shows this in simplified form. When the control lever is moved forward, point A moves forward. The control fork pivots about point B and the control valve moves (?) to the rear, releasing oil from the ram cylinder, while cutting off the supply to the pump. The implement then lowers under its own weight. When the lever is raised, point A moves to the rear, the fork pivots about point B, and the control valve moves forward (2) retaining oil in the ram cylinder and at the same time admitting oil to the hydraulic pump. This raises the implement.

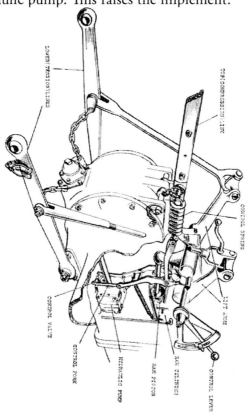

FIG. 24.—LAYOUT OF HYDRAULIC IMPLEMENT
CONTROL SYSTEM. (FERGUSON.)

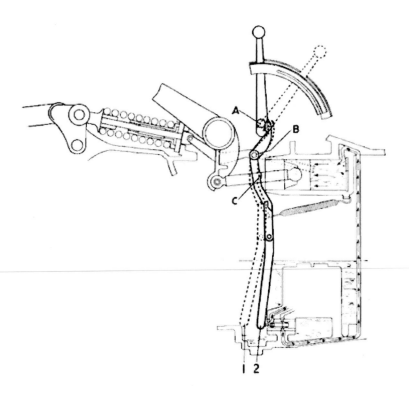

*FIG. 25.—FERGUSON SYSTEM, SHOWING HYDRAU-
LIC: LIFT.*

A, control lever boss; **B,** control fork pivot; **C,** lugs on
control fork; **1,** valve in lowering position; **2,** valve in lifting
position.

When the implement is fully raised, the ram piston
abuts against the lugs C and pushes the control valve to mid-
position. In this position, the control valve prevents any oil
being supplied to the pump, and also prevents the return of
oil from the lifting cylinder to the sump.

Depth Control. Depth control is achieved by utilizing the forces acting in the top link of the implement hitch. The greater the draught of the implement, the greater the force in the top link and the more the control spring is compressed.

The way in which the implement forces operate is by moving the position of the fulcrum about which the control fork rotates.

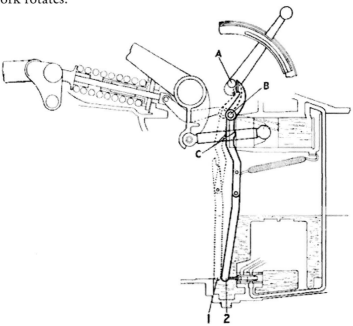

FIG. 26.—FERGUSON SYSTEM, SHOWING DEPTH CONTROL MECHANISM.

A, control lever boss; **B,** control fork pivot; **C,** lugs on control fork; **1,** valve in lowering position; **2,** valve in lifting position.

101

(Fig. 26.) Suppose that the implement is at work and the draught becomes excessive. Fulcrum B is pushed forward (i.e. to the right in the diagram). This causes an anti-clockwise movement of the control fork about the fulcrum. The result of this is that the control valve is pushed forward, (2) oil is admitted to the pump, and the implement is raised, until the force on the control spring is no longer excessive, the control fork returns to its normal position, and the control valve reverts to mid-position. Conversely, a decrease in the force in the top link causes the implement to be lowered.

If the action of the automatic control device is excessive, its effects can be regulated by re-setting the control lever.

Safety Device. If the implement strikes a serious obstruction, such that the control plunger is driven far forward, the lugs on control fork, C, strike the skirt of the lifting cylinder-, and the control fork, pivoting about this point, is rotated in a clockwise direction, thus releasing the oil from the lifting cylinder. The effect of this is that the weight of the implement, and the downward soil forces on it, are suddenly dropped from the tractor drive wheels, and wheel spin occurs, thus saving the tractor and implement from excessive strain.

Slave Cylinders. Fig. 27 shows how the "delayed action" lift on the rear tool-bar of an International "Farmall" is secured. The control lever, C, is shown in the neutral position, when the oil from the pump is simply by-passed direct to the reservoir. To raise the tool-bars the lever is pulled back to D, where it is held by the projection on the underside of the locking-lever, X. The by-pass is then shut off by piston, Y, and the oil pressure begins to build up in the hoses connected to the lifting-cylinders.

The rear lifting-cylinder has a delayed-action valve, W, which is so adjusted that the pressure required to force oil through it is greater than that required in the front cylinders to lift the front tool-bar. When the front tool-bar is fully lifted, the oil pressure immediately begins to rise, and when it has reached a predetermined level the valve W permits oil to start passing into the rear cylinder. With the rear tool-bar in its turn fully lifted, the oil pressure rises still further, and acting on the spring-loaded cap on which the end of the locking-lever rests, it pushes up the end of the locking-lever, thus releasing the control lever and allowing it to slide back to the neutral position. The piston Y thus uncovers the by-pass again, and the pump idles. Meantime, the pressure in the two lifting-cylinders is maintained by the non-return ball valves.

To lower the front tool-bar, the control lever is pushed forward to B, causing the upper ball valve to be pushed off its seating and allowing the oil from the front cylinder to pass back to the reservoir. If now the control lever is pushed farther forward to A, the cam plate, Z, coming into contact with the pump body at the top, pivots forward at the bottom and pushes the rear-cylinder non-return valve off its seating, thus releasing the oil and lowering the rear tool-bar.

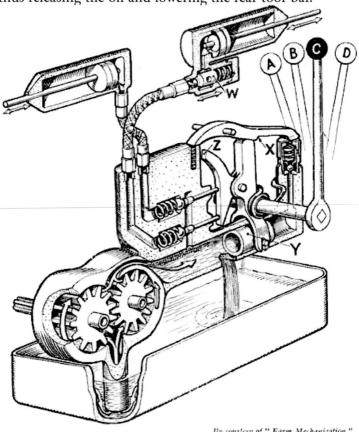

By courtesy of " Farm Mechanization "

FIG. 27.—DIAGRAM SHOWING DETAILS OF THE "FARMALL" LIFT FOR TWO TOOL-BARS. ONLY ONE OF THE TWO FRONT POWER CYLINDERS IS SHOWN.

Points of importance in any hydraulic lift system are as follows:—

A pump of sufficient capacity to lift the required weight at the required speed, with a ram designed to suit.

A high standard of engineering precision throughout, to ensure efficient operation and long life.

Efficient filtration of the oil.

Pump should be under load only when actually lifting. It should be automatically relieved of load when implement is both fully raised and fully lowered.

There must always be a safety relief valve to prevent the development of excessive oil pressures.

Ease and simplicity of control are essential.

FIG. 28.—MEDIUM-POWERED ALL-PURPOSE TRAC-TOR WITH 3-2 FURROW MOUNTED PLOUGH. (FORDSON.) NOTE WHEEL WEIGHTS ON TRACTOR, AND PLOUGH DEPTH CONTROL BY LAND WHEEL.

The Power-Lift Linkage, and Implement Control.

Applications of the hydraulic control system include raising and lowering of the following types of equipment:—

Front tool-bars, light bulldozer and front loaders. A hydraulic push-off device may be incorporated in manure loaders and hay stackers.

Under-belly or mid-mounted tool-bars of various types.

Rear tool-bars.

Mounted ploughs.

Mounted power-driven machines, e.g. potato diggers.

Trailed implements, tipping trailers, etc.

FIG. 29.—THREE-LINK HITCH FOR REAR TOOL-BAR. (FORDSON.)

The following types of linkage may be distinguished:—

(a) Tool-bar rigidly connected to the rear ends of a pair of long draw-links which pass beneath the rear axle and are pivotally connected at their forward ends to a bracket beneath the belly of the tractor. The pivot can usually be raised or lowered to control the pitch of the implement, and the lift is by means of chains or special links which allow the implement to "float". Depth control is partially achieved by land wheels on the implement.

(b) Tool-bar connected to the tractor by a parallel-motion linkage, with depth control by wheels on the implement.

(c) Tool-bar connected to the tractor by a 3-link system (Fig. 29) which gives a "semi-parallel" motion, with a virtual hitch point near the front axle, and depth control by means of implement land wheels. Most tool-bars arc allowed to float freely by using the lift rods in the position where telescopic action is provided.

(d) The Ferguson system, in which a 3-link hitch is employed, and the top link operates a spring-loaded control valve. For a given setting of the hand control lever the hydraulic system maintains a substantially constant draught on the implement. No implement depth-control wheels are used.

A preliminary study of systems *(a)*, *(c)* and *(d)* relative to depth control with ploughs, published by the National Institute of Agricultural Engineering, shows that system *(a)* may result in the most satisfactory work over undulating or uneven land. On the other hand, it is not so attractive in many other ways as systems *(c)* and *(d)*. The relative merits of depth control by hydraulic means or by ground wheels is still a subject of argument.

Method of Attaching Implements to 3-Link System. The attachment of heavy mounted implements having a depth wheel and cross-shaft control to tractors equipped with the

3-link system is an easy one-man operation provided that it is tackled in the correct sequence. This is as follows:

Back up squarely to the implement cross-shaft.

Let the top link settle as nearly as possible above the top-link bracket.

Stop the tractor, dismount, and adjust the length of the top link till the pin enters easily.

Attach the near-side link by adjusting the top link and the cross-shaft control.

Attach the off-side link by adjusting the top link, depth wheel, cross-shaft control and linkage levelling box.

There should be no need to move the implement during the hitching operation.

Tracklaying Tractors.

Tracklayers are designed to secure good adhesion and transmit high drawbar pulls in difficult conditions, where wheels fail to secure an adequate grip on the soil. They are particularly suitable for use on steep hillsides, on heavy land, on fen land where particularly deep cultivations are required, and in all conditions where running wheels over the land may harm the soil structure. For one or more of these reasons tracklayers are preferred for many types of field work, but they suffer from the disadvantages of high first cost and high costs of overhaul when repairs or replacements are needed. A further disadvantage compared with a rubber-tyred tractor is unsuitability for running on metalled roads. The chief users of large tracklayers (machines of 40 B.H.P. or more) are contractors who undertake heavy cultivations such as deep ploughing, subsoiling and mole draining. The larger sizes are also frequently used for bulldozing, etc. Such powerful machines arc now almost invariably fitted with diesel engines and are extremely expensive. They naturally need heavy equipment to suit their power, and this also can be very costly.

An early tracklayer was built in California in 1904. Since that time, steady improvements in design and in the materials of construction have occurred and the modern tracklayer is a very efficient type of tractor with respect to tractive power. The track may be looked upon as a continuous rail, made up of links and supported by the cleats. The tractor driving sprockets gear with the links of this rail, and the tractor rolls along it on the track rollers. The track is picked up after the tractor has passed over it and is relaid continuously. It provides a large area of contact with the ground and is eminently successful where adhesion is difficult or rolling resistance is high. A certain amount of power is, however, wasted in internal track friction. The weight of the tractor is carried on the track rollers and not on the driving sprocket, and the weight of the track itself is also partly removed from the driving sprocket by means of the upper track rollers.

The track itself generally consists of a series of links and ground-plates, coupled together by pin joints. It is flexible in both directions, so that obstacles are easily negotiated. Practically all heavy tracklayers are fitted with tracks of this type. The chief defect from an agricultural viewpoint is the wear that occurs at the pin joints, and the high initial cost.

A second type of track in fairly common use in Britain is the rubber-jointed type, where blocks of rubber are substituted for the pin joints. There is no pin movement with these tracks, internal deflection of the rubber blocks permitting the necessary flexibility. These tracks are not, however, flexible in both directions in the same way as the pin-jointed type. Indeed, they are built to a curve, and a considerable force is required to flatten the track. An advantage of this is that over rough ground the track rollers have a much smoother rail to run on than they do with the pin-jointed type. The disadvantage of these tracks is that rubber is not as strong as steel, and the tracks have to be made heavy. They are very satisfactory for small tractors (e.g. the Ransomes M.G. Cultivator, the Bristol, etc.), but are not a sound proposition for powerful machines of more than 6-7 tons or 50-60 b.h.p.

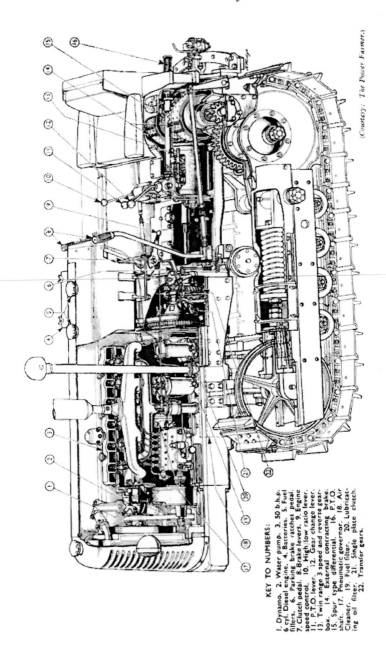

Courtesy: The Power Farmer.

KEY TO NUMBERS:

1. Dynamo. 2. Water pump. 3. 50 b.h.p. 6 cyl. Diesel engine. 4. Batteries. 5. Fuel filters. 6. Parking brake ratchet pedal. 7. Clutch pedal. 8. Brake levers. 9. Engine speed control. 10. High/low ratio lever. 11. P.T.O. lever. 12. Gear change lever. 13. Twin range 3-speed and reverse gear-box. 14. External contracting brake. 15. Spur type differential. 16. P.T.O. shaft. 17. Pneumatic governor. 18. Air Cleaner. 19. Fuel filter. 20. Lubricating oil filter. 21. Single plate clutch. 22. Transfer gears.

FIG. 30.—SECTIONAL DRAWING OF MEDIUM-POW-ERED TRACKLAYER. (DAVID BROWN TRACKMAS-TER.)

Experiments have been carried out over a long period with rubber tracks which have a steel-wire reinforcement. The advantages of this type are that adhesion is almost equal to that of a steel track in most agricultural conditions, and that the tractor can run along good roads without damage to either the road or the tractor. Disadvantages are a high manufacturing cost, and the fact that a scrubbing effect caused by turning in heavy-duty conditions causes excessive wear of the tread. Such tracks are seen to advantage when a very wide track is needed for traversing soft ground. For example, the Cuthbertson full-track "Water Buffalo", or half-track outfits equipped with tracks 24 in. wide or more, will pull a drainage plough across bogs which are completely impassable to ordinary tractors or vehicles. In the Cuthbertson track, wire-reinforced rubber pads are squeezed between cleat plates on the outside and driving plates on the inside. The cleat plates may be of steel, rubber, or rubber bonded to a steel base.

Fig. 31 shows the method commonly employed for transmitting the power to and steering the tracks of a tracklayer. The tractor is turned by cutting off the power from one track by means of the clutches C and CI, which are operated by the steering levers, S and S-I. For example, in turning to the right, the operator pulls on the right-hand steering lever, thus disengaging the power from that track. The extent of the turn is determined by the length of time the steering clutch lever is held back. Sharp turns may be made by use of the individual brakes, B and B-I, but these need only be used when making short turns, when running light, or when preventing the tractor from rolling down a steep hill.

The steering of some machines is achieved by the use of brakes and a differential.

Some manufacturers employ "controlled differential" steering, operation of which may be illustrated by reference to Fig. 32. When the tractor is going straight ahead, the whole differential assembly driven by the bevel gear No. 7, rotates as a unit without any internal motion. To make a turn to the right the right control lever is pulled. This holds stationary the right steering drum No. 10, and the drum gear No. 6, which are fastened together and which rotate on a bronze bushing on their shaft. As a consequence, pinion gear No. 4 and differential gear No. 2, connected to bevel gear No. 7, by means of shafts, begin to turn forward. Final drive gear No. 8 is slowed down because No. 2 gear is turning and rotating in the same direction.

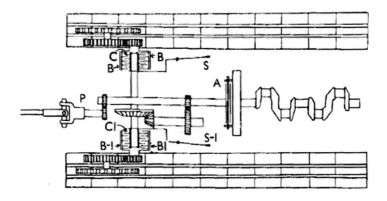

FIG. 31.—STEERING CLUTCHES AND BRAKES OF A TRACKLAYER (CATERPILLAR)

At the same time, differential gear No. 1, rotating around the final drive gear No. 9 in the opposite direction, is speeded up. This results in a positive drive to both tracks while turning. The disadvantage is that so long as the steering drum is held the urning circle is a fixed one, and to make a gradual turn the tractor has to do it in stages, going straight ahead for a time and then turning rather sharply. More perfect steering mechanisms have been developed for use in military tracklayers, but have not yet been adopted for farm tractors.

A slight disadvantage of tracklayers sometimes experienced in deep ploughing is that considerable side-draught may be unavoidable because it is necessary to run both tracks on the unploughed ground; but against this must be set the advantage of no compression of the furrow bottom. On wet, heavy land, tracklayers, on account of their low ground-pressure, can often be used when other types poach the soil so much that they cannot be profitably employed.

FIG. 32.—DIFFERENTIAL GEAR FOR TRACKLAYER. (CLETRAC.)

Half-Track Equipment.

The limitations of rubber-tyred wheels have caused a steady increase of interest in "halftrack" assemblies which may be fitted in place of the rear wheels of general-purpose tractors. The object is to provide an outfit which is capable of overcoming difficult conditions and will provide a substantial drawbar pull in circumstances where any type of wheel would be unsatisfactory.

One successful half-track unit (Fig. 9) operates on a girder-track principle. Like the rubber-jointed track, this type is flexible in only one direction. There is no relative motion of pins and bushes at the point where maximum stress occurs, and this appears to result in reduced wear compared with the fully flexible type of track. This type of girder track gives the effect of a big wheel (Fig. 33).

As with full-track machines, half-track, conversions should never be run with one track in the furrow when ploughing. On most types of soil such a practice will result in a track so treated having only half the life of one which is run on the land. When half-tracks are used for ploughing, suitable types of ploughs such as are designed for use with tracklayers should be chosen, and the swinging drawbar, which must pivot well forward, should be allowed to swing within reasonable limits.

Agricultural tracklayers and half-tracks are not designed for high-speed work, i.e. speeds of above 4 m.p.h. or so. One of the advantages of the half-track is that when such high speeds are required it is possible to change back to pneumatic tyres. Many-farmers find it convenient to fit half-tracks when starting autumn ploughing and to remove them when the heavy spring work is completed.

*FIG. 33.—GIRDER TRACK DISCONNECTED, SHOW-
ING THE "BIG-WHEEL" PRINCIPLE. (ROADLESS
TRACTION.)*

The girder type of half-track is available either with a standard track or with a skeleton track. The latter is intended, like skeleton wheels, for use on sticky clay soils or for work on seedbeds where the standard track causes excessive compaction of the soil surface. On sandy soils and light loams the skeleton track will dig in at high drawbar pulls, and the standard track is preferable for such conditions. In general, the skeleton track is preferable in conditions where skeleton type wheels give a better performance than standard types of steel wheels fitted with spade lugs.

It should be added that there are other types of half-track units in the development stage, at least one of which shows exceptional promise. It seems more than likely that more will be seen of half-track outfits of a roughly triangular shape, with the driving sprocket at the apex of the triangle and a simple bogie assembly carrying the weight of the tractor at the base.

Now that "all-purpose" tractors are regularly equipped with good steering brakes and a wide range of gearbox and engine speeds, the day of the half-track is due. It would not be surprising if simple half-track devices ultimately supersede full-track equipment for agricultural tractors up to about 40 b.h.p. Requirements of a fully successful half-track unit may be stated as follows:—

FIG. 34.—HALF-TRACK WITH WIDE, CLEAT PLATES FOR USE ON BOGGY LAND. (CUTHBERTSON.)

(i) The unit must be quickly and easily fitted or removed, so that changing from pneumatic tyres to half-track or the reverse may be undertaken as required.

(ii) The tractor must be as easily manœuvrable as the conventional full-track machine.

(iii) The outfit must be capable of operating efficiently on a wide range of surfaces and gradients.

(iv) The track must not pack the soil more than existing medium-powered full-track machines.

(v) Some adjustment of track width is desirable.

(vi) The track must be capable of transmitting fairly heavy drawbar pulls, but not necessarily to the point of stalling the engine at maximum rear axle torque.

It seems likely that all these requirements will be filled by outfits which are reasonable in both first cost and maintenance in comparison with conventional full-track machines.

Market-Garden Tractors.

There are several quite distinct types of tractors designed for use in market gardens and for horticultural crops on general farms, and in order to avoid confusion it is best to distinguish the main groups at once.

The smallest may best be described as motor hoes. They are two-wheeled or single-wheeled tractors with engines of up to about 3 h.p., designed for hoeing, grass-cutting, spraying, etc. Some include a plough among their equipment, but in most instances these very small tractors are not really intended for serious ploughing. These machines are extremely versatile and useful, and the smaller ones find steadily increasing use in large private gardens, as well as on commercial holdings.

The single-wheel type when equipped with hoes, cultivating tines, mowing cylinder or cutter bar is not difficult to balance, and has the advantage of needing no wheel adjustment when rows of differing widths are cultivated. When used for work like spraying or light transport, a 2-whceled bogie may be attached. These small machines are easy to operate, and are powerful enough to cultivate, spray, mow and trim hedges quite effectively.

The next important group comprises the 2-wheeled tractors with larger engines, of 3 to 8 hp., which can do quite satisfactory single-furrow ploughing as well as many other types of heavy cultivation including disc harrowing and rolling. Such tractors will also do multiple-row drilling and will pull a fairly heavy load on a trailer.

Out of these more powerful 2-wheelers have been developed a number of very small 3- or 4-wheelcd tractors, some of which are much like a conventional 4-wheeled machine, while others have the engine and transmission compactly arranged over the rear driving axles and an open frame which is supported at the front by a single wheel or a pair of wheels, and carries the tools. In some cases the tractor becomes in effect a carrier for all types of mounted tools including hoes, drills and spraying equipment, and is frequently called a self-propelled tool chassis. With these machines much attention has been paid to points of design which are important in securing really accurate work. Tractors of this type are now used on many of the larger market gardens, but are also being increasingly used as specialized drilling and hoeing units on general farms, especially where a big acreage of sugar beet is grown. One of the most widely used tractors of this general type is not designed for ploughing, but several manufacturers are aiming to develop machines which will operate a forward-mounted plough and will do deep cultivations, as well as inter-cultivations.

A type of market-garden tractor which is quite distinct is the baby tracklayer. One of these, the Ransomes M.G., has a single-cylinder 6 h.p. engine, and a maximum drawbar pull in normal working conditions of about 600-800 lb. It is particularly suitable for some kinds of light orchard work, being capable of working beneath very low trees.

This little tractor can do light ploughing with its own single-furrow plough, disc harrowing, cultivating, seeding and inter-row hoeing.

Another tracklayer (the Bristol), though considerably more powerful, is also of interest to many horticulturists on account of its small overall dimensions. This tractor has a 4-cylinder petrol/paraffin engine and is able to pull a 2-furrow plough in average working conditions.

*FIG. 35.—LIGHT 2-WHEELED TRACTOR EQUIPPED
WITH CULTIVATING TOOL-BAR. (BARFORD.)*

FIG. 36.—DRILLING 6 ROWS OF VEGETABLE SEED
WITH SELF-PROPELLED TOOL CHASSIS. (HUMBER-
SIDE ENG. CO.)

Another important type of tractor now widely used on
market gardens is the rotary cultivator or rotary hoe. Some
of the smallest sizes are multi-purpose tractors which include
a rotary hoe attachment along with a range of equipment
suitable for use in large private gardens. This small size, as a
rotary cultivator, is often considered too small for commercial
holdings. The most important size on market gardens is
the 2-wheeled special-purpose machine of about 6-8 h.p.
(Fig. 86). Such rotary cultivators arc capable of working to
a depth of about 9 inches, and on some holdings they are
invariably used for all seed-bed and plant-bed preparation,
taking the place of the plough and the whole range of

cultivation implements. Rotary cultivation is discussed in Chapter Seven.

FIG. 37.—SMALL TRACKLAYER WITH RUBBER-JOINTED TRACKS (BRISTOL.)

Walking Tractors: Constructional Features. Some of the points to study in choosing and using walking tractors are as follows:

For general work, the track width should be adjustable to permit working between rows, and there should be ample vertical clearance for straddling the rows of growing crops. Strong, well-designed tools are necessary, and lifting of the tools and control of the depth of working should be easy.

Methods of steering 2-wheeled tractors vary greatly. They include use of a differential gear and independent wheel brakes; or dog clutches and overrunning ratchets. The necessity for power steering naturally depends on the size of the tractor. Some of the very light fractional h.p. tractors which are used only for the lightest hoeing and weeding tasks are easily steered with nothing but pawl and ratchet gears; but the 3-8 h.p. machines used for ploughing and heavy cultivations require something better.

Gearboxes differ considerably, some having three forward gears and a reverse, others two forward and reverse, and others only one forward speed. Some makes employ an automatic centrifugal clutch which engages when the throttle is opened.

FIG. 38.—TWO-WHEELED TRACTOR AND PLOUGH.
(TRUSTY.)

The usefulness of these walking tractors generally depends to a great extent on the range of equipment available. Most makes have the implements designed to form an integral part of the machine, on the "unit" principle. Attachment of the various implements should be a simple and quick job, and when the implements are attached and the tractor is at work the outfit should be well balanced, so that the wheels have maximum adhesion and the tools can easily be lifted out of work.

Ease of steering is one of the chief problems in the use of walking tractors for hoeing between rows of plants. When a two-wheeled vehicle turns, it pivots about a point in line with the axle, and if hoes are rigidly attached behind the machine, they first move in the direction opposite to that in which the tractor is steered. If the hoes are carried at the front, it is not easy for the operator to see them and, moreover, the torque reaction of the engine tends to lift them out of the ground. One satisfactory method of attaching hoes is by means of a pivotal attachment, the pivot being placed ahead of the main axle and the hoes running at the rear. A difficulty in this is that the hoes may swing sideways owing to a difference of resistance at the two sides, but this may be controlled by exerting pressure on the hoe frame through a bar connected to the main steering handles.

*FIG. 39.—SMALL RIDING TRACTOR WITH MID-
MOUNTED TOOLS. (GARNER.)*

For ordinary farm work, when cost and performance are considered, the two-wheeled tractor compares unfavourably with the small four-wheeled models now obtainable. Their use in market gardens may increase, but it is unlikely that the gap between the single horse and the "two-plough" tractor on ordinary farms will be filled by two-wheeled or front-wheel drive machines.

Much thought has been given by horticulturists to the development of a tractor which will do everything from deep ploughing and subsoiling to the finest inter-row hoeing. While the development of such a tractor is not impossible, it is too much to hope that such a universal machine could be produced at a price which would suit a grower with a

5-acre holding. The best compromise for many such growers seems to be to accept the fact that once in a while it will be necessary to engage a contractor for deep cultivations. For the rest of the work a small 4-wheeled riding tractor with rear-mounted engine and a range of forward-mounted tools, including a reversible single-furrow plough, seems likely to meet the needs of many growers.

CHAPTER THREE

TILLAGE

"By what means Ploughs and Tillage itself came. at first to be invented, is uncertain; therefore we arc at liberty to guess."

Hone-Hoeing Husbandry. Jethro Tull, 1733.

Tillage may be described as the practice of modifying the state of the soil in order to provide conditions favourable to crop growth. It represents the most costly single item in the budget of an arable farmer and is a part of the business of farming which remains almost entirely an art. The main objects of tillage are: (1) the production of a suitable "tilth" or soil structure; (2) the control of soil moisture; (3) the destruction of weeds; (4) the burying or clearing of rubbish and the incorporation of fertilizers with the soil; and (5) the destruction or control of pests.

Soil Structure.

Agricultural soils consist mainly of a heterogeneous collection of mineral particles existing either singly or as small "crumbs" comprising several particles grouped together. Between the soil particles are spaces which may be filled by air or by water. When soils are left uncultivated for long periods, the soil particles become stuck together, the air spaces become small, the mass of soil becomes consolidated, and the friable structure of the soil is lost. Soils in such a condition do not provide a suitable environment for agricultural crops, and one of the chief uses of cultivation implements is to restore to such soils a "tilth" favourable to crop development.

The tilth required varies according to the crop to be grown, but an invariable requirement is the production of a certain amount of mould with a granular structure, which can easily be penetrated by plant roots and allows air and moisture to move easily within it. A good state of tilth cannot, however, generally be produced by implements alone; it depends for its formation on the natural weathering agencies. Weathering is especially important in the preparation of seed beds on heavy land. Ploughing leaves the soil in such a condition that alternations of weather such as frosts, rains and drying winds cause it to crumble into smaller pieces. The land becomes "mellow", and the various cultivation implements must be used in such a way as to assist this natural process as much as possible.

Certain soils containing a high proportion of coarse sand and little clay can frequently be easily worked to a tilth by the use of implements alone, without waiting for the action of the weather. On this type of land, use may be made of such machines as rotary cultivators and of trains of implements performing a number of different operations. Such soils may require the use of implements to consolidate the seed-bed and produce the firmness necessary for intimate contact between the plants and the soil.

The Control of Soil Moisture, Temperature and Aeration.

Tillage is concerned in many ways with the adjustment of the soil moisture content to the needs of the crop. Heavy soils are frequently deeply cultivated and left rough and open at the surface in order that rain may easily penetrate and quickly pass down into the drains. On the other hand, it may be desirable at seeding time to produce at the surface a finely divided crumb structure which will keep any moisture deposit in the upper layers. At other times, in preparing an early seed-bed on a wetter soil, it may be necessary to produce a rough, broken surface by use of a cultivator, in order to induce a partial drying of the surface layers. In the drought areas of America, one of the most successful farming practices recently developed is to leave the surface of the fallow land in the form of large basins, so that rain is collected instead of being allowed to run off the surface. Consolidation may sometimes have important effects on the movement of water through the capillary spaces in soils, but the effects of this have been shown to be less important than was at one time supposed. Many other instances of the control of soil moisture by tillage could be mentioned, and it should be sufficiently clear that important changes in the soil moisture may be produced or made possible whenever cultivation implements arc used.

Tillage and weathering modify the air spaces in the soil, as well as the sizes of the soil particles. Aeration, like many other results of tillage operations, is imperfectly understood. Oxygen is certainly required in the soil for the germination of seeds, the respiration of plant root systems and the activity of certain beneficial bacteria. It is also certain that it is good practice to leave land in a fairly "open" condition over winter, so that rain and air can percolate easily. In these circumstances, the soil becomes "sweet" and "mellow"; but it is quite impossible, in the present state of knowledge concerning such matters, to estimate how much of the improvement is due to aeration.

Soil temperature is closely bound up with such variables as moisture content and compactness. For example, soils which are wet in spring are cold because of the high specific heat of water compared with that of soil, the result being that the effects of a warm atmosphere are less rapidly felt on the wet soils because a great amount of heat is required to warm the water in them. Generally speaking, improvements in soil temperature conditions are brought about incidentally in the performance of tillage operations where the main objective is to modify texture and moisture content.

The Destruction of Weeds.

The destruction of weeds is one of the major objects of cultivations, and the methods of attack vary widely according to the type and condition of both the soil and the weeds. Annual weeds may generally be killed by cutting just below ground-level, by burying them completely, or by dragging them out and leaving them exposed on the surface of a dry soil. They are most easily destroyed in the seedling stage. Tillage has the further effect of making weed seeds germinate, and the seedlings may then be destroyed by subsequent cultivations.

One of the chief objects of stubble cleaning is to germinate weed seeds so that they may be killed by subsequent cultivations. This technique works well with chickweed, speedwells, cleavers, wild oats, black grass, poppy and charlock, to mention a few; but there are some weeds such as corn buttercup, fat hen and knot grass which often do not germinate until late in the year.

Perennial weeds are generally more difficult to eradicate than annuals; for many, such as docks, are not killed by burying or by pulling them out and leaving them on the surface for long periods. Many also, such as couch, thrive when cut into pieces by implements, the division into many parts serving merely to assist in their vegetative propagation. Special methods are therefore required for the eradication of perennial weeds, and these methods vary according to the particular species encountered. Bare and bastard fallows and autumn and spring cleaning operations may be mentioned in this connection. For details of these processes, reference may be made to many books dealing with agriculture. All that need be said here is that every farmer on heavy land should understand exactly what is meant by killing weeds "in the clod" and should know how to set about doing it.

Inter-cultivations between rows of growing crops are generally aimed at the destruction of weeds, and experiments both in this country and abroad show that keeping the soil free from weeds is the important factor, and that stirring the soil is relatively unimportant. Indeed, careful experiments have shown that with crops such as sugar beet, additional hoeings beyond the amount necessary to control weeds may reduce yields rather than increase them.

The Burying of Rubbish and Mixing of Fertilizers.

Crop remains and farmyard manure must be buried so that the manurial constituents contained in them may be made available to the crop plants by processes of humification and decay, and so that their presence on the surface may not interfere with cultivations. The plough, with its inverting action, is generally used for this operation, and one of the chief requirements of good ploughing in British conditions is that all rubbish should be buried. The necessity of burying all rubbish does not, however, apply to ploughing throughout the world, for there are areas in the United States, the Canadian Prairie Provinces and in other countries, where serious wind erosion may occur if all rubbish is buried. In such circumstances, one of the chief requirements in ploughing may be to leave as much rubbish as possible protruding from the soil surface.

Cultivators or harrows may be used for mixing with the soil artificial manures which have been broadcast on the surface. The optimum position of fertilizers with relation to the various crop plants has not yet been fully determined, and further reference is made to this question in Chapter Nine. Preliminary trials with a few crops have shown that fertilizers often do not give their maximum effect if distributed throughout the surface layer of soil.

The Control of Insect Pests.

Tillage may be used to combat the action of insect pests by direct and indirect methods. Direct effects include exposing the pests to the attacks of birds in ploughing and cultivating, and the limitation of the movement of certain pests in the soil by consolidating the surface. The action of pests may also be minimized by using tillage implements in such a way that the plants can grow away from the attacks without suffering serious damage. The fact that winter wheat sometimes resists wireworm or wheat bulb fly better where the seed bed is firm than where it is "fluffy" is probably due more to this indirect effect on the plants than to any direct effect of firmness on the activity of wireworms or the larvae of wheat bulb flies.

Changing Views on Tillage.

In recent years, many farmers have departed from the traditional methods of tillage. Modern engineering developments have provided them with new machines which have made it possible to telescope many of the tillage operations. For example, some farmers now plough, press the furrows, sow fertilizer, drill and cover wheat in one operation, with good results. Such developments make it necessary to enquire what are the fundamental requirements of tillages, and what are the minimum cultivations required to produce suitable seed beds for the various crops. It is possible that new developments in farm machinery will make it necessary to modify the generally accepted views briefly set out above concerning the essentials of tillages; for certain points, not mentioned above, which were considered fundamental until a few years ago, have been shown to have little significance in practice.

The wartime ploughing campaign did much to emphasize the importance of good cultivation, especially in the preparation of seed beds after old turf. Against a few experiments which suggest that cultivations may sometimes be overdone may be set scores of examples from practical farming where a little extra cultivation makes the difference between success and failure. There are some crops, such as wheat, that are relatively indifferent to cultivations, providing that a certain minimum standard is reached; but there arc others, such as barley, where a higher standard of cultivation is necessary for success.

No definite rules can be laid down as to what implements are necessary to produce the best results in different circumstances. The plough remains and is likely to remain the basic cultivation implement in Britain. Beyond this, all that can be said is that there are many types of cultivators, harrows and rolls, and that securing the best results depends on the farmer's ability to choose the right implement at the right time. It may be added that experience with such methods as sub-surface tillage and basin listing in the dry-farming regions of North America, though extremely interesting, has no bearing on the cultivation of British farm land.

Influence of Mechanization.

Whatever the object of tillage, mechanization makes it easier to accomplish. Adequate tractor power and suitable implements make it possible to undertake operations that would be out of the question without them. For example, so much advantage can be taken of short favourable periods in autumn or spring that a cleaning effect comparable with that of a bastard fallow may often be obtained in less than a month; heavy land can be burst up when it is dry, and need never be worked when too wet, with the result that soil structure becomes greatly improved; spring cultivations may be done so rapidly that moisture losses are negligible; and seed beds may be made just as fine or firm as necessary, with a minimum expenditure of time and trouble.

The scientific study of the soil in relation to cultivations, now being actively pursued in many countries, may in time produce data leading to modifications of implement design calculated to increase the efficiency of tillage.

One interesting aspect of tillage implement design is the way in which old ideas and methods crop up in new circumstances. The blade weeder, recently hailed as the salvation of dry-land farming in some parts of the North American prairies, is exactly like the broadshare or thistle blade known in Britain for generations; the Dutch harrow, recently introduced to British sugar beet growers, looks uncommonly like the type of harrow that was generally used in Britain over a century ago; and even some of the recently introduced power driven diggers and rotary cultivators are fundamentally the same as machines first tried out soon after steam engines were invented.

CHAPTER FOUR

DEVELOPMENT, DESIGN AND CONSTRUCTION OF THE PLOUGH

"The Plough is certainly the most valuable and the most extensively employed of all agricultural implements."

The Implements of Agriculture. J. A. Ransome, 1843.

Ploughing is the basic tillage operation. Its essential feature is that a layer of soil is separated from the underlying subsoil and is inverted, so that any vegetation or manure present on the surface is buried, and a layer of soil from below is brought to the surface, where it is exposed to the action of weathering agents and of other implements. The ploughed land is laid up in furrow slices, the type of slice depending upon the type of plough used and the nature of the soil. Ploughing is the most important operation of the arable farmer, not only because of the basic nature of the work but also from the standpoint of the power required.

Development and Design of the Plough.

The modern plough is the result of experience which has accumulated slowly over thousands of years, and it is of interest roughly to trace its development during historical times. There is little doubt that its most primitive prototype antedates history and was simply a pointed wooden stick. A very old illustration on a monument in Asia Minor shows animals being used to pull what is apparently the natural fork of a tree with some strengthening braces added, and it may be mentioned that similar implements are still used in the East. Several references to ploughs may be found in the Old Testament, and in 287 B.C. Theophrastus wrote: "The earth may seem cold, but if it is inverted, it becomes free, light and clear of weeds, so it can most easily afford nourishment." It is not, however, by any means certain that the ploughs in use at that time achieved this desirable inversion of the soil. Virgil, in the Georgics, described a plough which possessed a beam, a handle, and a body formed of two pieces of wood which met at an angle at the front and were shod with an iron share; but no serious attempt to invert the soil with this plough was apparent.

The development of the English plough in its early days is wrapped in mystery. A considerable amount of information concerning the Roman and Saxon ploughs is available, but there is no trace of the agricultural implements used by the Ancient Britons, either before or during the Roman occupation. The evidence available suggests that our present-day implement developed from the Saxon plough, and was not much influenced either by the Roman occupation or by anything that existed in Britain before the Romans came.13 The Saxon appears to have brought over his own plough, and information concerning this implement is derived from various manuscripts, tapestries and psalters extending from the eighth to the fourteenth centuries. The ploughs illustrated and described usually consisted of a beam carrying a vertical coulter to cut the side wall of the furrow; behind this was a "share beam" carrying the share, which cut the bottom of the furrow and raised the furrow slice from the subsoil. (The share beam was a long piece of wood, sliding on the ground and braced at its fore end to the beam by means of a part called the "sheath".) There was also a breast or mould-board* which turned the furrow slice aside and more or less inverted it. There were either one or two handles for guiding the plough, and generally a pair of large wheels supporting the front of the straight beam, though an illustration in the Louterell Psalter (fourteenth century) shows a plough with a curved beam and no wheels.

Progress over the period after the Norman Conquest and during the Dark and Middle Ages was slow and uneven. Fitzherbert's *Boke of Husbandry*, which appeared in 1523, and Tusser's *A Hundred Good Pointes of Husbandrie* (1557) show the ploughs of the sixteenth century to have been heavy and cumbrous. The share and coulter were more generally made of iron, and Fitzherbert was familiar with a complicated device for varying the depth and width of furrows.

Walter Blith, in *The English Improver Improved* (1653), described the "Hertfordshire" plough, which was a direct descendant of the primitive types mentioned above. This plough was in general use in the Midlands, and developed into the type described by Jethro Tull in his *Horse-Hoeing Husbandry* (1733) and by Thomas Hale in the *Compleat Body of Husbandry* (1756). The Hertfordshire plough was, at best, a clumsy implement, of heavy draught. Moreover, the design of the frame was unsatisfactory, the method of attaching the share beam to the beam being mechanically unsound. Blith was alive to the deficiencies of the Hertfordshire plough, and he also described an improved plough which he invented. It appears that Blith's plough was largely based on a type at that time being used in Norfolk, and that the design of the latter was influenced by a type which was in common use in the Netherlands. It was a light swing plough with a curved beam, and this came into fairly common use in East Anglia during the latter half of the seventeenth century.

Progress was by no means general throughout the country, for while 2-horse ploughing was the rule in East Anglia, clumsy implements existed along with the improved in other parts. In the "Large Letter" of R. Child to Samuel

Hartlib, the former says: "Some Kentish farmers used four, six, yea twelve horses or oxen on their ploughs"—and in Ireland, farmers still yoked horses by their tails!

The beginning of a new era in plough design was marked by the patenting in 1730 of the "Rotherham" plough. It was a small light swing plough, with the beam, handles and frame made of wood, and an iron hake, coulter and share. The curved wooden mouldboard usually had an iron plate attached to it. The method of attachment of the share and mouldboard showed an improvement over the arrangement employed in Blith's plough. Arbuthnot, a Norfolk farmer, further improved the Rotherham plough, and he wrote a remarkable paper on plough design that was included by Arthur Young in his *Eastern Tour*, published in 1771. James Small was another farmer who made improvements on the Rotherham plough, and his book, *The Plough*, published in 1784, places on record some of the important contributions that he made to plough design.

The close of the eighteenth and the dawn of the nineteenth century was a period of rapid development, during which the iron plough began to replace the wooden one, and the plough assumed a general shape differing little from that of the modern horse implement. Cast iron shares had been used since about 1785, and when, in 1800, William Plenty patented a metal frame to which other parts of the plough were bolted, and in 1803 Robert Ransome patented the method of making self-sharpening chilled cast iron shares, the way was prepared for those improvements in the details of design which have led up to the modern plough.

Ploughing matches, begun towards the end of the eighteenth century, assisted in the rapid development which then took place. During the early part of the nineteenth century, the differing requirements of autumn and spring ploughing brought about the development of distinct types of ploughs for these two processes. For the autumn work, a long-breasted plough with a gentle twist designed to set up the furrow slices was used, and this has developed into our modern lea plough. For spring work, a plough with a short, sharply curved breast was used. The action of this implement was to pulverize the soil, and its development has given us the modern digging type of plough body.

By the middle of the nineteenth century, a fairly complete grasp of the principles of plough construction as we understand them to-day was possessed in some quarters. These principles were set out by J. E. Ransome in 1865 in *Ploughs and Ploughing,* a paper which gives a clear account of the principle of designing breasts to conform to the manner in which a furrow slice naturally turns. The optimum shape of the furrow slice was also discussed, and this question was carried further by W. R. Bousfield in a paper read before the Institution of Mechanical Engineers in 1880. Bousfield showed that the fundamental form of mouldboard could, in theory, be developed from first principles, and then discussed how the fundamental form was modified in practice. In more recent times, E. A. White 14 has proceeded in the reverse direction, taking standard ploughs and expressing their shapes mathematically. He has shown that the surfaces of most plough breasts contain two sets of straight lines, and has developed equations expressing the relation between various points on the surfaces. Intensive research on plough design in the United States by M. L. Nichols and his colleagues15 has thrown some light on the complex relation between the plough and the soil. Careful studies have shown that Bousfield's simple geometrical explanation of the action of the breast is incomplete, and that the problems of plough design cannot be separated from studies of the soil reactions.

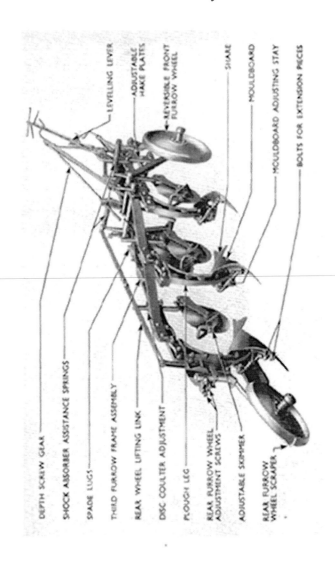

FIG. 40.—COMPONENTS OF A 2-3 FURROW
TRAILED TRACTOR PLOUGH FITTED WITH LEA
BODIES (FORDSON.)

An easily applicable scientific method of designing plough bodies would pave the way to a measure of standardization and the elimination of superfluous types. Wartime experience showed that far greater uniformity of design could be achieved with advantage. Though attempts to raise plough breast design to a scientific basis have achieved some slight success, it is an unquestionable fact that recent researches of this type have had practically no influence on the construction of modern ploughs. Ploughs of to-day are constructed largely according to such practical and empirical principles as were set out by J. E. Ransome in 1865.

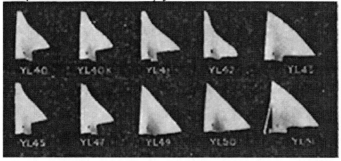

FIG. 41.—A FEW OF THE SHARE TYPES AVAILABLE FOR ONE MANUFACTURER'S GENERAL-PURPOSE PLOUGH BODIES, (RANSOMES.)

Plough Components.

The *beam* is the part to which the power is applied, and the other parts may be considered as attached, directly or indirectly, to it. The essential parts that actually do the. work on the soil are, with the exception of the coulter, carried on *the frame,* " leg" or "frog" which is bolted to the beam. A few tractor ploughs have straight beams (Fig. 40) and cast steel legs somewhat similar in type to those on most English horse ploughs. Most tractor ploughs, however, have curved beams and small frames consisting of simple steel stampings (Figs. 44 and 46).

The *share,* which makes the horizontal cut separating the furrow slice from the soil below, fits on to the front or "nose" of the frame. Plough shares, like breasts, are of very varied shapes. The share of the lea plough has a definite "neck" that has the effect of placing the cutting edge well in advance of the breast. This "lead" at the cutting edge assists in keeping the furrow slice unbroken, enabling it to rise gently up the breast without any pulverizing action. The cutting edge of the digging plough share is close to the forward part of the breast and often almost at right angles to the line of draught. Points and wings are usually separately renewable (Fig. 42).

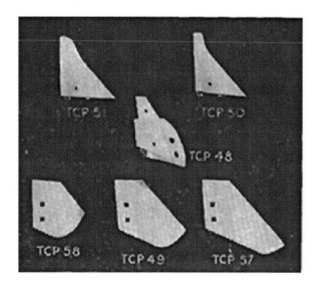

FIG. 42.—PARTS OF 3-PIECE SHARES FOR DIGGER PLOUGH. (RANSOMES.)

The vast majority of shares used in Britain are made of chilled cast iron, but malleable iron, cast alloy steel and forged steel are also used. The principle of the manufacture of chilled cast iron shares was discovered more or less by accident by Robert Ransome in 1803, and the importance of this discovery to agricultural progress cannot be overemphasized. Chilling of the under side and land side of the share makes these parts wear more slowly than the upper part and renders the share self-sharpening. When chilled cast iron shares become blunt they may be improved by grinding the upper edge. Many attempts have been made to manufacture self-sharpening steel shares, but complete success in this direction has not yet been achieved. When a steel share becomes blunt, it may be sharpened by a competent smith by beating out and re-hardening. Steel shares may be cast in one piece or made of two or three pieces of metal welded together. Soft-centre steel is employed by a few manufacturers. Whether the share be of crucible steel or soft-centre steel the method of sharpening is the same. The edge of the share is heated in a forge to a cherry red colour, and the upper side is hammered out, the lower side being laid flat on the anvil. Only a small piece should be heated at a time, work being started at the point and finished at the wing. Care must be taken to avoid spoiling the "suction".

Steel shares can be treated to give very long wear by using a welding outfit to apply a layer of stellite, a very hard compound, along the bottom side of the cutting edge. Steel shares are not generally used in Britain except in conditions where chilled cast iron will not withstand the shocks encountered, or where the soil contains little abrasive material and the shares will remain sharp for a long time. Most horse ploughs and small tractor ploughs use chilled cast iron shares; but steel shares are generally used on deep-digging tractor ploughs. This position may be contrasted with that in Germany, where steel shares are the rule, although the soils are generally sandy.

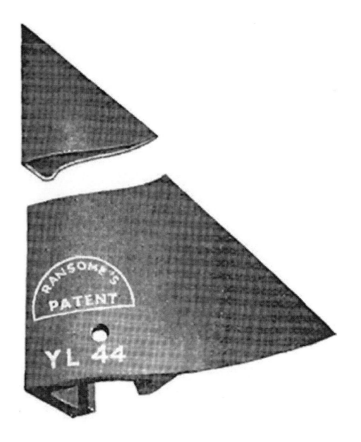

FIG. 43.—CHILLED CAST IRON PLOUGH SHARE,
(RANSOMES.)

On rocky or hard, stony land, the "bar point" share may be used. This consists of a steel bar which passes through the frame and forms the point of the share. When the point has become blunt, the bar may be turned over or reversed end for end. Under extreme conditions, use of the bar point share results in considerable economy owing to its comparatively long life.

The *breast* or mouldboard is the part that inverts the furrow slice in the case of the lea plough and pulverizes it in the digging type. It is bolted to the frame, the countersunk bolt heads fitting flush with the face of the breast. The lea plough breast is long, and turns the furrow slice very slowly, firmly pressing it into place against the previous one. It is convex from top to bottom throughout its length. In the digging body, the action of the breast is very different, for the furrow slice is pulverized by the use of a short breast which is concave both from top to bottom and from shin to tail. The furrow slice rises rapidly up the breast and is bent back upon itself and finally thrown aside by the abrupt curvature. Digging ploughs are used on suitable land when ploughing in spring for spring-sown crops, so that a seed bed may be obtained in as few operations as possible. Many farmers, however, use digging ploughs for deep autumn ploughing, partly because only digging ploughs can do deep work satisfactorily, and partly because the soil often weathers much better if thrown up very rough. The vertical cut is often made by a special renewable "shin" or "cutter" attached to the breast, but where a coulter is used, it is placed nearer to the breast than with the lea plough.

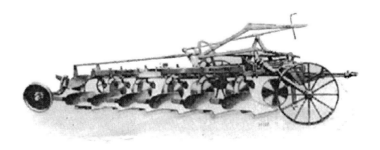

FIG. 44.—SIX-FURROW TRACTOR PLOUGH WITH DIGGER BODIES. (RANSOMES.)

The *landside* is the part that receives the side thrust due to the turning of the furrow slice. It is bolted to the left side of the frame and is generally made of hard cast iron, and of a suitable height and length to meet the requirements of the soil and plough type. On some horse ploughs, it may be over 2 ft. long, while on tractor ploughs, where the side thrust is partly taken up by the wheels, it is sometimes entirely dispensed with. The "slade" receives the downward pressure due to the weight of the plough and of the furrow slice. It is bolted to the bottom of the frame and is often part of one L-shaped casting which performs the dual purpose of landside and slade.

The *coulter* makes the vertical cut separating the furrow slice from the unploughed land. The knife coulter is fastened to the beam by a clamp, which permits adjustments to suit the work. The coulter acts as a knife, and the sharper it is, the better; but it must, of course, be thick enough to withstand wear and shocks.

The *disc* coulter is used where the knife type would cause difficulty by blocking, i.e. on most tractor ploughs. The disc coulter has a tendency to act as a wheel and lift the plough out of the ground, and can therefore only be used, in hard ground, on such ploughs as are adapted by their weight and design to counteract this tendency. The disc type is almost universally used on tractor ploughs, except in very hard ground, while the knife coulter is generally used on animal draught ploughs. Wavy-edge discs may be used with advantage where straw or other loose rubbish pushes along in front of plain discs.

The *skim* coulter is used to assist in the complete burial of manure, rubbish and crop remains. It may also be used where land is to be worked down to a seed bed immediately after ploughing, to cut away the corners of the furrow slices in order to facilitate the production of a firm bottom to the seed bed. The skimmer consists of a miniature plough, fixed rigidly to the beam in front of the knife or just behind the disc. A drag-chain is frequently used to assist in the burial of loose rubbish or farmyard manure.

The main constructional features of common types of tractor ploughs and various special ploughs are described in Chapter Six.

FIG. 45.—WAVY-EDGE DISC COULTER USED FOR PLOUGHING IN STRAW. (N.I.A.E. PHOTO.)

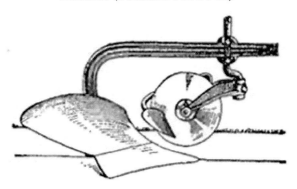

FIG. 46.—COMBINED SKIM AND DISC COULTER

REFERENCES

(13) Passmore, J. B., *The English Plough.* Oxford University Press, 1930.

(14) White, E. A., "The Study of the Plough Bottom." *Trans. Amer. Soc. Agric. Engnrs.,* vol. xii, p. 42.

(15) Nichols, M. L., *et al.,* "The Dynamic Properties of the Soil."

Agric. Eng., July 1931, August 1931, August 1932, and November 1932.

* "Breast" is the more usual English term; "mouldboard" is the usual American one.

CHAPTER FIVE

THE SETTING AND OPERATION OF PLOUGHS

"The edge of the share must be set a little below the level of the share beam."

The English Improver Improved. Walter Blith, 1653.

Correct setting of the plough leads to fuel economy and a better standard of work. When a plough is properly set, it should run steadily along, cutting a clean surface on the furrow wall and bottom. The plough body should run level, the heel of the landside just rubbing against the furrow wall, the heel of the slade sliding smoothly over the furrow bottom, and the share running level.

Bad setting of tractor ploughs passes unnoticed more readily than it does with horse ploughs because the tractor is less sensitive to bad setting than the horse; and there is often no extra strain on the tractor driver if the plough is badly set. It is, therefore, necessary for the tractor driver to study both plough and ploughing closely to obtain the best setting.

Since 1939 much attention has been devoted to tractor plough setting, andit has been demonstrated on many occasions that setting has an important influence on draught and fuel consumption. It has, moreover, been found essential to do good ploughing when old grassland is broken up, in order to save much time and trouble in the preparation of seed beds. The chief adjustments needing attention are the pitch of shares and bodies, the set of the coulters, the width of furrow, and the hitch between plough and tractor. These adjustments are dealt with in turn below.

The Share.

Correct setting of the share is essential. If a straight edge be laid from the point of the share to the heel of the landside, there should be a gap of about $\frac{1}{8}-\frac{3}{8}$in. (Fig. 47) between the straight edge and the share at the point where the share and the landside meet. This is called the "land" or "land suction" of the share. Similarly, the point of the share must be inclined downwards about $\frac{1}{8}-\frac{1}{2}$in., and this setting is called the "pitch" or "suck". Both suck and land requirements vary widely according to the type and condition of the soil. The softer the soil, the less suck and land are required to keep the plough running smoothly and cutting its full depth and width. On most types of plough bodies, where the share box fits on to the nose of the frame, there is no simple adjustment of the suck and land, and the best that can be done is to pack the inside of the share box. An understanding of the requirements with respect to suck and land is, of course, essential when steel shares are sharpened.

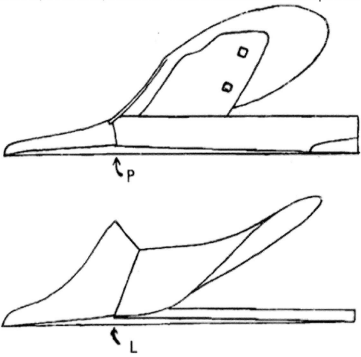

FIG. 47.—SETTING OF THE SHARE **P**, *PITCH OR "SUCK"*; **L**, *LAND.*

Faults in the construction, fitting or setting of the share cannot be remedied by adjustments elsewhere. If the share has insufficient suck, for example, it may be possible to keep the plough at work by raising the hitch at the hake; but if this is done, the plough will run on its nose, with the heel of the slade off the furrow bottom. In such circumstances the implement is difficult to handle, and bad work is done, the share gouging the furrow bottom and leaving the depth of work irregular. In a similar way, adjustments of the hitch in a horizontal direction will not remedy incorrect setting of the "land", which may cause the plough to run towards or away from the unploughed land.

On many modern tractor ploughs, the pitch of the complete body in relation to the beam may be varied by the use of set screws and a slotted frame, or by an eccentric device in the frame.

(Fig. 48). The "pitch measurement" is usually reckoned to be the vertical distance from the under-side of the share to the bottom of the beam, and when making adjustments to get all the bodies level, or checking a plough for strained beams, it is important to make the measurement at the correct point on the share. Some manufacturers advise fitting new shares and measuring from the share point. The pitch adjustment cannot be employed to correct for beams which have been severely distorted, and if, when all plough bodies are given the same pitch adjustment, one share point is found to be much lower than the others, it is usually an indication that the beam needs to be removed and re-set, or replaced.

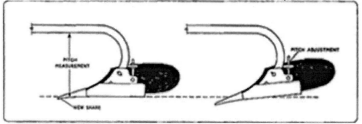

FIG. 48.—PITCH ADJUSTMENT. LEFT, BODY WITH CORRECT PITCH. RIGHT, TOO MUCH PITCH. (N.I.A.E. DRAWING.)

The Coulter.

The coulter, whether it be of the knife, disc or skim type, must be set correctly for the performance of good work. The knife coulter is usually set with its point slightly forward, so that it forms an angle of from 55 to 85 degrees with the slade; but where stumps or rocks are encountered, the point may be set backward so that the plough is thrown out of the ground if the coulter strikes an obstruction. For

normal work, the point of a knife coulter is set $\frac{1}{2}$-1 in. above

the point of the share and $\frac{1}{2}$ in. towards the unploughed

163

land. The coulter, like all other cutting tools, must be given a clearance behind the cutting edge, so that the side which is towards the unploughed land does not rub hard against the furrow wall. The clearance at the back edge should be from $\frac{1}{16}$ to $\frac{1}{8}$ in.

FIG. 49.—SETTING OF DISC AND SKIMMER FOR PLOUGHING OLD TURF

For average conditions the disc coulter should be set $\frac{3}{8}$-$\frac{5}{8}$ in. to the land side of the share point, and a vertical line dropped from the axle of the coulter should fall just behind the point.

Bad coulter setting is responsible for much bad tractor ploughing. Discs always tend to run narrow, and it is easy to show with the help of a dynamometer that setting the rear disc too narrow causes a considerable increase of draught (often 200-300 lb. with a 3-furrow plough) as well as bad work.

In hard going it is necessary to draw the disc back so as to give the share a "lead", to set the disc shallow, and to check up the setting to land every few hours.

For ploughing old turf the disc should be set well forward, to allow the turf to be cut and the skimmer to operate before the furrow slice is lifted by the share. In normal conditions the bottom of the disc may be set about 1 in. above the top of the share, while in very matted turf it may go deeper.

Modern disc coulters have a special fitting that allows them to be tilted. With the discs set over 2-3 notches, so that the lower edge undercuts the unploughed land, a crested furrow slice is produced; and this often gives a good result with rough old turf where skimmers will not function effectively.

Some farmers like to use "wing-shares", which have a vertical coulter attached to the land side. This takes the place of a coulter, and often works well where discs refuse to turn in wet, heavy land.

The Skimmer.

Skim coulters are of two types, viz. the skimmer with a straight shank attached direct to the plough beam, and the combined skim and disc coulter. Both types should be set with the point close to the disc to avoid blocking. In difficult conditions the independent skimmer often works best.

The skimmer is generally used to remove a small "neck" of soil from the left edge of the top side of the furrow slice, so that weeds arc prevented from continuing to grow through the cracks between adjacent slices. The most common fault in setting the skimmer is to make it cut more of the top of the slice than is necessary to achieve this object.

Skims are also often set with the point too far from the disc, or too far back on the disc. They should be set forward so that the cut is made before the furrow begins to rise. In general the cut should not be more than $\frac{1}{2}$ in. deep and 1 to 1$\frac{1}{2}$ in. wide and the piece cut should drop into the bottom of the furrow so that it does not prevent the furrow slices packing together.

With single-furrow deep-digging ploughs a rear-attached skimmer has many advantages. It has plenty of room to work without blocking, and the skimmed piece can be made much bigger. The skimmed part is pressed into the bottom of the furrow by the tractor wheel on its next round, and the tractor runs more level because the furrow is partly filled.

The Breast.

Many tractor drivers who cannot get the furrow slices to turn over attempt to press them together by setting out the tail of the breast. This generally only makes matters worse, and often results in a considerable increase in draught. If the breast is at fault the trouble is more likely to be cured by drawing it in; but when furrows will not turn a common reason is that the depth is too great in relation to the width.

In 1939 there was a widespread prejudice against wide furrow slices; but experience of ploughing turf soon showed

that it is necessary to have the width $1\frac{1}{2}$ times as great as the depth in order to do satisfactory work.

Principles of Hitching Trailed Ploughs.

With the plough body and coulter correctly fitted and set, it is necessary to adjust the hitch and the wheels to produce the required depth and width of work.

When the depth of a trailed plough is to be altered the correct sequence of operations is as follows:

Set the depth control screw to its new position. This will raise or lower both sides of the plough equally.

Set the levelling lever to tilt the plough so that the front furrow matches up with the previous work.

On the next bout, re-adjust the levelling lever to bring the plough back to level at the new working depth.

A complete understanding of the hitching problems of ploughs can only be obtained if the principles are understood. All the forces acting upon a plough at work may be considered as balanced about a theoretical point situated between the breast and the landside, called the "centre of resistance". With a plough body of ordinary proportions this point is generally about a foot back from the point of the share, 2 in. in from the edge of the landside, and 2 in. up from the bottom of the slade. If a single chain could be attached at this point, the plough could be pulled straight ahead at a uniform depth and width. The centre of resistance of a 2-furrow plough is midway between the centres of each furrow considered separately, and that of a 3-furrow implement is at the centre of resistance of the middle furrow. The position of the centre of resistance may thus be roughly calculated for any plough, but it should be observed that this position is not rigidly fixed. In hard ploughing, where more pressure may be thrown on the breast, the centre may move over from the landside until it is near the middle of the plough body.

The "line of draught" is an imaginary line running from where flexibility starts at the power end of the hitch to where it ends in the solid attachment of the implement to the centre of resistance. The best hitch is obtained when the line of draught passes through both the "centre of power" and the centre of resistance. The centre of power with a single horse is midway between the hame hooks, provided that no part of the harness holds the traces out of a straight line. For an ordinary tractor with rear driving wheels and on level ground it is near a point midway between the rear wheels and slightly ahead of their centres, while with 4-wheel drive and track-laying machines it lies farther forward.

Vertical Adjustments. If the point of hitch does not lie on a line joining the centre of resistance and centre of power it will tend to move into that line. For example, if the hitch at the hake is too low, the front of the plough will be pulled up out of its work (Fig. 50). In a tractor outfit too much pressure is exerted on the plough wheels if the hitch at the hake is too high, and this causes bad work and excessive wear of the wheel bearings. If the rear landside or rear wheel presses hard on the bottom of the furrow the hitch should be raised at the hake plates; and if the landside runs well clear of the furrow bottom, the hitch should be lowered. In general, the hitch from plough to tractor should slope slightly upwards. The vertical hitch adjustment is, of course, always influenced by the length of the hitch as well as by the hake setting. With horses care should be taken to have the traces approximately at right angles to the horses' hames, or sore shoulders may result.

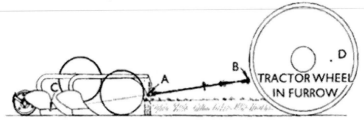

FIG. 50.—CORRECT VERTICAL HITCH FOR A TRAC-
TOR PLOUGH
A, *HITCH AT THE HAKE PLATES;* **B,** *TRACTOR*
DRAWBAR; **C,** *CENTRE OF RESISTANCE;* **D,** *CENTRE*
OF POWER.

Horizontal Adjustments. The horizontal adjustment of the hitch of tractor ploughs frequently presents some difficulty to farmers. Where possible the centre of power, the line of draught, and the centre of resistance should be in a line parallel to the direction of travel, but in some

170

circumstances, and especially when deep-ploughing with a tracklaying tractor which does not run in the furrow, this is impossible, and it is necessary to make the best hitch possible under the circumstances. Fig. 51 illustrates such an outfit at work. The centre of resistance of the plough lies much nearer the ploughed land than the line drawn through the centre of power parallel to the direction of travel. This arrangement inevitably leads to what is called "side-draught", a term which refers to the tendency of the two centres, when the plough is at work, to move sideways and come into line. Power consumed in overcoming side-draught is wasted, and the centre of resistance should be placed as directly behind the centre of power as possible.

When, as in Fig. 51, it is impossible to eliminate side-draught completely, the hitch must be made so as to distribute it between the tractor and the plough according to the amount that each can withstand. If the hitch is offset on the rear of the tractor until it is in the ideal position for the plough, the tractor takes all the side-draught; but unless it is heavy and the draught of the plough light, the front of the tractor will swing over towards the ploughed land, making it difficult to steer the outfit.

*FIG. 51.—ILLUSTRATING AN INSTANCE OF INEVI-
TABLE SIDE-DRAUGHT*

With a hitch that is ideal for the tractor, on the other hand, all the side thrust is thrown on the plough, which tends to swing round into line behind the centre of power, making good work impossible. The practical position lies between these two extremes, where the side-draught is distributed between the plough and the tractor.

With tracklaying machines, which frequently operate with the plough offset, it is important to employ a swinging drawbar which is pivoted at the mean centre of lateral resistance and to allow the drawbar to swing freely. The following may be indications of incorrect hitching in a. tractor outfit: —

The tractor running crab-wise or skidding into the furrow.

The plough running crab-wise, with the. land wheel skidding; excessive pressure on plough wheels, landsidc or slade; wheels, landside or slade "floating".

A broken or crumbled furrow wall or bottom; first furrow of multi-furrow plough not cutting its proper width.

With most small 2-furrow or 3-furrow tractor ploughs pulled by a tractor running with the off-side wheels in the furrow, the following method of securing a correct horizontal hitch may be tried.

The plough should first be drawn into its work and adjusted so that the width of the front furrow is the same as that of the others. If the hitch does not now correspond approximately with that shown in Fig. 52, the tractor should be stopped, the hitch uncoupled without moving tractor or plough, and the hitch re-coupled with the front end in the centre hole and the main drawbar parallel to the furrow. The hitch adjustment should be put in the centre of its travel and the end of the crossbar connected to the farthest hole on the hake-bar that it will reach.

The outfit should then be tried and the plough carefully watched to see if it "crabs". If it swings to the left at the rear, the drawbar should be moved to the right on the tractor draw-plate, and the main bar again lined up parallel to the furrow. If, on the other hand, the plough "crabs" towards the ploughed land, the hitch on the tractor should be moved to the left and again lined up.

The Setting of Tractor-Mounted Ploughs.

While the principles of setting tractor-mounted ploughs are much the same as those applicable to trailed models, there are important differences in the methods employed for altering the depth and width of furrows and for levelling. Almost all British mounted ploughs are attached to the tractor by the 3-link system, but ploughs so mounted may be divided into two major groups, viz. those in which depth of work is controlled by a land wheel, and those in which it is regulated through the tractor's hydraulic control system.

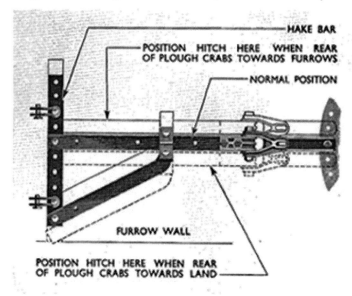

HAKE BAR

POSITION HITCH HERE WHEN REAR OF PLOUGH CRABS TOWARDS FURROWS

NORMAL POSITION

FURROW WALL

POSITION HITCH HERE WHEN REAR OF PLOUGH CRABS TOWARDS LAND

FIG. 52.—HORIZONTAL ADJUSTMENT OF HITCH OF SMALL 2-3 FURROW TRACTOR PLOUGH.

Depth Of Work. The depth regulating wheel on a mounted plough works in the same manner as on a trailed plough. On ploughs which have automatic hydraulic depth control, the depth of work pre-selected by setting the hand control lever is varied or maintained by the forces acting through the top link in conjunction with the hydraulic control mechanism. At a given setting, if the plough tends to go too deep, the additional draught increases the force in the top link, and this causes the control fork and control valve to move and so raise the plough until the draught is restored to the original setting, as explained in Chapter Two.

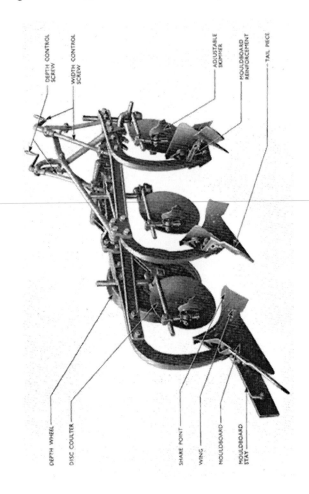

FIG. 53.—COMPONTENTS OF A 2-3 FURROW DI-RECT-MOUNTED PLOUGH WITH DEPTH WHEEL AND SEMI-DIGGER BODIES (FORD-RANSOME.)

On many tractors, the 3-link system can be attached to the tractor at two different points, according to the type and size of the tractor wheel equipment. The upper hitch pins and holes are for use with the smaller steel wheels, and the lower for use with pneumatic tyres. It is important that the linkage be attached to the correct points, as alteration of the height of the hitch-points above the ground has a considerable effect on plough penetration.

The pitch of the complete plough may be increased by shortening the top link, and this adjustment has the same kind of effect as raising the hitch at the hake of a trailed plough. If a mounted plough will not penetrate, it is often because it needs new shares. In general, the implement should be used with a minimum of pitch according to ground conditions; and in cases of failure to penetrate, a check should be made of the pitch of individual bodies, before shortening the top link.

Furrow Width. The furrow width of a single-furrow mounted plough, or the width of the first furrow of a multi-furrow one, may be varied by adjusting the setting of the tractor wheels, and also by adjustment of the plough cross-shaft. The importance of correct tractor wheel setting should not be overlooked. It is a mistake to try to remedy incorrect tractor wheel setting by adjustment of the cross-shaft, the correct function of which is to vary furrow width *after* the tractor wheels are properly set.

The Cross Shaft. All tractor ploughs employing the 3-link system are attached to the ends of the lower links by pins which form the ends of a cranked cross-shaft. Rotation of the cross-shaft alters the position of the pins relative to the plough frame. When the plough is hitched to the tractor, the pins are fixed, and the first effect of rotating the cross-shaft is to alter the angle of the plough relative to the direction of travel. When this has been done, and the plough again moves forward, the linkage and the plough move over to the right or left relative to the tractor, until the correct line of draught is substantially restored. On some makes of plough this adjustment is made by means of a screw gear, while on others it is necessary to loosen U-bolts which attach the cross-shaft to the plough beams, and rotate the cross-shaft by means of a spanner, not forgetting to tighten up the U-bolts again before starting work. A small angular movement of the cross-shaft will result in a. substantial sideways movement of the plough.

FIG. 54.—GROSS-SHAFT ADJUSTMENT. CLOCKWISE ROTATION OF SHAFT MOVES PLOUGH TORIGHT, AND VICE-VERSA.(N.I.A.E. DRAWING.)

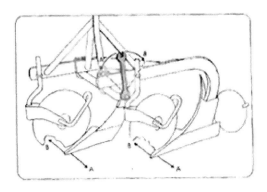

The levelling of mounted ploughs is achieved by lengthening or shortening one or both of the lift arms in the 3-link system, this being achieved by means of a screw gear which is easily accessible to the tractor driver.

Reversible Ploughs. The principles of setting reversible ploughs are, of course, exactly similar to those for right-hand ploughs and call for little special mention. With a well-designed plough, hitch adjustments made for right-hand work should not need alteration when the left-hand body is at work; but it is absolutely essential that tractor wheel settings should be correct for the plough and the work being done, otherwise satisfactory setting is almost impossible. Tractor rear wheel tyre pressures must be uniform, as must those of pneumatic tyred depth wheels on the plough.

Setting Tractor-Mounted Disc Ploughs. The setting of tractor-mounted disc ploughs equipped with a depth-regulating wheel and a rear wheel follows standard mounted

179

mouldboard plough practice except as regards setting of the rear wheel. The rear wheel setting is vitally important on any disc plough, and on a mounted disc plough it must be adjusted in conjunction with the cross-shaft setting. The inclined rear wheel normally has two adjustments, viz. (1) lateral and (2) angular. The lateral adjustment must be suited to the furrow width cut, and will vary with soil conditions. The angular adjustment is used to steer the plough towards or away from the ploughed land, and must be used in conjunction with the cross-shaft adjustment. The angle is correctly set when fine adjustment of the cross-shaft easily alters front furrow width. Adjustment of the cross-shaft should enable front furrow width to be kept correct when working across slopes without further alteration of the rear wheel setting.

When opening lands with a disc plough the "top" should be split, and the double-split system is usually preferable, as this leaves a level field with all the land ploughed.

The Draught of Ploughs.

The draught varies according to the construction of the implement and the type and condition of the soil ploughed. With ploughs of good design and construction the resistance of the soil per square inch of furrow section varies from about 5 lb. per sq. in. on light, loose, sandy land to about 16 lb.

per sq. in. on light, loose, sandy land to about 16 lb. per sq. on hard, heavy clays. Thus, whereas one horse may be able to plough a 9 in. by 6 in. furrow in the first case, four may be required in the second, and two or three for intermediate soils. Similarly, a tractor of 15 drawbar h.p. which can easily

pull three furrows 10 in. wide and 7 in. deep at $2\frac{1}{2}$ m.p.h. on a light soil may be unable to pull two furrows at the same speed on a heavy soil. It is an advantage to have ploughs which are easily convertible from three furrows to two; for while it is wasteful of power to pull two furrows where three could easily be pulled, it is worse management to attempt to pull three furrows where conditions make this just impossible, owing to a too heavy draught or wheel slip. Whether any particular tractor will pull 2, 3 or x furrows depends on the drawbar power available, the ploughing resistance, wheel grip and items such as gradients. A medium-powered tractor in good order may be expected to exert a pull of 2,000 lb. at about 3 m.p.h. It should, therefore, pull 3 furrows 10 in. wide and 6 in. deep at this speed on any land having a ploughing resistance of not more than $\frac{2,000}{3 \times 10 \times 6}$ equals 11 lb. per sq. in.

It is, therefore, advisable to exercise some care and even to make some preliminary tests before selecting new implements. It should perhaps be added that no field has

181

absolutely uniform soil, and there are often wide variations in the draught from point to point in the same field.

Some tracklayers that normally pull 3-furrow or 4-furrow ploughs could often, with advantage, pull 5 or 6 furrows if a suitable large plough were available. Where fields are large it may be worth while to pull two ploughs in tandem, by using a two-plough hitch such as that shown in Fig. 55. Where the two ploughs are not of the same size, the larger one must run first.

It should be added that good ploughing is often more important than a full load. If the addition of an extra furrow or two necessitates shallow ploughing, it is generally preferable to cover less ground and do the job thoroughly.

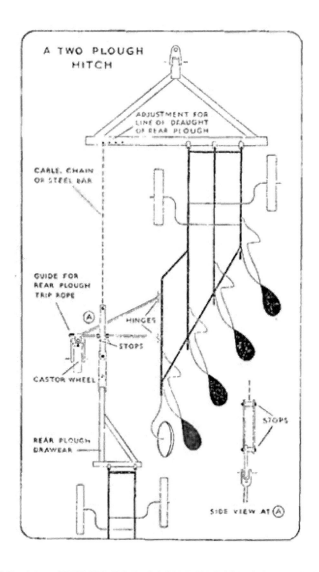

FIG. 55.—HITCH FOR COUPLING TWO TRACTOR PLOUGHS (N.I.A.E)

Methods of Tractor Ploughing.

The normal method of working with the horse plough is outlined in many textbooks of agriculture and need not be discussed. In tractor- ploughing, the procedure is slightly modified because the outfit is generally unable to make short turns. The headlands must be about 8 yds. wide for an outfit of medium size. This width just allows the tractor to turn in a semicircle, but not to split or gather work which is less than about 8 yds. wide without doing figures of eight. It is, therefore, necessary to work on two ridges as shown in Fig. 56. With ridges set up at A, B, G. etc., as in horse work, zones 1 are first ploughed by working on ridges A and B as illustrated. When at least 8 yds. have been ploughed in this manner on ridge A, zones 2 are ploughed by gathering the work around A. The work is then quite straightforward, zones 3, 4, 5, etc., being ploughed out successively in the manner illustrated. There is less idle running with this method than with any other.

A system of tractor ploughing that is popular in some districts for deep work, especially in the Fens, where the fields are well shaped and must be left level, is "square ploughing". A small land is first ploughed in the centre of the field. This is made of such a length and width that, when completed, the distances from its sides and ends to the adjacent edges of the field are all equal. Any "gores" must, of course, be worked out on this land. The outfit then proceeds round and round in a clockwise direction, the procedure at the corners being to lift the plough when it is just clear of the corner, turn the tractor in a small loop anticlockwise, and drop the plough in to work again just before it reaches the ploughed land on the next side. This procedure can result in very good work; for all the land is turned once and only once, and open furrows, which often cause much difficulty after deep ploughing, are eliminated. It is much more neat and thorough than the "round-about" system, in which the outfit starts either at the centre or at the edge of the field, and works round and round without lifting the plough at the corners. Square ploughing suffers from the disadvantage that it takes longer than the normal method of ridge ploughing, and it would be folly to use the square ploughing method for land that needs water furrows.

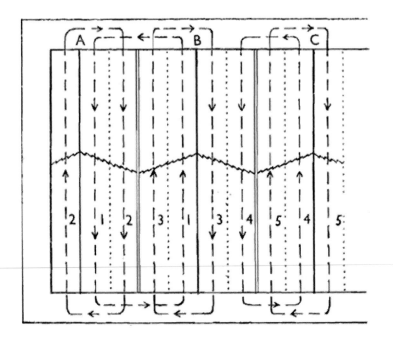

FIG. 56.—*NORMAL METHOD OF TRACTOR*
PLOUGHING **A, B, C,** *RIDGES.*

To keep the fields level it is necessary sometimes to plough from the edge to the middle. The best method then is to draw out "headlands" along the diagonals, lift the plough at the turns, and plough the diagonal headlands out last. If a field is of irregular shape and has one sharp corner, the plough can be kept in at the gentle turns and lifted at the sharp one, a headland being run from the sharp corner to the centre of the field, and this headland being ploughed out last.

Setting Up Tops And Making Finishes With A Tractor Plough. The details of good tractor ploughing can only be learned by practice; but one method of starting and finishing that is generally satisfactory may be briefly detailed. The method described is one of many, and other ways may be better for particular circumstances.

Even experts need to mark out the land accurately and to use sighting sticks. To start a top on old turf the front body or bodies may be raised clear of the ground, the rear disc set just below the top edge of the share and a shallow skim 2 in. deep cut with the. rear furrow only. Then, turning right at the end of the ridge, the plough may be set, shallow and level so that the first furrow falls across the original one and half covers it. Turning right again, the plough is again set level and shallow, and the hitch lever adjusted so that the first furrow just meets the first of the last bout.

This procedure leaves a small strip of unploughed land; but it is so well covered that grass or rubbish cannot grow through, and the top is so neat that subsequent cultivations are unlikely to expose the unploughed strip.

To avoid constant alterations of plough settings, all ridges should be set up before ploughing commences.

In making the finish it is essential to correct any irregularities in the width of the unploughed land before the last few bouts. The last two rounds must be ploughed shallower, and on the last bout the strip left should be one furrow width for a 2-furrow plough, two for a 3-furrow, and so on. For the last stroke the rear disc should be set to run below the bottom of the share, to prevent the plough crabbing; and the plough should be set so that the front body takes a full furrow and the rear one brings up a shallow mould from the bottom of the last furrow.

FIG. 57.—ILLUSTRATING THE ACTION OF A LEA PLOUGH

Styles of Ploughing.

The size and shape of the furrow slices are determined by the construction of the plough body and by its setting. The ordinary lea plough turns a rectangular slice which remains unbroken in medium or heavy soils. The usual aim with this type of plough is to set up the slices so that they lie at an angle of 45 degrees to the horizontal and leave equal areas of soil exposed on each side of the apices of the slices in the finished work. In order to do this it is necessary to turn a slice in which the width is greater than the depth in the proportion 10: 7. If the width is relatively less, the slices will stand more upright and there is a possibility of their falling back into the open furrow; if it is relatively greater, the slices will lie too flat.

At various times in the past there has been considerable controversy concerning the merits of different shapes of furrow slices for lea work, especially with regard to the "crested" style. Crested work is produced by the use of a share having "up-wing", so that the angle between the coulter and the cutting edge is acute. A similar effect is produced by setting the coulter over to "under-cut" the land.

FIG. 58.—ILLUSTRATING THE ACTION OF A DIG-GING PLOUGH

The crested slice exposes about 10 per cent, more surface to the action of the weather, but the advantages are to some extent discounted by the fact that in ploughing to the same average depth as with the rectangular style more work must be done to plough the same area, and the rate of working is reduced.

The common style of ploughing, when the aim is the rapid production of a tilth, is the pulverized work produced by the digging type of body. In this work the shape of the slice cut is generally rectangular, and the width/depth ratio can be varied over a wide range. An "under-cut" slice is popular for digging work in some districts. This is produced by using a body which has the shin and coulter set over to make an acute angle with the horizontal share. In some circumstances employment of an under-cut body may lead to more effective burying of rubbish.

The Essentials of Good Ploughing.

The rapid development of the plough towards the end of the eighteenth century coincided with the introduction of ploughing matches, at which the skill of ploughmen and the merits of various types and makes of plough could be compared.

The following are among the more important requirements of good ploughing for British conditions:—

The "top" or ridge should be set up with no unploughed ground beneath it, at approximately the same height as the rest of the work.

All furrows should be straight and of uniform depth and width. The furrow wall and bottom should be clean-cut.

All vegetation, manure, etc., should be completely covered.

The land should be left in the condition required. This may be either well set up with the slices unbroken and firmly

Farm Machinery

pressed together or completely pulverized to full plough depth.

5. The furrow ends should be finished off evenly with no packed or unploughed soil, and the lands should be finished off in a manner suited to the style and object of ploughing.

REFERENCES

(16) *Tractor Ploughing.* H.M.S.O. (1952).

CHAPTER SIX

THE PRINCIPAL TYPES OF TRACTOR PLOUGH

"Ploughs of various other shapes and fashions have from one imagined improvement to another multiplied to an almost endless variety."

The Implements of Agriculture. J. A. Ransome, 1843.

Tractor plough design has now settled down to a few well-established types. The usual type of trailed plough has a rigid framework, consisting of from one to six steel beams, carried on three wheels. The connection of the frame to the furrow wheel and to the land wheel is through cranked axles which can partially rotate in brackets attached to the frame. A rear furrow wheel is usually present, but may be dispensed with on certain small single-furrow and double-furrow types. Adjustment of the depth of working is easily effected by levers or screw mechanisms which control the positions of the cranked axles.

The hitch to the conventional three-wheeled tractor plough is "full-floating" or flexible in a vertical direction, and rigid in the horizontal plane. The weight of the plough and the downward force due to the lifting of the furrow slices are thus carried by the three plough wheels. When the rear furrow wheel is omitted, it is necessary to make the hitch semi-flexible or "semi-floating", so that when the plough is lifted it is drawn behind the tractor on its two wheels like a two-wheeled cart.

Lift Mechanisms. An essential feature of the tractor plough is a self-lift mechanism for raising it at the ends of the held and quickly returning it to its work after turning. The usual method is to have some type of positive clutch which, when engaged by the operator, rigidly connects the land wheel to its cranked axle. Rotation of the land wheel then rotates the crank about the bracket connecting it to the frame. A strap connecting the land wheel axle to the furrow wheel axle causes the latter to move in a similar way. When the plough is being lifted the cranks rotate until the axles are almost vertically beneath the brackets. The clutch is then automatically disengaged and the plough is locked in the lifted position. Another pull on the clutch lever releases the locking device and the plough is allowed to fall back into its work.

One of the simplest types of self-lift is the rack-and-pinion type. A pinion wheel is bolted to the land wheel and a curved rack is attached to the plough frame. With the plough in work, a pull on the lifting lever engages the rack with the pinion, and as the wheel turns, the rack and the frame to which it is attached are driven upwards by the pinion. By the time the end of the rack leaves the pinion a locking device has fixed the plough in the lifted position. A second pull on the same lever releases the locking device and allows the plough to fall back into work. A more common type of self-lift is one in which the clutch mechanism is enclosed in a drum attached to the land wheel. On ploughs of three or more furrows the implement becomes so long that a simple lift working only on the main wheels is insufficient to lift the rear bodies clear of the ground. The use of a rear-wheel lift overcomes this difficulty. A strap from a lever fixed to the land wheel axle passes back to a system of levers whose intermotion causes the frame to move vertically up with relation to the rear wheel. The rear wheel should castor when the plough is lifted and should be locked when it is in work.

Trailed ploughs may be lifted out of work by means of a hydraulic lift operated by the tractor. The normal method is to employ the tractor's built-in hydraulic lift pump, which delivers oil under pressure to a standard "slave" cylinder fixed to the plough, delivery and return of the oil being through flexible rubber tubing with rapid self-sealing couplings. British Standard No. 1773: 1951 gives details of the dimensions and operating characteristics of such lifts. They have the advantage of being more positive than any mechanism which depends on the forward movement of the implement's wheels for its operation.

The Wheels. The land wheel runs parallel to the plough beams and is vertical when the plough is level, but the furrow wheel axle is generally bent in such a way that when the plough is level the top of the wheel is set over towards the ploughed land, making an angle of about 5 degrees with the vertical. The forward edge of the wheel is also set nearer to the beams than the rear edge, and this setting helps to keep the plough up to the work. The wheels should be adjusted for depth so that when the plough is running level they are just carrying the weight of the plough. The rear wheel seen in plan, is set to run either parallel to the furrow or slightly away from the unploughed land. In elevation the top of the rear wheel is generally inclined away from the unploughed land at an angle of about 30 degrees. The object of this setting is to enable it to take up most of the side thrust due to the pressure of the furrow slices on the breasts. Where the rear wheel is adjustable it should be set so that the landslide and slade of the rear body are just touching the furrow wall and bottom.

FIG. 59.—TWO-FURROW DEEP-DIGGING TRACTOR PLOUGH. (RANSOMES.)

Heavy Tractor Ploughs for Special Work.

There is a wide range of special ploughs available for very deep work or difficult conditions.

Ploughs such as the Ransomes "Jumbo-trac", which will turn furrows 16-18 in. deep, are regularly used on some farms; and during the war period "Prairie Buster" or "Grub Breaker" ploughs, designed for difficult conditions in the North American prairies, were advantageously employed for breaking up rough land. The British "Bracre" turns a furrow 22 inches wide and 1 foot or more deep. There is a strong knife coulter joined to the point of the share, and land where a growth of thorn bushes or gorse makes it impossible to use ordinary ploughs can be satisfactorily dealt with by these implements.

The "Turnall" plough is essentially a tractor swing plough, which is adjusted simply by altering the pitch. It is designed for use with a light tractor, and has proved efficient at ploughing rough ground. When this plough strikes an obstruction it is able to swing sideways, and thereby avoids serious damage.

Direct-Coupled Tractor Ploughs.

Most direct-coupled ploughs, when at work, have the depth controlled by levers or screws that operate on a land wheel of some kind (Fig. 53); but some, e.g. the Ferguson, are controlled entirely by adjustment of the linkage mechanism connecting them to the tractor. Such ploughs are excellent in many ways. There would appear to be a limit to the size of plough that can be handled satisfactorily in this way.

"Reversible" Or "One-Way" Tractor Ploughs. In France and many other European countries "reversible" ploughs have been in common use for many years. Serviceable implements of this type were available for horse work, so when tractor ploughs were being developed it was natural that many manufacturers should develop one-way models.

One-way ploughs save the trouble of setting up and closing ridges, eliminate the troublesome open furrows associated with our normal methods of ploughing, save time in turning, and reduce the running along headlands. Level seed-beds are a great advantage for root crops such as sugar beet, and even when grass and corn crops are grown the disappearance of the open furrows leads to easier operation of many implements and machines, and often reduces machine breakages and allows an increased speed of working.

Disadvantages are the extra cost and weight of the implement:, and the fact that setting the implement to do good work takes longer and is more difficult.

The main types of one-way tractor ploughs employed in Britain are as follows:

Trailed ploughs suitable; for use with medium-powered track-layers. (Mainly deep-digging single-furrow and medium-depth 2-furrow models. 3-4 furrow trailed ploughs suitable for shallow to medium-depth work arc being developed.)

Semi-mounted deep-digging single-furrow and medium-depth 2-furrow ploughs for use with medium-powered tractors.

Direct-mounted ploughs, mainly deep-digging single-furrow and medium-depth-furrow, for use with medium-powered tractors. Several manufacturers are engaged in the production of direct-mounted reversible ploughs, and these, on tractors fitted with independent rear-wheel brakes, permit rapid turns on the headlands and result in a minimum of idle running time. In addition to the above, there are a few examples of types which arc much more common on the Continent of Europe, e.g. balance ploughs suitable for use with very large tracklayers, and similar to the multiple-furrow ploughs that used to be commonly employed with steam cable tackle.

FIG. 60.—SEMI-MOUNTED 2-FURROW REVERSIBLE PLOUGH. (BONNEL.)

Subsoiling Ploughs.

Subsoil ploughs are designed to break up the lower layers of soil without bringing them to the surface. Many trials have been carried out with the object of determining the value of subsoiling, but the results have been of little value owing to a lack of studies of the exact condition of the soil before and after the operation. Tests of the soil condition must be the basis on which the possible effect of subsoiling is judged. It may be stated, in general terms, that subsoiling will probably have a beneficial action where there is any "pan". On certain types of land a pan is formed just below ploughing depth by the action of the plough slade on the furrow bottom, and this may seriously interfere with plant growth. On other soils, "iron" pans may be formed in the lower layers of the soil by chemical action. On heavy land, bursting up the lower layers of the soil in dry weather opens up the soil and subsequently has a beneficial effect on drainage.

FIG. 61.—*SINGLE-FURROW DIRECTED-MOUNT-ED REVERSIBLE PLOUGH WITH DEPTH WHEELS (FORD-RANSOME)*

One method of subsoiling to is attach a subsoiling tine behind the body of an ordinary plough or to replace one of the furrows of a double-furrow plough by a subsoiling tine.

A second method of sub-soiling at the same time as ploughing when using a medium-powered tractor is to attach a subsoiling tine or tines behind the bodies of a mounted plough. On some ploughs a subsoiler may be fitted in front of the plough body, in such a manner that the tractor wheel does not run along the furrow bottom after the subsoiler has done its work.

FIG. 62.—*TWO-FORROW TRACTOR PLOUGH CON-
VERTED TO SINGLE FURROW WITH SUBSOILDER
(RANSOMES.)*

Fig. 64 shows a deep-digging tractor plough with subsoiling tines working behind the bodies. Such an implement is capable of ploughing 12 to 18 in. deep and subsoiling a further 6 in., but requires a very powerful tracklayer to operate it. A third method is to use a special tractor subsoiler. This may be a mounted implement designed for use with a medium-powered tractor, as seen in Fig. 65, or may be a trailed implement designed for very deep work with a powerful tractor.

FIG. 63.—2-FURROW MOUNTED PLOUGH (FORD-RANSOME) WITH SUBSOILING ATTACHMENT FITTED BEHIND EACH BODY. (CENCO.)

The trailed subsoiler consists of a very strong frame with two wheels, mounted on a single cranked axle and carrying one. or two subsoiling tines. This implement, which

204

is fitted with a self-lift, will break the subsoil to a depth of 20 in. There are some farms where regular subsoiling would pay; and it is surprising that few farmers either appreciate the value of subsoiling or know anything of the range of equipment available for carrying out the work.

In circumstances where subsoiling is only rarely necessary, it may be most economical to have it done by contractors, using heavy tracklayers.

FIG. 64.—TWO-FURROW DEEP-DIGGING PLOUGH WITH SUBSOIL TINE BEHIND EACH BODY. (RAN-SOMES.)

FIG. 65.—DIRECT-MOUNTED SUBSOILER. (FERGU-SON.)

FIG. 66.—SELF-LIFT TRACTOR SUB-SOILER. (RAN-SOMES.)

Disc Ploughs.

Disc ploughs bear little resemblance to m o u l d-b o a r d ploughs, for a large revolving concave steel disc replaces the share, coulter and breast. The disc is set at an angle to the line of travel, and it turns a furrow slice to one side with a scooping action. The usual size of the discs is about 24 in., and these will turn a furrow 10 to 12 in. in width. The disc is mounted on a heavy frame, and the large amount of side-thrust due to the pressure of the soil is carried by the three wheels, the two furrow ones being inclined in a suitable direction to withstand it. The wheels are often weighted to assist in the penetration of hard land, since the discs themselves have little "suction". (See Fig. 9.)

The more vertical the disc is set, the deeper it penetrates and the better it covers rubbish. Correct setting of the scraper will assist in inverting the furrow and burying rubbish. The greater the horizontal angle between the disc and the direction of travel, the wider the slice cut and the less the tendency of the disc to revolve. Disc ploughs suitable for mounting on the hydraulic lift linkage of medium-powered tractors have now been developed—largely in response to the needs of farmers in countries where mouldboard ploughs are seldom or never employed. Disc ploughs are well adapted to ploughing extremely hard soils. They also deal with certain types of sticky soil better than the mouldboard plough, and will work in long, loose rubbish where no mouldboard plough will run. Disc ploughs are widely used in America and in South Africa, but find little favour in England. They cannot satisfactorily replace the mouldboard plough for ordinary purposes in this country owing to their failure to bury rubbish completely or to cut all the ground. They can, however, be applied to certain special jobs, and are particularly useful for the rapid breaking up of stubbles where there is a large amount of rubbish, as when straw is left on the field by a combine harvester.

FIG. 67.—TRACTOR-MOUNTED 3-FURROW DISC PLOUGH. (HUDSON.)

"Stump-jump" disc ploughs of Australian origin have been used successfully for very rough work such as preparing land for reseeding where there are tree stumps, tree roots or large rocks to contend with. The stump-jump disc plough will stand up to work that would ruin any mouldboard plough, however strongly constructed.

FIG. 68.—STUMP-JUMP DISC PLOUGH. (SUNSHINE)

FIG. 69.—*POLY-DISC DISTRIBUTING NITROG-
ENOUS FERTILIZER AND WORKING IN STRAW LEFT
BY COMBINE HARVESTER, (RANSOMES.)*

The "harrow plough" or "poly-disc" has an action
intermediate between that of the true disc plough and the
disc harrow. It is commonly employed, with a seed-box
attached, for ploughing and seeding in one operation in the
drier parts of the Canadian Prairie Provinces; but for British
conditions its inability to bury rubbish effectively restricts
its use to such operations as stubble cleaning.

The general tendency in the Canadian prairies and
similar-regions of the United States is towards wider
implements and shallower work. Many years ago, in the

days of the early settlers, mouldboard ploughs were used. These were succeeded in turn by the disc plough and the "poly-disc", and the most recent development is the "discer", usually a very wide implement, with moderately small discs mounted in gangs of six and capable of independent vertical movement on the main frame. This implement normally works only about 3 in. deep, and usually has a seed-box attached. The power lift raises the discs and shuts off the flow of seed, but does not raise the whole frame of the implement as it does in the case of the poly-disc. The "discer" is now the only cultivation implement of any consequence on some Canadian farms.

Potato-lifting ploughs, sugar beet lifters and mole-draining ploughs are dealt with in Chapters Fourteen and Fifteen.

Maintenance of Ploughs.

Some tractor drivers forget that ploughs have a number of bearings that need regular lubrication. There are grease nipples on disc coulters and wheel bearings that need frequent attention. It should, however, be mentioned that self-lifts occasionally receive too much grease; and when this happens they may refuse to function until some of the excess grease is removed.

In hard ground it is necessary to tighten nuts and bolts frequently.

Rusty breasts and coulters are often responsible for delays and unnecessary wear. It takes only a minute or two to smear the breast with waste oil, and this should be done whenever the plough is left for more than a day. Many good farmers insist on having it done every night. An anti-rust compound should be used if it is known that the plough will stand for a long time.

CHAPTER SEVEN

IMPLEMENTS FOR THE PREPARATION OF SEED-BEDS

"It is not ploughing, it is not digging, it is not harrowing, raking, hoeing, rolling, scarifying, clod-crushing, scuffling, grubbing, ridging, casting, gathering that we want: all these are the time-honoured, time-bothered means to a certain result. That result is—a seed-bed."

Talpa, or Chronicles of a Clay Farm. Wren Hoskyns, 1845.

Cultivators and harrows are implements whose main use is breaking up clods. For example, they are used after ploughs for breaking tip the furrow slices and working the soil to a tilth in the preparation of seed beds. Other uses include the destruction of weeds, mixing of fertilizers with the soil and covering of seeds. There is no essential difference between cultivators and harrows, but, generally speaking, cultivators are used to deal with heavy work and large clods, and harrows are later employed to continue the work of preparing a fine tilth. The terms "cultivator" and "harrow" are, however, loosely used; for some of the tractor grassland harrows mentioned below could be as correctly termed "cultivators".

The tines used on the more common types of horse harrow are straight and are rigidly fixed in a perpendicular position in a "zigzag" frame. The depth of penetration of such implements is small; it depends upon the firmness of the soil, the size and shape of the tines and points, and the weight of the implement. Straight tines with the points inclined forward tend to penetrate deeper than vertically mounted ones. Such tines loosen the soil and lift weeds to the surface, while vertical tines compress the lower layers and do not pull out so much rubbish.

Curved tines are used on both harrows and cultivators, the simplest form having a vertical stem and a curved point. This type is widely used on drag harrows and scufflers of all kinds. With well-designed points, deep penetration can be secured. The sickle-shaped tine, either as a spring tine or spring-mounted, is a common type for cultivators. Cultivator tines may be subject to severe stresses, and for satisfactory service they must be strongly made in order to stand up to the work. Spring tines are frequently said to produce a better pulverizing effect and to have a lighter draught than rigid ones. There is little reliable evidence on this point, and in general there is little to choose between spring-mounted and rigid tines, the important considerations being the shape of the point or share, the effective working depth, and the strength of the tines.

FIG. 70.—"EQUITINE," SELF-LIFT TRACTOR CULTI-VATOR. (RANSOMES.)

Some types of spring-loaded tines employ a pair of heavy extension springs to protect the tine and the implement against breakages when working among tree roots, etc., or in very hard land, and these are becoming popular on tractor-mounted cultivators.

One type of cultivator (e.g. Ransomes' "Equitine") employs the principle of "compensating" tines. The tines are so constructed that they can swing individually in their mountings, but they are connected to each other by a system of chains and levers. When one of the tines strikes a serious obstruction or is in danger of choking, it can clear itself by swinging backwards, and is drawn back into position by the other tines pulling on the chain, as soon as the obstruction is cleared.

The shares or points used on cultivators are very varied, ranging from reversible points for average work to strong narrow points for deep work and broad A shares for shallow work. There are also special blades for cutting thistles, and knives for cutting matted turf (Fig. 71). The points and shares are generally made of steel, but some of the very heavy cultivators have massive chilled cast iron points (Fig. 75).

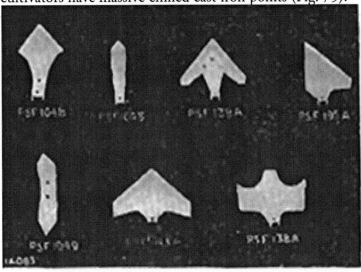

FIG. 71.—TYPES OF CULTIVATOR SHARES. (RAN-SOMES.)

Cultivators.

It was not until the middle of the nineteenth century that any serious attempt was made to cultivate the ground deeply by the use of tined implements, and the cultivators used through most of the century had roughly constructed shares and were very heavy in order to secure adequate penetration. Some of these heavy implements are still in use and perform useful work in such operations as broadsharing stubbles, when fitted with modern shares. The fore-runner of the modern cultivator, with its light frame and sickle-shaped spring-mounted tines fitted with well-designed points, was developed towards the end of the nineteenth century, though, of course, many improvements and modifications have since been made in details of construction and in such directions as the development of strong tractor implements.

The cultivator usually has two or three rows of tines which are "staggered". The object of mounting tines in two or more rows is to provide a clearance between adjacent tines through which clods and rubbish can freely pass. When tines are mounted too close together they quickly block in difficult conditions. Steel points of various shapes may be bolted to the tines. For general purposes the reversible double-ended type of point, fitted to spring-mounted tines, is very suitable.

Tractor cultivators may be subject to severe stresses and must be strongly constructed. Modern trailed implements have rigid braced steel frames, carried on steel wheels with cranked axles. The wheels are sometimes fitted with dust-proof roller bearings, but more usually a plain renewable cast iron bush is used. Depth adjustment is effected by a screw mechanism which can be reached from the driver's seat. A simple and efficient self-lift must be fitted, and it is an advantage if this operates from either wheel. On the larger cultivators, a special hitch, controlled by the self-lilt, ensures that the lifting of the implement is "parallel", i.e. that the front of the frame rises at the hitch as the rear is raised by partial rotation of the cranked axles.

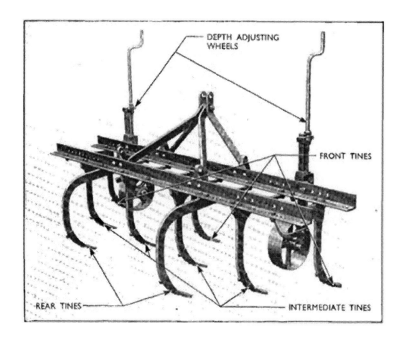

FIG. 72.—REAR VIEW OF GENERAL PURPOSE TRAC-
TOR-MOUNTED CULTIVATOR WITH RIGID TINES
ARRANGED IN THREE ROWS. (FORDSON.)

Tractor cultivators mounted direct on to the rear of
the tractor and operated by a power lift are now largely
superseding trailed implements. The work done by a good
general purpose directly-coupled cultivator is as good as that
done by trailed machines. The tractor-mounted unit has the
advantages of simplicity, cheapness, and a positive lift in the
most difficult conditions.

FIG. 73.—DIRECT-COUPLED TRACTOR CULTIVA-
TOR WITH SPRING-LOADED TINES AND THREE-
LINK HITCH. (FERGUSON.)

The heaviest tractor cultivators, such as that illustrated in Fig. 75, may be used for bursting up unploughed land. Such an implement requires a tracklayer of about 40 drawbar h.p. or more to pull it, and has an effect equal to that of the heavy cultivators used with steam cable tackle. It is very desirable to burst up heavy land when it is dry, and this kind of equipment enables the job to be done thoroughly.

FIG. 74.—DIRECT-MOUNTED CULTIVATOR WITH HEAVY SPRING-LOADED TINES SUTIABLE FOR DEEP WORK. (FORDSON)

FIG. 75.—VERY HEAVY TRACTOR CULTIVATOR SUITABLE FOR BURSTING UP HEAVY LAND WITH-OUT PLOUGHING. (RANSOMES.)

221

The Broadshare.

The broadshare is a heavy type of cultivator fitted with a share or shares up to 18 in. wide. Owing to its weight and strength it is possible to work this implement in very hard ground at a depth of 2 to 3 in, This feature makes it a useful implement for stubble cleaning.

In some districts a popular cultivator fitting is a heavy steel blade, 6 ft. long or so, about 3 in. wide, $\frac{1}{2}$ in. thick and sharpened at the forward edge. This is bolted on to the cultivator tines and set to run about 2-3 in. below ground. Its chief use is for cutting thistles and other weeds on land which cannot be worked deep.

Harrows.

Harrows are employed for a great variety of purposes, such as the preparation of seed beds, the covering of seeds, the destruction of weeds and the aeration of pastures. Many types and sizes are used for carrying out these widely-differing functions. The most common and simple type is that with a zigzag iron frame and rigid tines. In this class may be grouped seeds harrows, two-horse harrows, three-horse harrows and drag harrows.

FIG. 76.—ZIGZAG TRACTOR HARROW AT WORK

Seeds harrows are light one-horse implements with straight, closely spaced tines, about 4 in. long. Owing to the zigzag construction of the frame the tines are staggered so that those at the rear do not follow in the tracks of the front ones. These implements are especially employed for the final touches in the preparation of fine seed beds and for covering seeds after the drill. One horse can usually pull two or three sections, making a total width of up to 8 ft., but the draught naturally depends upon the condition of the soil. Sets from 20 to 30 ft. wide are available for tractor work, and are easily within the capacity of small machines. When they are used for covering seeds, harrows are frequently hitched directly behind the drill.

Two-horse and three-horse harrows are heavier than seeds harrows and are used for such purposes as the preparation of seed beds, mixing of fertilizers with the soil, and spring cultivation of autumn-sown corn. Drag (or duckfoot) harrows are a heavy type of zigzag harrow with long, curved teeth. The points of the teeth may be either chisel or "duckfoot" shaped. The teeth are up to 10 in. long, and in some conditions the draught is extremely heavy. These implements do much the same work as cultivators, but have less tendency to bring clods to the surface. They are, therefore, often used in preference to cultivators when deep working is not essential and it is desired to obtain a seed bed quickly.

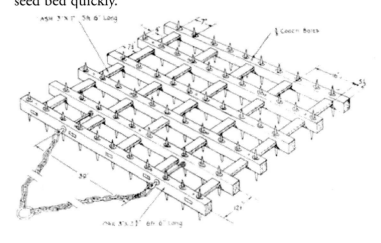

FIG. 77.—WOODEN-FRAMED SPIKED HARROW,
DUTCH PATTERN. (N.I.A.E. DRAWING.)

Heavy Wooden-framed Harrow.

Fig. 77 illustrates a heavy wooden-framed spiked harrow which sugar-beet growers are hading useful for the preparation of firm, level seed-beds. The frame consists of six main beams, preferably of oak, about 3 in. by $3\frac{3}{4}$ in. by 6 ft. 6 in. long, and held together by smaller cross members. The tines, which are 7 in. apart and staggered, are driven into drilled holes in the main beams and project 3 in. on one side and 5 in. on the other. The tines may be mounted at a slight angle, so that by choice of length of tine and angle four different effects arc obtainable. The harrow is pulled straight, with the main baulks at right angles to the direction of travel. The harrow levels off the surface and at the same time breaks down small clods. Such harrows are much used by Dutch beet growers.

Spring-tined Harrows.

The spring-tined harrow is really a light cultivator which can be adjusted to produce very variable effects. The tines are fitted to axles which can be partially rotated by means of a lever. With the implement working very shallow, the points are almost vertical and the action is that of a light harrow that does not penetrate to the lower layers. When set in an intermediate position the points arc inclined, and the action is that of a light cultivator. At full depth the points are almost horizontal and the soil is deeply stirred. This type of harrow is very popular on many small European farms because it can do the work of both cultivators and harrows on light soils. Its use on British farms is extending.

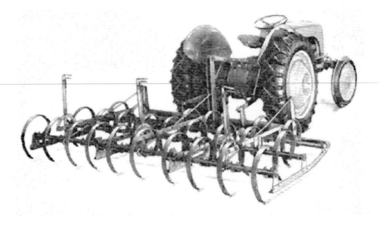

FIG. 78.—TRACTOR-MOUNTED FOLDING SPRING-TINED HARROW. (FERGUSON.)

Spring-toothed Weeders.

Spring-toothed weeders or finger weeders have been well known in America for many years, but it is only recently that they have been introduced into Britain. They consist of a number of long spring-steel teeth which have the edges bent back to form a U or finger shape at the bottom. This implement is effective mainly on free-working soils, and its proper function is the killing of seedling weeds either before or just after emergence of the crop. It may be used in suitable soil conditions on crops such as sugar beet, as well as on corn, peas and potatoes. It is often superior in action to a light, zig-zag harrow owing to the flexibility of the individual teeth and to the fact that there is little or no tendency to push along the surface soil. On some wide implements the tines may be attached to a tool-bar in independent sections about 18 in. wide, which are free to follow ground irregularities.

FIG. 79.—TRACTOR-MOUNTED FOLDING SPRING-TOOTHED WEEDER. (FERGUSON.)

The spring-toothed weeder is an extremely valuable implement when the surface tilth is good, but is ineffective

227

where the surface soil has set to a hard cap. As with zig-zag harrows, some damage to the crop will usually be inevitable if weeds are to be effectively controlled. Only experience can teach how and when to use the implement, but crop damage that looks serious when the job is done often proves to be negligible after a few days.

Chain Harrows.

Chain harrows have no rigid frame; the original type consists simply of a network of chain links, and is useful for such work as rolling up weeds which have been dragged out of arable land by other harrows. Spiked link harrows in which the teeth are formed by an extension of the link are more generally useful. The most valuable types combine the advantages of flexibility and of teeth that will effectively penetrate into the soil. Such harrows (e.g. the Aitkenhead, Fig. 80) may often be used with good effect on pastures as well as on arable land. On the heavy types of flexible harrow, the knives may be adjusted so that different effects are obtained by turning them over and using the other side. Some harrows have turf blades on one side and arable land tines on the other.

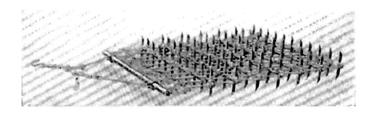

FIG. 80.—SPIKED FLEXIBLE HARROW. (AITKEN-HEAD.)

Self-cleaning Tractor Harrows.

On farms where most of the horses have been displaced by tractors, harrows especially adapted to tractor work are required. A tractor harrow should provide the tractor with an economic load; but with the lighter implements this necessitates a considerable working width and special hitches, and a better solution may be found in carrying out two or more operations in tandem. Where it is necessary to work in rubbish, it is almost essential that the implement should be self-cleaning. Another requirement is that the implement should be easily transported from field to field.

Wilder's Pitch-Pole harrow has two sets of teeth mounted at 180 degrees to one another on the same "axles", set in a rectangular frame, so that when one set of teeth is at work the other set projects into the air. A pull on a trip cord operated from the driver's seat causes the axles to rotate through a semicircle, bringing the other set of teeth into operation and depositing the rubbish. The teeth are automatically locked in the new position until another pull on the trip cord enables them to rotate again. The depth of working is controlled by means of levers which regulate the positions of the wheels relative to the harrow frame. These harrows work well where there is not an excessive quantity of rubbish, and are popular implements for the severe cultivation of permanent pastures.

Tines suitable for fairly deep arable cultivations are available, and some farmers use the Pitch-Pole mainly or exclusively as an arable land cultivator.

*FIG. 81.—TRACTOR-MOUNTED "PITCH-POLE"
HARROW. (WILDER.)*

With tractor harrows, as with many other tractor implements, the general adoption of the hydraulic lift facilitates the design of satisfactory implements. The power-lift harrow frame (Fig. 82) solves the problems of transporting the harrows and of raising them clear of the ground to deposit rubbish. Fig. 83 shows a tractor-mounted harrow with easily adjustable tines folded for transport.

FIG. 82.—POWER LIFT HARROW FRAME, WITH SADDLE-BACK HARROWS. (PATRICK AND WILKINSON.)

FIG. 83.—*TRACTOR-MOUNTED HARROW WITH*
ADJUSTABLE TINES, SHOWN IN TRANSPORT POSI-
TION. (FERGUSON.)

It might be thought that the development of power-
lifted tool-bars would remove the need for special self-
cleaning harrows such as the pitch-pole. In fact, however,
the hydraulic lift is too slow, scattering the rubbish collected,
and leaving too much land uncultivated unless the tractor
stops each time the implement is cleared. It is, of course,
possible to combine the advantages of the hydraulic lift
with rapid mechanical clearing of the implement, as in the
tractor-mounted pitch-pole (Fig. 81).

Reciprocating Harrows.

Harrows of conventional zig-zag type often produce a disappointingly slight effect in one journey across a field, and at the same time they provide such a light load that a medium-powered tractor is employed at only a small fraction of its full capacity, even with a wide implement. A recently-developed P.T.O.-driven tractor-mounted reciprocating harrow is therefore of considerable interest in that a light, narrow machine can produce an effect equal to that of several ordinary harrowings and rollings. Like the rotary cultivator, the reciprocating harrow differs fundamentally from the many harrows which have been developed from implements originally used for animal draught.

The Horstman Cultarrow may be used with any medium-powered tractor equipped with 3-point linkage. It consists of two rows of straight harrow teeth mounted on bars which are linked together by three centrally pivoted levers, and reciprocate sideways when the P.T.O. is engaged. Depth control is by land wheels at both sides of the harrow. Many of the joints on the harrow are equipped with flexible rubber bushes similar to those employed on the springs of some motor vehicles. These joints do not need lubrication.

The reciprocating principle is also employed on a mid-mounted German harrow which is driven by the pitman normally used to drive a mid-mounted mower. This implement is designed to work as the tractor is ploughing, and harrows the furrows ploughed on the previous bout.

It is still a little early to assess the full potentialities of reciprocating harrows. The complication of several joints working near the soil is a disadvantage, but if the flexible joints prove to be sufficiently robust and long-lived, as seems likely, this type of implement may ultimately displace the traditional type to a great extent.

Disc Harrows.

Disc harrows have a number of saucer-shaped discs mounted on one, two or more axles, which may be set at a variable angle to the line of draught. The discs are generally from 12 to 20 in. in diameter. They rotate as the harrow is pulled along, and their action on the soil is not unlike that of small digging-type ploughs. The precise action, however, depends on the size of discs, the depth of work, and especially upon the angle at which the disc gangs are set relative to the line of travel. If the disc gangs are set perpendicular to the line of draught, penetration is shallow, but the surface of the soil is pulverized and the lower layers are compressed. The penetration of disc harrows is achieved by virtue of the weight of the implement, and may be increased by the addition of extra weight to the pans attached to the frame for that purpose. When maximum penetration is desired, the gangs should be set with the forward edges of the discs parallel to the direction of travel.

When disc harrows are used with light-medium tractors it is advantageous to reduce the load on the tractor when turning. On some tractors this is achieved by connecting the hydraulic lift mechanism to a device which alters the angle of the discs. Raising the lift control lever rapidly straightens the discs, while a fine adjustment of the working angle may be achieved by the control lever setting. Other light types of disc harrows may be directly mounted on the hydraulic lift, and raised clear of the ground at the headlands or for transport.

The bearings are an important feature of the construction of disc harrows. It is necessary to counteract the end-thrust of the gangs due to the thrust of the soil on the insides of the discs, and this may be achieved by the use of chilled iron thrust bearings, tapered roller bearings or ball thrust bearings. On many implements, the journal bearings are of hardwood soaked in oil, and the end-thrust of each gang is resisted by that of the opposing gang by placing large "bumper washers" on the innermost disc of each gang. The chief wearing parts are the bearings and the discs themselves. The discs should be made of properly heat-treated high-carbon steel. It is essential that tractor implements be made of high-quality materials. They generally have two pairs of disc gangs, the front pair being set to throw the soil outwards and the rear pair to throw it inwards.

In British conditions the main use of disc harrows is in the preparation of seed beds in conditions where tined implements are either ineffective or cannot be employed because of pulling buried turf, rubbish or manure out of the soil. They have a consolidating effect on the lower part of the furrow slices that may be very valuable in certain circumstances. The draught is heavy, and while horses were the only source of power for cultivations, the disc harrow could never make much progress; but with tractor power widely available, this implement has achieved increased popularity both for seed bed preparations and for such operations as stubble cleaning.

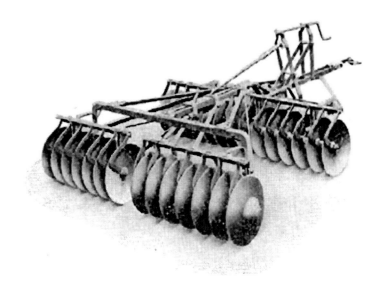

FIG. 84.—TRACTOR TANDEM DISC HARROW. (RAN-SOMES.)

Disc harrows are ideal implements for the preparation of seed beds after ploughing old turf; and much of the success of the wartime ploughing campaign may be attributed to the fact that a majority of the fields were disced. In many instances success or failure was shown to depend on the number of times the land was worked with discs. Many farmers have tried using disc harrows on old turf before ploughing. In the right conditions, and with suitable heavy discs, this method of working is sound, for the turf can be well pulverized with 4-6 discings in different directions, and when subsequently ploughed and disced again a deep, firm tilth is obtained.

In some instances seed beds both for corn crops and for direct re-seeding have been prepared from old turf by the use of disc harrows alone. This method has its limitations, and cannot be recommended for general adoption.

Disc harrows have a few disadvantages, the chief being a rather higher cost and rate of depreciation than most tined implements, a certain amount of difficulty with transport, and the fact that where there is couch ("twitch") the discs tend to cut it up and make it more difficult to pull out. In general, it may be stated that where a tined implement will do work as good as that done by discs the tined implement should be used. Disc harrows do so many jobs better than tined implements that some farmers are apt to use them without considering whether a tined implement would do the job as well or better.

Off-Set Disc Harrows. On some farms there are many uses for a disc harrow which can be off-set from the tractor, to work beneath overhanging trees. In Britain there is rather less demand for such an implement than hitherto, owing to the trend towards grass orchards. The off-set disc usually has two gangs, set in opposite directions (Fig. 290). Both gangs can be used directly behind the tractor, and with a heavy implement, deep work can be done at such a setting. If it is necessary to work land 4 ft. beyond the. tractor wheels the first gang is off-set about half this distance and the rear gang is set. to work shallower in order to keep the side thrust within reasonable limits.

Rolls.

The main objects of rolling are to consolidate the soil, to crush clods and to smooth the surface. Consolidation is necessary on some soils to give plants a firm root-hold and to ensure continuity between the top soil and subsoil. There is, at present, no reliable scientific basis for deciding when rolling is necessary. Little evidence is available concerning the degree of consolidation produced by rolls under varying conditions, or the bearing of consolidation on such questions as the moisture relationships of soils. The effects of varying degrees of consolidation on plant growth are not yet understood. In these circumstances it is necessary to be guided in the use of the roll by practical experience.

The effects produced by the roll both with respect to consolidation and to clod-crushing are largely dependent upon the soil moisture content. Rolls cannot be used when the soil is really wet, but clods will not break down easily when they are very dry. The ideal condition for clod-crushing is when the clods have been thoroughly wetted and have started to dry. They then frequently crumble at a touch.

On light land, rolls are used in the preparation of both autumn and spring seed beds. Crops do not thrive where light land is left too "fluffy", and wheat, in particular, suffers if it is sown in a loose light soil or if the surface is made very loose by the action of frosts. On heavy land the roll is seldom used in autumn except when old turf is ploughed, since it is undesirable to produce a smooth surface which may "run together" or "cap" during the wet weather. When old turf is ploughed, rolling or pressing is generally the next operation, and it is important to do this immediately, while the furrow slices are still moist, in order to get the best results.

Spring rolling of autumn-sown corn is a common practice, and during the grassland ploughing campaign great emphasis was placed on such rolling as a safeguard against the attacks of wire-worms. While one or two rollings in the spring are very desirable, especially on crops sown after old turf, it should be observed that the proper time to consolidate the lower layers of the soil is before seeding, and not when the crop is growing. Spring rolling generally consolidates only the top two or three inches of soil, and has little effect on the lower layers.

It should be understood that where rolling assists crops against wireworms it does so more by providing better conditions for plant growth than by any direct effects on the wireworms. Some farmers roll oats almost every day, on seeing damage by wireworms. The mechanical damage to the oats resulting from such treatment is considerable. There is little doubt that the best plan is to give one or two really heavy rollings and then give the crop a chance to grow by leaving it alone for a time.

One useful effect of spring rolling is the smoothing, which causes the binder to run better at harvest. Care must be taken not to roll heavy land when it is wet down below, owing to the danger of the soil becoming "puddled" or "poached".

The chief implemental factors which influence the work done by rolls are the weight per unit length, the diameter, and the type of surface presented to the soil. Different types of rolls show wide variations in these factors. If other factors remain constant the greater the weight of the roll, the greater are the degree of consolidation and the draught. The greater the diameter of a roll for a given weight, the lower are the draught and the consolidation. Rolls of small diameter sink into the soil farther than large ones, and the chief disadvantage of this is the tendency of the roll to push a wave of soil in front of it. This action may have disastrous effects on a growing crop. Another disadvantage of rolls of small diameter is the difficulty of turning at the headlands. The most common diameters range from 20 to 30 in., and the weight per foot of length generally lies between 1 and 4 cwt. All kinds of surfaces are employed, the plain and the ribbed or "Cambridge" types being most common.

The chief wearing parts of rolls are the bearings, which may be of metal, with renewable bronze bushes, or of hardwood soaked in oil. Hardwood boxes are very satisfactory if kept well lubricated. They have a low draught, a reasonably long life, and are easily and cheaply replaced. On heavy horse rolls the shafts are adjustable so that one, two or three horses may be used. In flat country the roll usually has a "swing" frame, with a castor wheel at the front. This has the advantage of removing all downward thrust from the horse's back. In hilly country, however, shafts are essential, and it may be necessary to add brakes. The draught varies widely according to the surface on which the roll is used. A roll that is easily pulled by a pony on a hard level pasture may require three good horses on soft arable land. It is, therefore, impossible to give useful figures for the draught of these implements.

Flat Rolls consist of steel or cast iron cylinders made in two or three sections to facilitate turning. The normal weight is about 10 cwt. for a width of 7 ft. The CAMBRIDGE or "ring" roll has a number of ribbed cast iron segments mounted loose on an axle. The usual width of the rings is 3 in., but 2-in. rings are sometimes used. Each ring or section is raised at its centre to a narrow rim, so that the implement leaves a characteristic pattern of small ridges and grooves on the soil surface. The rings are free to turn independently on the axle, and are mounted so that there is a certain amount of side-play to assist in keeping them free from soil. The bush of each ring, which is in contact with the axle, is chilled to make it resistant to wear. No attempt is made to lubricate this part, but the axle itself can rotate in lubricated bearings. The Cambridge roll is more effective as a clod-crusher and more generally useful than the flat roll. It is usually heavier, weighing about $2\frac{1}{2}$ cwt. per foot of width. It has many special uses, such as the breaking of surface crusts, and the preparation of a suitable surface for the broadcasting of grass seeds. Many types of clod-crushing rolls with spiked or serrated surfaces of various kinds have been used in the past, and in some districts are still in use. They generally weigh more and are heavier in draught than the Cambridge roll and are sometimes effective clod-crushers; but none appears to have sufficient all-round merit to warrant its general use.

Tractor Rolls. When tractor power is used for rolling, the ordinary horse swing roll is sometimes used; but this provides the tractor with much less than its optimum load, and unless other implements can be used in tandem, it is wasteful of power to use the tractor in this way. For efficient operation a set of special tractor rolls is required. The hitch should be constructed so that the rolls can easily be arranged in tandem for transport. In normal conditions a set of rolls 20 to 30 ft. wide is required to provide a full load for a small tractor. Little is known concerning the effect of tractor speed on the nature of the work done. Since the draught is generally light the tractor is often run in high gear. This may be advantageous where the object is clod-crushing, but the consolidating effect is not quite so great at high speeds as at low.

The "Roll Pack" is a very heavy implement used for crushing clods and especially for the consolidation of the lower layers of the soil. It consists of sections assembled like those of the Cambridge roll, but very different in shape. They are 5 to 6 in. wide and taper to a rim with a V section, the height of the V being about 6 in. The diameter of the sections is generally about 1 to $1\frac{1}{2}$ ft. to the bottoms of the grooves and an extra foot to the tips of the rims. Two axles are frequently used, the sections of the rear axle running between those of the front. These rolls are chiefly used on very sandy soils in other countries.

FIG. 85.—A HEAVY FURROW PRESS BEHIND A
FOUR-FURROW PLOUGH

The Furrow Press is used, after ploughing, to break down the lower parts of the furrow slices and ensure that there is no discontinuity between the top soil and subsoil caused by hollow spaces beneath the furrow slices. It is a heavy implement with wedge-shaped press wheels about 3 ft. in diameter, mounted on a spindle at such a spacing that adjacent wheels run in the grooves between adjacent furrow slices. The press is commonly used in the preparation of seed beds for wheat after leys, especially on light land.

The Planker Or Scrubber is sometimes used for breaking clods on heavy land in the spring. It is made by fastening together several planks of dimensions approximately 6 ft. by

10 in. by 2 in., so that they overlap about 2 in. The working corners of the implement may then be shod with iron bars. This is a useful implement on heavy land when the surface is dry but the lower layers are too wet to support the weight of a roll.

The Float is an implement that may be used for smoothing seed beds for root crops, especially sugar beet. It is widely used in some parts of Europe, and has been employed with success in Norfolk, on fields where sugar beet is to be "cross-blocked".

It consists essentially of a framework carrying about three boards which are placed at right angles to the direction of travel and are spaced about 2-3 ft. apart. The boards are set at a slight angle (often adjustable), so that they pare the soil surface rather like a plane, picking up soil from the high places and dropping it in the hollows.

Rotary Cultivators.

The possibilities of rotary cultivation as a method of preparing seed beds have engaged the attention of agriculturists ever since mechanical power became available to farmers. The idea of rotary cultivation is an attractive one, but until recent years attempts to produce a satisfactory machine had met with indifferent results. Steam power was used towards the end of the nineteenth century to drive various types of milling and digging mechanisms, but none of these achieved much success. The internal combustion engine has many advantages over the steam engine as a source of power for rotary cultivation, and rotary cultivators now occupy an important position in the equipment of many horticulturists and market gardeners. The most popular type of machine for market gardens has an engine of 5-8 h.p., carried on two wheels, which are driven through clutches and suitable reduction gearing. Power is transmitted from the engine to a "miller" shaft, mounted parallel to the axle of the travelling wheels. On the miller shaft are mounted either small curved hooks or sturdy hoes which successively enter the ground as the machine moves forward. The soil is thrown backwards and upwards from the machine in a loose, "fluffy" mass that generally needs some consolidation before small seeds are sown. The loose tilth produced by the rotary cultivator is suitable for much horticultural work.

The tilth produced by the rotary hoe is not necessarily very fine, and may be varied somewhat by choice of gear, a high forward speed giving a coarser tilth. Small tractors are not, however, able to provide sufficient power to operate them at high forward speeds if working depth is to be maintained.

FIG. 86.—2-CYLINDER 8 H.P. WALKING TYPE RO-
TARY CULTIVATOR WITH OFFSET ROTOR. (ROTARY
HOES.)

With sturdily constructed rotary hoes breakages seldom cause much trouble, even in severe working conditions. The reliability of the machines is well demonstrated by the fact that they are frequently set to work on building sites where all kinds of obstructions are encountered. Modern machines are incomparably more durable than some of the earlier spring-tined rotary cultivators.

Tractor-mounted rotary hoes are now available for all the leading medium-powered British tractors, as well as for some kinds of more powerful tracklayers. The machines fitted to medium-powered tractors equipped with hydraulic lifts may be raised clear of the ground on the power lift linkage. Depth is controlled by a wheel fitted at the rear of the tiller, and on some models the tilth is controllable by varying rotor speeds. Width of cut ranges from about 36 to 48 inches and depth of work is up to about 10 inches. The necessity to provide a degree of rigidity in the drive gears renders the attachments somewhat less easy to put on and remove from the tractor than most P.T.O.-driven machines.

Farmer & Stockbreeder

FIG. 87.—SMALL ROTARY CULTIVATOR

These machines are now used for a variety of farm jobs including the first working of stubbles after the combine harvester. It is not yet possible to assess to what extent they are likely to be used in future for seed-bed preparation in general farming, but their use for such purposes is increasing, and farmers who have them for other purposes find the machines particularly useful for such odd jobs as weed control on the headlands of root fields.

Rotary cultivators often reduce the number of cultivations necessary for producing a tilth, and can sometimes be used to prepare seed beds in one operation. It is, however, only in the best conditions that one operation suffices for the production of a satisfactory seed bed.

There is no reliable evidence either for or against the claim often made that rotary cultivation is the most "efficient" method of applying power to cultivation operations. It is true that when a tractor is pulling a plough a proportion of the power is wasted in its transmission from the engine to the implement drawbar. It is also true that rotary cultivators waste less power in this way. In spite of this it is doubtful whether, except in special circumstances, an ordinary seed bpd can at present be produced with the expenditure of less work by rotary cultivation than by the traditional methods. The traditional methods have been developed in such a way that the natural weathering agents do much of the work in the preparation of a tilth, and though it may often be useful to have machines which can prepare seed beds in spite of the weather, it is not always desirable to attempt to do this. Nevertheless, the steady increase in use of rotary cultivators throughout the world is an indication of their usefulness for many kinds of cultivation.

FIG. 88.—TRACTOR-MOUNTED ROTARY CULTIVA-TOR. (ROTARY HOES LTD.)

CHAPTER EIGHT

SEED DRILLS, PLANTING MACHINES AND IMPLEMENTS FOR INTER-CULTIVATIONS

"And in Essex, they use to have a chylde, to go in the
forowe before the horses or oxen with a bagge or a hopper
full of corne; and he taketh his hande-full of corne, and by
lyttel and lyttel casteth it in the sayde forowe. Me semeth,
that chylde oughte to have moche dyscrection."

Boke of Husbandry. Fitzherbert, 1523.

Seed may be sown by broadcasting, dibbling, ploughing
it in, or drilling. For most crops drilling is the modern
method of seeding. It has advantages over the other methods
in economy, efficiency, and adaptability to the modern
methods of farming. One of its chief merits is that it renders
inter-cultivations possible.

A rude kind of drill has been used in some countries
from a very remote period; the cultivators of China, Japan
and Arabia have drilled or dibbled in their seed from time
immemorial. In this country Sir Hugh Piatt, in *The newe
and admirable arte of setting come,* published in 1601, wrote:

"Happily some sillie wench, having a few cornes of wheate, mixed with some other seed, and being carelesse of the worke shee had in hand, might now and then instead of a raddish or carret seede, let falle a wheate corne into the ground, which after branding itself into manie eares, and yeelding so great encrease, gave just occasion of some farther triall." A little later, Gabriel Platte described a rough drilling machine, and John Worlidge, in his *Systema Agriculturae,* published in 1669, advocated the use, not only of the seed drill but of the manure drill. At a later period (1730-40) Jethro Tull devoted his energy to the introduction of the drill "more especially as it admitted the use of the horse hoe". He says, in his *Horse Hoeing Husbandry* (1733), "The Drill is the Engine that plants our Corn and other Seeds in Rows; It makes the Chanels, sows the seed in them, and covers them, at the same time, with great Exactness and Expedition". Tull's drill bore little resemblance to modern machines, but by 1782 James Cooke had produced a machine in which may be detected the counterparts of most of the essentials of the modern "Suffolk" drill. The development of drills after this was rapid. Early in the nineteenth century the firm of Smyth, of Peasenhall, started to produce drills that were the forerunners of their Suffolk drill which is widely used in the Eastern Counties to-day, while as early as 1839 Messrs. Hornsby took out a patent for a "drop" or spacing drill. In recent years American influence has been responsible for the introduction of light all-metal drills differing in many respects from the Suffolk type, and these have come into favour along with tractor farming.

The *desiderata* of drilling are, briefly, that the seed should be deposited in the desired amounts at a uniform depth and spacing, the depth and spacing to be variable at will. It is generally considered that the ideal arrangement is for seed to be deposited in straight rows and at regular, adjustable intervals in the row. This ideal is attempted in "spacing" drills; but in most drills, all that is attempted is to control rate per acre and to get the average density of distribution the same in every row. Trials conducted by the Royal Agricultural Society at Doncaster in 1912 showed that there was a wide variation between the amount of seed sown by individual coulters in even the best of the drills. Careful scientific investigation later showed that there were enormous differences between the numbers of seeds sown on adjacent foot-lengths of the same drill row. These irregularities in action maybe demonstrated in some modern drills, in spite of improvements in certain features.

It is possible that the performance of drills with respect to uniformity of distribution, both between different rows and within individual rows, might be improved by the general use of machined parts in the feed mechanisms, but it is doubtful whether the greater cost of such machines could be justified. Even if it were possible to obtain absolute uniformity of seeding, it has yet to be shown that this would lead to increased yields. Preliminary fundamental studies on the relation between spacing and yield have so far failed to show that absolute uniformity of spacing is necessary. There can, however, be no doubt that yield is lowered, if large patches of land receive too little or too much seed.

General Design of the Drill.

The modern general-purpose drill may be used for sowing all kinds of seeds from the size of beans to that of clover, at various seed rates and depths, and in rows at various distances apart. The drill has a braced frame and a seed box extending across its width. The seed is delivered from this box by a feed mechanism driven by gearing from one or both land wheels. A clutch is included for putting the feed mechanism in and out of gear. The seed passes down seed tubes to the coulters, which cut grooves in the soil. The seed rate is adjusted by alterations of the gearing or of the feed mechanism, and the spacing between rows and the depth of sowing are adjusted by the setting of the coulters. The more common types of feed mechanisms and coulters are described below.

FIG. 89.—TWENTY-EIGHT ROW SELF-LIFT TRACTOR DRILL WITH FORGE-FEED AND DISO COULTERS. (MASSEY-HARRIS.)

Feed Mechanisms.

The CUP FEED is the type most common in British drills, and is still one of the most generally useful types for mixed farming. The seed hopper is divided into two parts, an upper compartment, the grain box, and a lower, the feed box. These are connected by shutters. The seed is placed in the grain box, and the shutters between this and the feed box are adjusted so that there is always an adequate supply of seed in the feed box, but not an excess. The feed box is divided into a number of chambers in which the delivery mechanism operates. The latter consists of a spindle carrying a series of discs with a ring of cups attached to the periphery of each; the spindle and its attachments is called the seed barrel. As it rotates, one disc, with its set of cups, rotates in each compartment of the feed box. Each cup picks up a few seeds and drops them into the small hoppers leading to the coulters. The seed barrel is usually reversible, the cups being made with two faces, one for large seeds and one for small. Two seed barrels with double-faced cups suffice for all normal requirements. On one barrel, the larger cups are used for beans, peas and oats, and the smaller ones for wheat and barley. On the other barrel, the larger cups are suitable for roots, while the other side may be used for such small seeds as clover, mustard and kale. The cup feed, like nearly all feed mechanisms, delivers the seed to the seed tubes in small batches; but by the time the seed has fallen down the flexible seed-tube and reached the soil the flow is more or less continuous.

FIG. 90.—CUP-FEED MECHANISM. (SMYTH.)

There are several useful modifications of the simple cup feed. In some foreign drills (e.g. the Melichar) the size of the cups may be varied by adjustment of a wheel at the end of the seed barrel. On the Melichar this adjustment is a very fine one, and the drill can be set to sow almost any crop at any desired rate.

Another useful modification is the disc feed, used on some grass and clover drills (e.g. the Coultas and Rainforth). The discs on the seed barrel are about ⅛in. thick and the cups by which the seed is delivered consist of pockets cut in the circumference.

The seed rate in cup-feed drills is controlled by the size of the cups and the rate at which the seed barrel revolves. The rate of rotation may be varied by fitting pinions of different sizes to the seed barrel; the larger the pinion fitted, the lower the rate of seeding. The different sizes of pinion wheels are accommodated by the use of an idler pinion between them and the pinion connected to the land wheel. The makers provide tables showing the proper pinions and cups to use for sowing various quantities of different kinds of seed. Where there is doubt, the seed rate may be roughly adjusted beforehand by jacking up the drill and collecting the seed delivered when the drive wheel is turned a given number of revolutions. Under these conditions the shaking which affects the seed rate in the field is absent, and a more reliable method is to push the drill along for a known distance and measure the quantity of seed delivered into bags placed over the coulters. With either method a simple calculation shows the seed rate per acre.

The seed box must be kept horizontal when the drill is operating up and down hill, or the seed rate will fluctuate owing to the irregular flow of grain through the shutters, and the change in the relative positions of the seed hoppers and the centre of the seed barrel. The box is kept horizontal by operation of a crank which usually has attached to it a pinion; the pinion engages with a vertical rack at the front of the box.

The outstanding advantage of the cup feed is its adaptability. It can be adjusted to sow all kinds of seeds, and the use of spring flaps or tilting hoppers at the tops of the seed tubes facilitates adjustments involving the change of seed barrels or shutting off various coulters. The mechanism cannot injure the seed, and when the work is finished the seed box may be easily emptied and cleaned.

The chief disadvantage is the irregularity of feed caused by rough conditions, when the seed may be jolted out of the cups before reaching the normal position above the seed hoppers.

The External Forge-Feed. A fluted roller was used as the feed mechanism on some of the earliest English drills. The mechanism of the modern external force-feed drill consists essentially of a fluted roller which rotates just below the seed box and draws seed from the bottom of the box into hoppers at the tops of the seed tubes. The seed box is a single compartment, and the fluted rollers rotate in feed runs attached to the bottom of it. The rollers are fluted over only half their length, and can be moved laterally so that either the plain or the fluted portion, or a part of each, is in contact with the seed. This provides a simple regulation of the seed rate; for no seed is delivered by the smooth part of the roller. The rollers are driven by a spindle which receives its motion from the land wheels. On some machines the rollers are connected to the spindle by individual dog clutches, and coulters may be shut off as required by disengaging these clutches. Modern machines are sometimes fitted with adjustable spring-loaded baffle plates in the feed runs. The use of these causes less damage to large seeds and permits adjustments for drilling various types of seed and for varying the seed rate.

FIG. 91.—EXTERNAT. FORGE-FEED MECHANISM.
(MASSEY-HARRIS.)

The force-feed drill is steadier over clods than cup-feed types, and works more uniformly and with less attention on hilly land. It has also the advantages of simplicity and cheapness; but it is not as generally adaptable as the cup-feed type, though it is claimed for many of the modern machines that they will sow any kind of seed from beans to clover.

The Internal Double-Run Force Feed. The internal force-feed mechanism has been developed in the United States, and is especially efficient in the sowing of cereals. The feed mechanism consists of a spindle carrying a series of flanged discs, the insides of the flanges being slightly corrugated. Each feed disc is housed in a casting which fits in the bottom of the seed box. As the wheels rotate, the seed is drawn by the corrugations past baffles to a point outside the box, where it falls into the seed tubes. On one side of the flanges, both the serrations and the clearance between them and the baffle are smaller than on the other side. Either side may be used at will by adjusting the position of a hinged flap in the hopper, which covers either the coarse or the fine side of the feed run. When sowing oats the coarse side is required, while for wheat the fine side generally gives a sufficiently high seed rate.

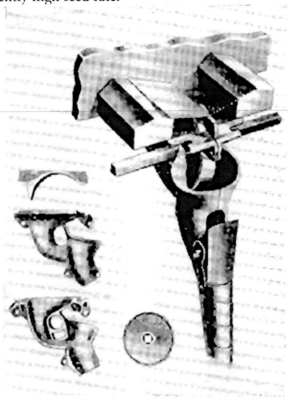

FIG. 92.—INTERNAL DOUBLE-RUN FORCE-FEED
MECHANISM. (MASSEY-HARRIS.)

This type of mechanism is one of the most efficient available for sowing wheat or barley. It has the advantages of all force-feed drills, and gives a more continuous flow of seed than the external-feed type. Like other force-feed drills it suffers from the disadvantage that cleaning out is not easy, but this is partly offset by the ability to sow down to the last pound of seed. Its main disadvantage is that it is less adaptable than either the cup feed or the external force-feed type, being really suitable for sowing few crops apart from cereals.

The internal force-feed mechanism is used for the grain feed on many combined seed and fertilizer drills. Combine drilling is discussed in Chapter Nine.

Other Types of Feep. The feed mechanism of most of the drills at present employed in Britain is of one of the three types described above, but it should be stated that various other types of drill are in existence, especially in other parts of the world. A brush feed, similar to that used in the seed barrow, is used on many root drills, while a "plate" feed is used on many American drills used for sowing maize, sugar beet, etc. The plate feed consists essentially of a flat disc, with grooves cut in the edge, rotating in the bottom of a cylindrical hopper. One or more seeds drop into the grooves in the disc, and are carried round until they are above a hole, through which they drop to the seed tubes. Various types of plates are used for different jobs. In some instances the grooves are of such a size and shape that a single seed is sown. For a full description of this type of feed mechanism, reference should be made to American textbooks. In Germany and many other European countries, a common type of feed consists of a studded roller. This has many advantages, one of the chief of which is a wide adaptability. A few drills of this type also have the advantage of a special device which facilitates calibration and cleaning out of the drill.

Drill Coulters.

The Suffolk type of coulter is still most commonly used in Britain, but the single-disc type is rapidly gaining popularity. In the Suffolk, a renewable chilled iron shoe cuts a groove in the soil and the seed is delivered by the seed tube into the groove. In good conditions this coulter works satisfactorily, the soil falling back on the seed and covering it well. It opens a narrow furrow and does not easily choke with weeds. Its disadvantages arc that it will not penetrate the soil in hard conditions and blocks easily on sticky land.

In the single-disc coulter a saucer-shaped hardened steel disc cuts the furrow, and the seed is delivered to a "boot" attached to the convex side of the disc, just below and behind its centre. The great advantage of the disc type is that it will work satisfactorily in all kinds of unfavourable conditions. Where the ground is hard a fair penetration can be secured; where there is rubbish the discs will cut through it and not collect it; where the tilth is poor, the discs effect an improvement by their pulverizing action; and where it is sticky, discs will keep clean better than other types of coulter. The single-disc coulter thus has many attractive features. Its disadvantages are the initial expense, the fact that it does not place all the seed at the same depth, and the constant need for attention to the lubrication of the chilled iron bearings.

The double-disc coulter consists of two plain discs set at a small angle to one another. The seed is delivered between them. They work well in rubbish, place all the seed at the same depth, and cover it well; but they are rarely used owing to the large number of wearing parts and the high cost. The hoe coulter resembles a cultivator share, with a half-round or V section. It is generally reversible and is made of hardened steel. This coulter is not much used in Britain. Its chief advantage is its simplicity and ease of penetration, but its depth of work is not uniform and it causes trouble when working in rubbish.

The coulters are attached to levers which hinge on a crossbar at the front of the drill. They are thus enabled to follow inequalities in the surface and to ride over obstructions. On the common Suffolk type, weights may be attached to the ends of the levers to increase the penetration. Chains attached to the levers and passing over a windlass permit control of the maximum penetration and enable the coulters to be raised at the ends. Press irons, by which the coulters may be forced into the ground, may be used either instead of or in addition to the weights. On modern drills, pressure is generally applied to the coulters through springs, the pressure being controlled by a lever at the back of the drill. The coulter levers on corn drills are usually arranged in two ranks, alternate coulters being in the front and rear ranks. This "staggered" arrangement is designed to secure freedom from blocking when working in rubbish or large clods.

Drill Gearing.

Many improvements have been made in the design and construction of the gearing in modern drills. The old type of drive, with a gear wheel attached to one of the land wheels, is being largely replaced by a pawl and ratchet drive from both wheels. This ensures that the feed mechanism is driven when the drill is turning corners in either direction, or is swinging over a rough surface. In many drills it is no longer necessary to change pinions in order to regulate the seed rate.

On most external force-feed drills a single control lever is used, and on many internal force-feed machines seed rate is regulated by the use of two mechanisms, viz. an epicyclic two-speed gear, combined with a multiple bevel gear disc *(vide* Appendix Two). Adjustment of the seed rate is then secured by means of two levers controlling these gears.

On modern drills a simple land measure, consisting of a dial driven from the axle by worm gearing, registers the acreage seeded. Roller bearings are used in the axle boxes to reduce friction, and the whole drill is made light, easy running and strong. On heavy soils, however, some farmers still prefer the old Suffolk type on account of its simplicity, steadiness and general ruggedness.

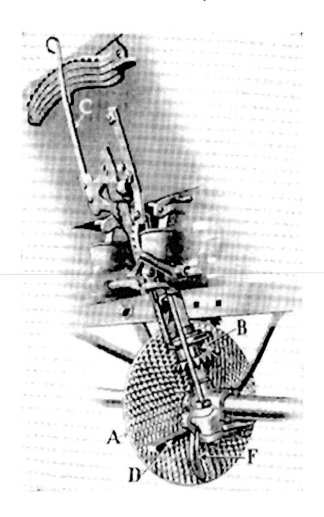

*FIG. 93.—MULTIPLE BEVEL GEAR FOR CONTROL-
LING SEED RATE, (MASSEY-HARRIS.)*

A, face gear; **B,** bevel pinion; **C,** lever for regulating seed
rate; **F,** lever for putting mechanism out of gear.

Steerage.

The majority of the drills used in this country have only two land wheels, but in some districts drills arc regularly fitted with a two-wheeled "steerage" fore-carriage. The advantages of this for horse work are a reduction of "swinging", the possibility of steadier and more accurate work on rough land, and the removal of downward thrust from the shaft horse's back. The axle bar of the fore-carriage usually pivots on a hinge attached to a stub pole, fixed centrally on the front of the drill. A steerage fore-carriage with stub-axles and Ackermann or automobile type steerage is in general use on the Continent of Europe.

The steerage drill has many advantages under adverse conditions, but it has the disadvantages of extra initial expense, heavier draught, and sometimes extra cost of operation. In good conditions work may be as well done with a pole or a shafts fitting, but on rough, late-autumn seed beds and for market garden crops the steerage forecarriage often proves its worth.

Tractor Drills.

Drills specially constructed for tractor operation are normally fitted with adjustable markers, and with a self-lift.

The self-lift operates in the same way as that on a plough or cultivator, the tractor driver pulling a cord to engage a clutch, which raises the coulters and puts the sowing mechanism out of gear.

Semi-Mounted Tragtor Drills. Some tractor drills are carried on and driven by their own two wheels, but are connected to the hydraulically operated 3-link system of the towing tractor.

FIG. 94.—SEMI-MOUNTED TRACTOR DRILL. (FER-GUSON.)

The hitch is so devised that raising of the drill coulters and shutting off the seed delivery is simply effected by operation of the tractor's hydraulic lift control lever. This is a great advantage, eliminating the uncertainty and slow action typical of many mechanical drill lifts. The lift can, of course, be operated without the necessity of moving forwards. Headlands can be narrower, and the drilling can

Farm Machinery

be started or stopped to a more accurate line than with most trailed power-lifted drills.

Tractor-Mounted Drills. Several firms now manufacture their lighter types of drills for direct mounting on the tractor's 3-link hydraulic lift system. With such drills the wheels are used to drive the mechanism, and putting the drill out of gear is simply effected by lifting the whole machine bodily off the ground. As soon as the drive wheels are clear of the ground the drive to the feed mechanism automatically stops.

It seems likely that this method of mounting will be applied to most of the light drills of the future, while the semi-mounted method is likely to be increasingly employed for the heavier corn and combine drills.

Setting Markers. Setting of the markers for tractor work often causes difficulty, and is sometimes done incorrectly. The rule is simple. The distance of the marker from the outside coulter must be equal to the distance between the front wheel track and the outside coulter on the same side, plus one coulter width. If this rule is followed it does not matter whether the drill is not off-set or not; and the same rule may be applied to the use of markers for any job. With ridging, for example, the distance from centre of outside ridge to marker will be equal to the distance from front wheel to centre of ridge on the same side, plus the distance between ridges.

271

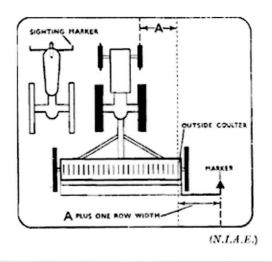

FIG. 95.—ADJUSTMENT OF MARKERS

Operation of the Drill.

The power required for drilling varies between wide limits. Whereas a 7-ft. drill fitted with shafts, and drilling four rows of sugar beet is easy work for one horse, the same drill, fitted with a fore-carriage and twelve coulters, drilling wheat on wet, heavy land may be a heavy load for four horses. The first operation after the drill has been "set" should, in difficult conditions, be the drilling of the headlands. If the headlands are left until last it is difficult to cover the seed owing to the effect of trampling. When the drill is turned or backed the coulters should always be raised to avoid the risk of bending the drag levers or the coulters themselves. When a drill is being stopped or started it should be remembered that there is always a lag between the operation of the feed mechanism and the arrival of the seed at the coulters. In cup-feed drills this lag is generally equivalent to about a yard along the field. The feed mechanism must, therefore, be put in gear just before the position where it is desired to start sowing; and if the drill stops in the middle of the field it should either be backed before restarting or a handful of grain should be broadcast where the stop occurred.

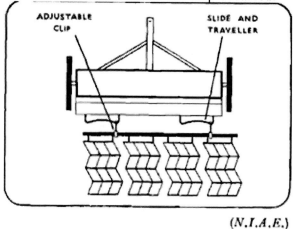

(*N.I.A.E.*)

*FIG. 96.—METHOD OF ATTACHING HARROWS
BEHIND DRILL. (A SINGLE, CENTRALLY ATTACHED
SLIDE IS OFTEN USED.)*

The amount of drilling done by tractor power steadily increases, on all types of land. Even on sticky clay land light tractors, fitted with skeleton type wheels, may be seen drilling efficiently where formerly a team of four horses put up an indifferent performance.

When tractors are used for drilling it is advisable to equip the drill with a strong foot-board, so that the follower can ride in comfort and attend to his job efficiently. The foot-board should be braced so that harrows can be attached to it for "heeling in". The hitch for the harrows should be made as shown in Figure 96, so that the harrows are slightly off-set and do not cover the outside wheel mark. With a well-made hitch the harrows automatically swing over into the right position as the drill turns.

Root Drills.

Roots grown "on the flat" may be sown by corn drills with some of the coulters out of action. The 12-coulter cup-feed corn drill, with coulters 7 in. apart, is immediately ready for drilling roots 21 in. apart if all coulters except the second, fifth, eighth and eleventh are removed. But in some districts roots are normally grown "on the ridge", and special drills are required for this purpose. These are generally 2-row machines (Fig. 97), with a concave roller preceding each coulter and a small pressing roller following it. The concave roller serves to keep the coulter running along the centre of the ridge. The feed mechanism in these drills is generally of the simple cup-feed type. Tractor-mounted ridge drills are now available.

Figure 98 shows a 4-row root drill that is popular for sowing sugar beet on light soils in the West Midlands. The land is worked to a level surface and a fine tilth, and the coulters are set to drop the seed near the surface as a pair of sweeps form a small ridge of fine soil. This drill may be fitted with side hoes for working the crop later.

FIG. 97.—TWO-ROW RIDGE DRILL. (GOWER.)

Unit Drills for Root and Vegetable Crops.

While it is possible to make quite a good job of drilling almost any kind of root crop or the commonly grown vegetable crops with a good general-purpose drill, many farmers and market gardeners find it more convenient and efficient to use a unit type drill which has been specially designed for such crops. There is now available a wide range of unit drills for use with almost all kinds of tractor-toolbars, as well as units for use with specialized market garden tractors (Fig. 36).

Unit drills may be broadly classified into two main types, viz. (1) those in which the feed mechanism consists essentially of a hole and an agitator to keep the seed flowing, and (2) those in which there is a positive feed mechanism. The latter type may be further subdivided into those in which the mechanism sows several seeds at a time and relies on the drop down a long seed-tube to provide scatter, and those with mechanisms designed to sow single seeds, with a short seed tube to ensure that uniformity of spacing achieved by the sowing mechanism is not nullified by a long drop to the soil.

FIG. 98.—FOUR-ROW COMBINED RIDGING AND DRILLING MACHINE. (GOWER.)

[N.I.A.E. photo]

FIG. 99.—SIX-ROW REAR-MOUNTED DRILL UNIT WITH DRIVE FROM TRAILING WHEEL. (HUMBER-SIDE ENG. CO.)

With the agitator type (e.g. Planet, Bean, Wild, etc.), seed rate is governed by the size of hole used, and the flow of most seeds is little influenced by the speed of the wavy-disc agitator. With a given setting the seed flows from the hopper at a more or less constant speed regardless of the rate of forward travel. This means in practice that the faster the drill travels at a given setting, the lower the seed-rate; and it is important to control the tractor speed carefully if a uniform sowing rate is to be achieved.

Sugar Beet Drilling.

"Single-seed" unit drills are of great interest both to farmers and horticulturists. There has been much development work both in Britain and abroad on drills for sowing sugar beet seed thinly and uniformly, the aim being to reduce the labour required for hoeing and singling, and to produce better stands of sugar beet. Since the work on sugar beet illustrates a problem which is common to many crops, a brief account of recent developments is given.

The problem of eliminating hand work from the singling of sugar beet is complicated by the fact that the beet "seed" is really a seed cluster, generally containing one, two or three viable seeds. With a perfect spacing drill and natural beet seed, the problem of singling remains.

As a result of research on this problem in Germany and the United States, as well as in Britain, a process for splitting the seed clusters into single seeds was developed; and machines for this purpose were manufactured and used on an extensive scale in the United States. The machine for "shearing" the seed consisted essentially of a silicon carbide stone that rotated close to a steel shear plate. With seed at a suitable moisture content and the machine carefully adjusted, seed clusters could be separated into individual seeds without very much damage to germination.

With the sheared seed, which had to be carefully sieved and graded before sowing, a suitable drill, and mechanical cross-blocking, it was found possible in the United States to obtain satisfactory stands of beet without employing any hand labour for singling. Bainer 17 quotes the results obtained in a field in Colorado, where with sheared seed spaced an inch apart, and cross-blocking carried out with hoes set to cut 3 in. and leave a gap of 1 in., only 2.45 man hours per acre were needed to operate the machine and later chop out a few surplus beet. In comparison, the customary method of drilling, hand blocking and singling, using normal seed, required 27.2 man hours per acre. Moreover, there was no difference in yield, both crops producing 12 tons (2,000 lb.) per acre.

Such results showed great possibilities of saving labour in the use of treated seed and suitable drills, but the results of careful trials in this country were not so encouraging. It has been found that saving of labour by the use of suitably treated seed and cross-blocking or other mechanical gapping is possible; but the method of shearing the seed described was never entirely satisfactory and has been replaced by a more gentle rubbing process. Research continues into methods of "milling" the seed and various other treatments prior to sowing. The problem would, of course, be simplified if plant breeders could produce a strain of sugar beet with single seeds.

Spacing-Drills.

In spacing-drills, the object is to sow seed singly or in small bunches at equally spaced intervals along the row. In this country interest in spacing-drills is mainly in connection with root crops. The chief objects of spacing are the establishment of a better "plant", reduction of the seed rate, and reduction of the labour required for "chopping out" and singling. Many spacing drills have been invented at various times, but none, with the exception of those used for sowing maize in America, has yet been sufficiently successful to come into general use. Of those which have worked at all, many have failed to achieve much semblance of spacing, while others have deposited the clusters of seeds too close together, so that singling is extremely difficult.

The principles on which the design of a successful spacing drill depend are being gradually established, partly as a result of repeated attempts by inventors to resolve the difficulties single-handed, and partly by means of studies at research stations. New workers in this field should make a point of studying all the trials which have already been made before attempting to put their own ideas to the test. This is the kind of development on which research institutes such as the National Institute of Agricultural Engineering are well fitted to assist.

The increased use of tractor machines in root crops should be considered in connection with the development of spacing drills. Where cross cultivations are to be carried out the ordinary spacing drill is of no use. In the sowing of maize the use of "check-row" planters is well established. These machines sow maize seeds with almost perfect precision, spacing the seeds accurately in the row, and also planting them exactly opposite one another in adjacent rows so that the crop is in rows in two directions at right angles.

This method is not applicable to crops requiring narrow spacings, e.g. sugar beet, and even for maize its use is confined to specially favourable conditions. For such crops as sugar beet the most sensible objective seems to be to employ a drill which sows a single seed at regular short intervals (e.g. i inch) along the row and to chop out the surplus plants either by "down-the-row" thinners or by "cross-blocking".

"Single-Seed" Drills.

The principle of operation of most "single-seed" drills is to employ a seed ring or disc with a series of slots which are just big enough to hold one seed from a carefully graded sample. The seed is carried round in the slot and is positively ejected when it arrives above the furrow opener. It is essential to have a very short drop from the seeding mechanism to the ground, so that the seeds remain regularly spaced after they have fallen into the soil. Fig. 100 illustrates one kind of mechanism working on this principle. Other types of mechanism are being or have been developed, and it seems likely that such drills will be used for many other crops where accurate drilling is important.

Some vegetable seeds, owing to the peculiar characteristics of their seed coats, present special problems, and for all seeds with rough or irregular seed coats the process of "pelleting" is attractive. Some existing drill mechanisms work very efficiently with uniform pelleted seed, but more research is still needed on the pelleting process to produce pelleting materials which are hard enough to stand up to

handling, yet do not interfere with germination and are economic in price.

Press Drills.

A method of sowing corn employed in some districts is with a seed box attached to a plough press. The seeding mechanism, mostly of simple force-feed type, is driven from the land wheel. The seed is dropped into the grooves made by the press, and harrows follow to cover it.

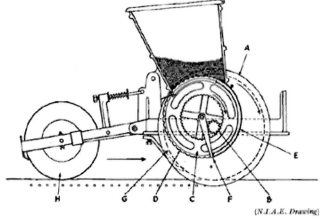

[N.I.A.E. Drawing]

FIG. 100.—SECTION OF SINGLE-SEED DRILL UNIT.
(MILTON.)

A, double-disc furrow opener; B, driving pinion; C, seed wheel; D, seed cells; E, seed retainer; F, ejector finger; G, depth bands; H, press wheels.

283

Press drills are often seen to greatest advantage in late autumn, when corn or beans can often be sown well by this method at a time when normal methods of preparing seed beds would be impossible.

Grass Seed Drills.

The essential feature of a satisfactory grass seed drill is a coulter spacing of not more than $3\frac{1}{2}$ inches. The main type of feed mechanism employed on grass seed drills is a modified cup-feed. The seed barrel carries discs which may range in thickness from about $\frac{1}{4}$in. to $\frac{3}{8}$in., and these discs have cups in the periphery. Regulation of seed rate is mainly by changing seed barrels and gear wheels, but some adjustment is also obtainable by a slide which regulates the flow of seed from the main seed box into the seed compartments.

Broadcasting Machines.

Broadcasting machines are frequently preferred to drills for the sowing of grass and clover seeds, though special drills with closely spaced coulters may do better work in some circumstances. Broadcasting machines generally have a long seed box with a brush feed. This may be mounted on a pair of wheels, with shafts for a horse or a tractor drawbar, or it may be. on a barrow frame for hand work. Seed is brushed from the box and falls direct to the surface of the soil. The seed barrow is the most commonly used broadcasting machine of this type. The seed is pushed from the box through a series of circular apertures, the size of which can be readily adjusted for different seeds and seed rates. The brush feed works well while the brushes remain in good condition, but is not reliable when they are worn. They are, however, easily and cheaply replaceable.

The Bean grass seed drill (Fig. 101) is designed to be attached behind a Cambridge roll by bolting it to the roll frame, and is driven by a trailing wheel. It has a fluted roller feed which is suitable for sowing seeds mixtures but not for pure crops of very small free-flowing seeds such as clover, timothy or seeds of "brassica" crops. The drive is easily disengaged from the tractor seat by pulling a cord attached to a lever which raises the trailing wheel clear of the ground. Sowing width is 8 ft., and sowing rate is varied by an adjustment which moves the fluted roller longitudinally within the seed box.

A small broadcasting device called the "Fiddle" is preferred by many farmers, especially for hillside work. This machine, consisting of a hopper and a mechanical distributor, is strapped to a man's shoulders. As the sower walks along he draws to and fro a "bow", which causes a small ribbed platform to rotate and fling the seeds out to the sides. Heavy seeds are, naturally, often thrown farther than light ones.

Many farmers have found that they can sow grass and clover seeds quite efficiently with their fertilizer distributors. While some distributors are unsuitable for this work, most modern ones that are in good order may be used.

The advantage of broadcasting over drilling for grass and clover seeds is that in good conditions a more complete covering of the ground is achieved. But in certain conditions, especially in dry districts and on heavy land, it may be better to place the seeds on moist soil just below the surface by drilling with small, closely spaced coulters.

FIG. 101.—GRASS SEED BROADCASTER FITTED TO
CAMBRIDGE ROLL. (HUMBERSIDE ENG. CO.)

Potato Planters.

Before 1939 potato planting machines had been little used in Britain, and there appeared to be little or no demand for planters from the chief potato-growing areas, though a number of machines had been available for many years. In the United States, on the other hand, mechanical planting was the normal method in the principal potato-growing areas, and American machines had been developed to a fairly high level of efficiency.

When it became necessary to expand the acreage under potatoes, and to grow them on a large scale in districts where there was no established technique, it was not surprising to find farmers seeking methods of mechanizing the planting operation. A few American machines were imported; and a number of machines similar to pre-existing British ones were used; but the most striking development was the invention of a number of efficient new machines, many of which were quite unlike any previously used. A brief description of the main types of planters in use to-day is given below.

(i) Implements Used to Facilitate Hand Planting. In Soft land some farmers use a device consisting of three large wheels with blunt spikes on the rims. This implement is used after the ridger, and it makes depressions at regular intervals in the bottoms of the baulks, into which the potatoes are set.

A dibbling implement with star-shaped wheels, which digs out pockets of soil at regular intervals, either on level ground or in the bottoms of baulks, is used in Germany. It has been tried in Britain but is not in general use. Both the above types ensure regular spacing. They also enable the workers to drop the potatoes into the holes, instead of having to place each one in position.

The dibber wheels may be combined with the ridger and a fertilizer distributor, and the whole mounted on a tractor toolbar. This combination has the advantage of getting three jobs done in one trip across the field, and avoids compaction of the soil in the bottoms of the ridges. The dibber wheels provide the drive for the fertilizer distributing mechanism.

(ii) Planting Attachments for Ploughs. Innumerable ingenious devices have been invented to facilitate dropping potatoes into the furrow when ploughing. Some consist of little more than a hopper and a seat, while others are more or less automatic. None of these devices is sufficiently widely used to merit special description.

(iii) Sledge-Type Planters. With the simple sledge-type planter a V-shaped share opens a furrow, and two operators, working alternately, drop the potatoes directly into the groove. Covering discs follow to complete the work. The tractor must run slowly for satisfactory work. A clicker wheel which beats a regular time may be used to assist in getting even spacing.

This type has been largely displaced by the toolbar-mounted hand-fed type.

(iv) Toolbar-Mounted Hand-Fed Planters. One of the most common types of planter is a simple 2-row machine designed for mounting on a standard 3-row ridging toolbar. It normally works off the flat on a well-prepared seed-bed. It consists essentially of a hopper or a rest for chitting trays to hold the potatoes, and two conveniently-placed chutes which end in simple V-shaped furrow openers. The two operators drop the potatoes as regularly as they can into the planting chutes, and the potatoes fall directly into the groove made by the furrow openers. The furrow openers slightly precede the ridging bodies, which cover the sets as soon as they fall and so prevent them from rolling. There is usually a clicker wheel and bell, which may be set to ring at adjustable intervals corresponding to the commonly required spacing distances. These are cheap and simple machines, with no moving parts except the clicker wheel. Their efficient use naturally depends much on the ability and industry of the operators.

FIG. 102.—THREE-ROW TRACTOR-MOUNTED SEMI-AUTOMATIC POTATO PLANTER WITH FERTILIZER ATTACHMENT. (ROBOT.)

(v) Semi-Automatic Planters. Many British planters are of the semi-automatic type, in which the potatoes have to be placed in cups by operators sitting on the machine. The Bruff planter is a two-row machine, with conveyor belts running fore and aft on both sides. Row widths are adjustable from 24 to 30 in. Chitted seed can be planted without much damage. Either two or four operators are needed, according to the tractor speed. With the Robot planter (Fig. 102), the operators place the sets in shallow hinged cups on the edge of a horizontal wheel. The wheel slowly turns, with the cups horizontal; but when the cups arrive above the chute they hinge downwards, dropping the set into the previously opened furrow. Covering discs or mould-boards can be set to make a good ridge behind. This type can also be used for chitted seed.

The "Coverwell" has horizontal planting wheels with bottomless compartments. The operators place a potato in each compartment, and as the wheel rotates each potato in turn drops into the delivery chute.

The "Proudlock" employs vertical planting wheels with compartments in the periphery

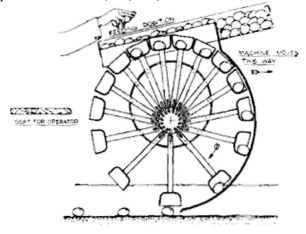

FIG. 103.—DIAGRAM SHOWING MODE OF OPERA-
TION OF THE PACKMAN SEMI-MECHANICAL PO-
TATO PLANTER.

The Packman planter (Fig. 103) employs an ingenious rotor system which brings the cups together for easy feeding at the top of the machine and places the potatoes accurately in the soil. Chitted seed of almost any shape and size may be planted without damage. As with some other semi-mechanical machines which are suitable for planting either chitted or unchitted seed, arrangements are made for either tray-carrying devices or fitting of a hopper.

The speed at which hand-fed planters can work is governed by the fact that operators on most machines cannot handle more than about 120 potatoes per minute; and at 18-inch spacing this results in a forward speed of not over 2 m.p.h. The rate of work is, of course, influenced by the operator's working position and by the exact movements that have to be performed, and there is a considerable amount of variation between different machines in this respect. There is also a good deal of variation between different machines in the accuracy of spacing, this depending on whether there is appreciable wheel slip and on whether the potatoes are allowed to roll in the furrow bottom. Preliminary experiments by N.I.A.E. suggest that absolute accuracy of spacing may be relatively unimportant, but there will obviously be a reduction in yield of ware potatoes where careless work results in doubles and misses.

Many of the semi-automatic planters are obtainable in a variety of forms—e.g. 2- or 3-row, tractor-mounted or trailed, and with or without fertilizer attachment.

(vi)Automatic Planters. With planters of the "assisted feed" type, there is an automatic feed mechanism, which picks the seed out of a hopper and then passes it in some kind of conveyor in front of the operator, who can correct any misses or doubles. This type of machine is capable of very good work with well-graded unchitted seed.

The "Smallford" is a 3-row planter which requires only one operator. As the machine is drawn forward the potatoes from the hopper fall on to an agitator which leads them to an endless conveyor having a series of cups arranged in groups of three. After passing the operator the potatoes fall into a box compartment at the rear of the machine. The potatoes in the centre cups fall straight to the ground, while those in the two outside rows are carried outwards by paddles to their appropriate chutes. This machine can work at about 2 $\frac{1}{2}$ m.p.h. and can plant up to 2 acres per hour.

The Dollé planter is a 2-row machine having vertically mounted revolving wheels with cups arranged radially at the periphery. There is an ingenious mechanical device for dropping a potato from an auxiliary supply when a cup fails to pick out a potato, and the chief function of the single operator is to keep a few potatoes in this auxiliary feed mechanism.

The feed mechanism on some American automatic planters (e.g. the Iron Age) consists of a picker wheel carrying a number of arms, each of which carries near the end two or more sharp, slender steel picks. When the arm passes through the hopper the picks protrude and pick up a tuber; and when the arm arrives above the seed spout the action of a cam causes the picks to be withdrawn, and allows the tuber to fall down to the ground.

These planters practically always operate with cut seed, which is planted in wide rows (3 ft. to 3 ft. 6 in.) at very narrow spacing (often as little as 6 in.). There are frequent misses and doubles, but with the very small spacing distance little attention is paid to this point. The fertilizer is almost invariably placed in a band about 2 in. away from the seed and at the same level, and the whole operation of planting and fertilizer sowing is usually carried out at a high forward speed (about 5 m.p.h.).

Fertilizer Placement for Potatoes.

Many experiments in Britain have shown that the practice of first drawing out the ridges and then broadcasting the fertilizer over them is basically sound, and that there can in some seasons be an appreciable reduction of yield if the fertilizer is first broadcast over the flat surface and the potatoes are planted in the bottoms of deep ridges drawn afterwards. Drilling all the fertilizer in a concentrated band with the seed sometimes gives good results, but is unsafe, since it may cause serious scorching and reduction of yield in a dry spring. Side placement 2 in. from the seed is a method that is both safe and efficient, but most British machines equipped with fertilizer attachments lack the necessary coulters to achieve such placement. Placement attachments are being developed for use in conjunction with some types of planters, and this may provide the ultimate solution; but further research is needed to determine the most efficient ways of using hand-fed planters in conjunction with ordinary fertilizer distributors. At present it appears that reasonably satisfactory results may often be achieved if the fertilizer is well worked into the soil, the potatoes are planted fairly shallow, and the covering bodies are set to draw the surface soil near the seed, deep ridging being carried out as a subsequent operation.

Economics of the Use of Potato Planters.

There can be no doubt that even the simple types of potato planter can achieve appreciable savings of labour required for potato planting.

Compared with sugar beet singling, and potato or beet harvesting, however, potato planting requires little labour. A gang of a dozen Fen women will plant 10 to 12 acres a day of unchitted seed, or 7 to g of chitted; and so long as good labour is available farmers in such districts are unlikely to be greatly interested in mechanical planters. It seems possible, however, that the days of large labour gangs are numbered, and it would not be surprising if many farmers who have no use for potato planters to-day may be unable to do without them in a few years' time. High wages will accelerate such changes.

With good mechanical planters yields are as good as, or better than, with hand planting. Advantages include conservation of soil moisture and avoidance of consolidation of the land beneath the tubers. Another advantage of some types is that the fertilizer may be placed more or less where desired in relation to the seed. It is clear that optimum placement must depend on soil conditions, and especially on soil moisture. There can be no doubt that there is an optimum method of fertilizer placement for every condition, and a machine capable of producing that placement might easily justify itself by an appreciable increase in yield.

Transplanting Machines.

Mechanical planting machines may be broadly grouped into two types, viz. "hand-fed" and "mechanical". In the first type the operators sit on low seats on the machine and drop the plants by hand direct into a furrow of suitable depth and width, made by a hollow coulter. A pair of press wheels usually follows and compresses the soil around the plants.

The early types of hand-fed planters were almost all designed as sledge-type machines for attachment direct to the tractor drawbar or to specially provided drawbars, and lifting at the ends was somewhat clumsy. A more recent trend is towards units designed for mounting on standard types of rear-mounted hydraulically lifted tractor tool-bars (Fig. 6). Such units have the advantages of simplicity and of being able to utilize the lifting and regulating mechanisms already provided with tool-bars. Hand-fed planters must normally be pulled very slowly, speeds as low as f m.p.h. often being called for. The forward speed at which any planter can work naturally depends on the spacing of the plants and on the number of operators who can work on one row. With hand-fed machines a common procedure is to have a 2-row outfit with one man planting each row and a third handing the plants to them. With mechanical planters, two or sometimes more people can work on each row. Operators place the plants in trays or between fingers on an endless conveyor, and from this point the plants are dealt with entirely mechanically. The conveyor takes them down to an open furrow, where they are released in the correct position and surrounded by soil. The chief advantage of this method over the hand-fed one is that the forward speed of the outfit need not be as low, and that unskilled labour may be employed after a very short training.

FIG. 104.—*PLANTING SPRING CABBAGES WITH TWO MECHANICAL TRANSPLANTERS. (ROBOT.)*

There is an intermediate type (e.g. the "Multiplanter"— Fig. 105) which employs a very simple conveyor consisting of a pair of sorbo rubber discs. The seedlings are placed, with the roots uppermost, between the rubber discs where they are apart at the top; and as the discs rotate, rollers squeeze them together, so gripping the plants by the leaves and carrying them round to the soil. Other rollers force the discs apart and release the plants when the root is in the groove formed by the furrow opener, and press wheels follow as with directly hand-fed types. There is no positive spacing mechanism with this type of planter.

An example of the mechanical type, which can be used for the transplanting of such crops as cabbages, sprouts, strawberry plants, leeks, etc., is the "Robot" mechanical planter. This machine has an endless chain carrying pairs of fingers, which open as they pass over the top of the machine. Workers sitting on the machine place plants between the fingers, and these later automatically close and carry the plants down to the ground, releasing them in an upright position in a furrow at the same time as earth is pressed around them by two inclined wheels. A device for watering each plant just before the soil is pressed around it may be fitted for use in dry ground. This machine is adjustable within wide limits and works at a high speed.

FIG. 105.—TRANSPLANTER WHICH EMPLOYS SOR-BO RUBBER DISCS TO CONVEY THE SEEDLINGS

INTO THE SOIL. (RUSSELL.)

It is now a well-established fact that in adverse planting conditions the work done by mechanical transplanters is often appreciably better than that done by good hand workers, even if no watering device is used. Where watering is necessary, a device on a transplanter which puts the water exactly where it will do most good—i.e. under the ground and in contact with the plant roots—is a great advantage.

The use of transplanting machines steadily increases, but many growers prefer to plant some crops by hand—e.g. brussels sprouts, because this enables them to be put in "on the square" and later cultivated both ways.

Implements for Inter-cultivations.

The drilling of seeds in rows enables cultivations to be carried out during the growth of the crop. Jethro Tull (1674-1740) did much by writing and by example to popularize the use of drills and horse hoes, and he emphasized the benefits conferred on the crop by repeated hoeings. His *Horse Hoeing Husbandry*, published in 1733, was largely responsible for bringing about a general change from broadcasting to drilling. Recent research suggests that hoeing in excess of the amount required to kill weeds may be not only wasted effort but also actually detrimental to the crop. Doubt has been cast upon the value of a surface mulch in reducing moisture losses, except in special circumstances. In any event, the main object of inter-cultivations is the destruction of annual weeds, and few British farmers now carry out more hoeings than the minimum required to achieve this object. In some crops, e.g. corn, inter-cultivation can only be practised during a limited period in the life of the crop. In others, such as sugar beet and potatoes, cultivation may be continued until the crops approach maturity; these are therefore called "cleaning crops".

The Single-row Horse Hoe.

The single-row horse hoe or scuffler is a simple implement that is chiefly used for deep inter-cultivations such as are required for potatoes. It consists essentially of a strong, narrow frame to which various types of tine may be attached. A pair of handles is at the rear, and at the front is a single wheel by which the depth of work may be adjusted. The "expanding" type has a frame which may be adjusted to work in various widths of row. On some implements adjustability is achieved by having the side members of the frame hinged at the front, and fixing them to cut any desired width at the back by means of a clamping screw acting on a rear cross member. This type of adjustment is unsatisfactory for certain shares which must always face squarely up to the work, and the more costly "parallel expansion" type is more generally useful. Adjustment of the width makes it possible to work very close to the rows in crops such as beans and roots.

Various types of tine may be used in these implements. For the initial inter-cultivation of potatoes, all tines may have deep-working grubber points fitted. For beans, the front tine may have an A share, and two L shares may be attached at the rear. The draught depends mainly upon the depth of working. Apart from the initial grubbing in potatoes, which is often two-horse work, the single-row implement normally requires one horse.

Multi-row tool-bar grubbers are now available for tractor' work.

The Steerage Horse Hoe.

The common type of multi-row horse hoe consists essentially of a two-wheeled fore-carriage and a trailing frame which carries the hoes and may be steered independently. The wheels may be set in any position on the axle to suit varying row widths. The trailing frame usually comprises two parallel bars to which various numbers and sizes of hoc may be attached to suit the different crops. This type of implement used to be extensively used for hoeing corn, but the practice, which began with Jethro Tull and was general in the last century, is now rare. The chief application of the steerage hoe is to root crops.

The success of hoeing in a crop such as sugar beet depends largely on ability to perform the first operations while the weed seedlings are still very young, and without harm to the young crop plants. It is partly for this reason that roots are always drilled "on the ridge" in some districts. With accurate ridge drilling and hoeing the first side hoeing may be carried out quite close to the row, before it is possible to see the rows at all clearly. With work on the flat the rows are generally first "marked out" by running narrow A blades up the centres as soon as the young plants appear. Then, as soon as the rows can be clearly seen, a close hoeing either with special L shares or with disc attachments that prevent the soil being pushed on to the young plants should be carried out.

The "Lever" type horse hoe has the shares attached to the ends of levers similar to those which carry the coulters of corn drills, each share having independent movement in a vertical direction. The raising of the hoes and regulation of the depth of work is effected much as with corn drill coulters, and steerage is effected by sliding sideways the bar to which the levers are attached. Regulation of the depth of each hoe is sometimes steadied by the use of a depth wheel or a sliding shoe attached to each lever. The lever hoe is well adapted to use on uneven or stony ground. With the decline in the hoeing of corn this implement almost disappeared, but the lever principle is being employed again on many new tractor-mounted hoes.

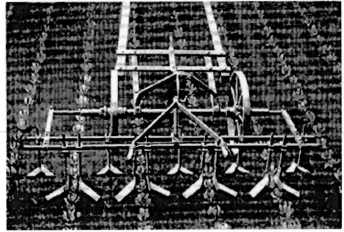

FIG. 106.—STEERAGE HORSE HOE. (F. RANDELE.)

The horse hoes used on the Continent of Europe are generally extremely well constructed implements, with the hoes almost invariably mounted on independently self-regulating brackets or levers. It should be realized, however, that independent mounting has disadvantages as well as advantages, for the hoes may fail to enter the ground when a hard patch is encountered.

The steerage hoe requires a boy to lead the horse and a man to guide the hoes. The hoe must generally be arranged to take either a lull or half width of the drill; for even with the best of drilling, adjacent bouts are not exactly parallel at all parts of the field, and plants will be cut up if the hoe covers rows which are not part of a single drill bout.

Tractor Tool-bars for Inter-cultivations.

A movement to apply tractor power to inter-cultivations began soon after the 1914-18 war and spread rapidly in the United States, chiefly owing to the immediate success of the International "Farmall" row-crop tractor. This tractor appeared on the market in 1923, in a form not differing in most essentials from that of many row-crop tractors sold to-day. The main features of the row-crop tractor itself are discussed in Chapter Two.

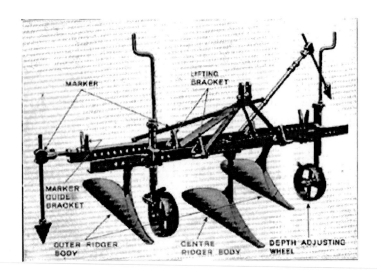

FIG. 107.—RIDGING BODIES ON REAR TOOL-BAR.
(FORDSON.)

It should be understood that the tasks which tractor tool-bars are now called upon to perform are so diverse that no one type of tool-bar can be best for all. It may assist consideration of the subject to list some of the operations. They include general-purpose cultivating; ridging; splitting ridges; inter-row cultivating with deep tines; drilling; hoeing root and vegetable crops; lifting sugar beet, and so on. Most of these operations are best done by means of a strong, rear-mounted tool-bar which is coupled to the tractor by a linkage that permits only very limited sideways motion and cannot be independently steered. Splitting ridges, on the other hand, is easiest done by a mid-mounted or, better, a forward-mounted tool-bar that is really robust; while the drilling and hoeing of root crops in the most efficient manner calls for specialized equipment with requirements quite different from those for the other operations. The farmer who has only one tractor and wishes to mechanize these operations has, therefore, a difficult task in selecting his equipment.

Ridgers.

Ridging ploughs are chiefly used in the cultivation of root crops, and especially of the potato crop. They are used for drawing out ridges in which the potatoes are planted, for "splitting" the ridges, and for "moulding up" when cultivations are complete. In some districts, notably the north of England, roots are often grown, on the ridge.

*FIG. 108.—ROW-CROP TRACTOR WITH RIDGING
BODIES BETWEEN THE WHEELS. (OLIVER.)*

There arc several distinct types of tractor ridging outfits,
which may be classified as follows:—(i) the independent
three-row ridger pulled by tractor instead of by horses; this
is, at best, a makeshift, and can only be used satisfactorily
for drawing the ridges on level ground. (2) The rear-attached
tool-bar fitted with ridging bodies. This is quite efficient at
drawing out the ridges, but splitting back is difficult; some
farmers get over this difficulty splitting two ridges at a time.
(3) Tool-bar attached amidships or in front. This is quite
efficient, the disadvantage being that the tractor wheels run

along the bottom of the baulk after it is made and pan it down. Splitting with these is easy.

FIG. 109.—FORWARD-MOUNTED TOOL-BAR FOR SPLITTING POTATO RIDGES. (SUN ENG. CO.)

FIG. 110.—REAR-MOUNTED STEERAGE HOE WITH DISCS.(FERGUSON.)

Hoeing Tool-bars.

Tractor hoeing outfits include the following main types: —(I) Rear-attached tool-bars with no independent steerage. While these may be successfully used for cultivating potatoes, they are unsuitable for accurate hoeing. (2) Rear-attached tool-bars which are independently steered (Figs, 110 and 111). These can do excellent work, but the output per man is not high, since two men normally hoe only four to six rows at a time. (3) Mid-mounted hoes, which respond immediately to steering of the tractor itself (Fig. 112). These can be quite satisfactory. Hoe blades must be used to eliminate the rear wheel tracks. (4) Tool-bars mounted in front of the front wheels. With rear wheel-track eliminators this makes a long outfit, and the hoes are too far away for the driver to have a really good view of them. (5) There are, in addition, the self-propelled tool-bars, which can do very accurate work, and may be satisfactorily used where there is a sufficient amount of drilling and hoeing to justify purchase of such an outfit.

FIG. 111.—REAR-MOUNTED STEERAGE TOOL-BAR WITH INDEPENDENTLY FLEXIBLE HOES. (FORD-SON.)

*FIG. 113.—FRONT-MOUNTED TOOL-BAR WITH
INDEPENDENTLY FLEXIBLE HOES, AND REAR-
MOUNTED HARROWS. (STANDEN.)*

FIG. 112.—HOEING SUGAR BEET, MID-MOUNTED
TOOL-BAR WITH INDEPENDENTLY FLEXIBLE
GROUP OF HOES FOR EACH ROW. (STANHAY.)

It should be noted that there are important differences in construction between different machines in all the above main groups. Thus, some have the hoes mounted on a parallel-linkage framework, while others have them independently mounted at the ends of levers, like the coulters of a drill. Some have springs to press the hoes into the ground, while others employ weight.

FIG. 114.—INDEPENDENTLY FLEXIBLE HOES, AD-JUSTABLE WITHOUT USE OF SPANNERS. (HUMBER-SIDE ENG. GO.)

Certain essential requirements apply equally to all kinds of drilling and hoeing outfits. The tools must be easily put on and removed from the tractor. The hoes must be easily visible from the driving seat. The hoes should float easily in a vertical direction, so as to accommodate themselves to uneven land, but must be rigid in a sideways direction, so that they do not wobble, and so that they respond immediately to steering.

Where there is sufficient hoeing and drilling to justify the expenditure, there is no task that was formerly performed by horses that cannot be carried out satisfactorily by one or other of the outfits now available.

Whatever type of tool-bar is used, successful work in root crops depends on careful preparation of the seed bed and drilling.

Where the crop is to be cross-blocked it is advisable to have the work set out "on the square", so that the crop can be hoed in both directions without changing the setting. This means that drilling must be in narrow rows, 14-15 in. apart, in order to obtain sufficient plants when the crop is cross-blocked at this distance.

The size of gap to be left in cross-blocking depends on the plant. The narrowest possible setting should be used, without leaving any blanks. In general, a gap of about 3 in. is left. Good cross-blocking is found to reduce the cost of singling by about 20 per cent. If the use of milled seed and the single-seeder drill is developed, it seems likely that the practice of cross-blocking may be more generally adopted.

Gapping or Thinning Machines.

The "Kent" gapper is a machine which has for many years been used on a limited scale for chopping out root crops. It consists of a pair of wheels which span the row and drive a spinner mounted on a horizontal axis. The spinner carries two or more hoe blades, which chop out the crop much as a man would, except that once the machine has been adjusted the action is entirely uninfluenced by the state of the braird. The machine is sometimes useful to sugar beet growers in minimizing the harm done if a crop gets too forward before singling.

The "Dixie" thinner is a similar machine to the gapper, the chief difference being that it was originally developed in U.S.A. for use in cotton, where the need is for a large number of small hoe cuts instead of a few wide ones. Along with the development of single-seed drills in U.S.A., tiny hoes specially suitable for use in sugar beet have been fitted to the Dixie thinner, and a new technique of using the machines with the object of eliminating altogether the labour required for singling has been developed. Experiments on these lines in Britain have begun and show reasonable promise.

Inter-Plant Hoeing.

Though well-constructed hoes on fine soil can be run quite close to rows of plants, the idea of hoeing between plants, except by "cross-blocking" or "check-row" spacing, was regarded as quite impracticable until a few years ago. There are now, however, machines which can be used to hoe between plants as well as between the rows. One type of inter-plant hoe consists of two wheels running on either side of the row, and carrying two knives. The tips of the knives cross at a point coinciding with the row of plants, and if the machine were drawn along with the knives crossing thus, the whole row of plants would be cut out; but a mechanism is provided whereby, on arriving at a plant, the two knives can be made to diverge from one another and to close again when the plant has been passed. In this way the intervening ground can be hoed without damage to the plants.

On one machine the operator holds a control lever fitted with a twist grip. A lateral movement of the lever steers the hoe, a vertical movement regulates the depth of work, and a foot pedal operates the mechanism which opens the knives, allowing plants to pass between. With a special tractor running at about $\frac{3}{4}$m.p.h., excellent work can be done. A more recent development, indicative of the great possibilities still open to mechanization, is the electronically controlled inter-plant hoe.

This was originally designed to employ a photo-electric cell, but more recent models have a sensitive feeler which touches the crop plants. The hoeing mechanism is an electrically driven spinner mounted on a horizontal axis and equipped with a series of small blades. When the feeler mechanism touches a plant, an electrical relay causes the hoe blades to pivot rearwards so that they are raised clear of the soil and the plant. The mechanisms are complicated, but progress to date justifies the opinion that satisfactory automatic inter-plant hoes will become available in the not too distant future.

REFERENCES

(17) Bainer, R., "New Developments in Sugar Beet Production." *Agric, Eng.*, August 1943, vol. xxiv, no. 8.

(18) Cooke, G. W., Fertilizer Placement, Experiments. *Agric. Eng. Record.* Summer, 1949.

CHAPTER NINE

MANURE DISTRIBUTORS

"Remember, that it is not only the first cost of all manures which makes them expensive, but the comparative LABOUR saved in their application, which must also be taken into the account, when the cultivator is estimating their value."

Farmers' Encyclopaedia. Cuthbert Johnson.

An essential feature of all permanent systems of farming is the return to the soil of the nourishment taken from it by the crops that are removed. This is commonly done by the application of farmyard manure, liquid manure and artificial or organic fertilizers. The distribution of each of these categories of manure presents its own problems, which are discussed separately below.

Fertilizer Distributors.

Artificial manures have been in general use for a long time, but it is only recently that fundamental research work has been aimed at a determination of the ideal *position* in which the fertilizer should be placed in relation to the plant. The need for more knowledge on this subject is pressing. In the past it has been usual either to mix the fertilizer with the soil as uniformly as possible by working it into the seed bed, or to broadcast it in the form of a "top dressing". The ideal placing of the manure varies for different crops, soils and manures. While working it into the seed bed may be suitable for some crops, some do better with a top dressing, some with the fertilizer ploughed in, and others if it is drilled with the seed. The need for work on this question is emphasized by recent experiments which indicate that the best position of the fertilizer for some plants, sown in rows, is a few inches away from the row of plants and a few inches below the surface of the soil. In many parts of Britain, very good results have been obtained with combine drills, which sow corn and fertilizer together (p. 214).

Most fertilizer distributors are broadcasting machines. The object in the past has been to secure a uniform broadcasting of the manure over the surface of the soil. Uniformity of distribution may be especially important where small amounts of concentrated fertilizers are applied as a top dressing, but recent experimental work on fertilizer placement suggests that the importance of a very fine and uniform distribution of manure applied to seed beds may have been over-emphasized. In any event, in order to achieve a good distribution, whether the aim is to distribute the manure over the surface or to arrange it in some pattern, it is essential that the fertilizer be in good condition. Many of those in common use are apt to set in hard lumps, and the best of distributors will not do really satisfactory work unless these lumps are removed, either by grinding or by screening them out. Some distributors get rid of the lumps without difficulty, while others become blocked. Freedom from trouble in this respect is a desirable feature in a distributor, but it sometimes encourages the sowing of manure that is in poor condition.

A fertilizer distributor for general use should, until more is known concerning fertilizer placement, be capable of uniformly distributing quantities varying from 1 to 30 cwt. per acre, with easy adjustment of the rate of sowing.

It should deal satisfactorily with fertilizer in poor condition, should be simple in construction, easy to operate and easy to clean. There should be few moving parts in contact with the manure, and these parts should be resistant to corrosion

FIG. 115.—DISTRIBUTING FERTILIZER IN POTATO RIDGES WITH REGIPROCATING-PLATE MACHINE FITTED WITH DEFLECTORS. (BAMFORDS.)

. The variety of types of distributor available is great, but there are few machines which comply in every respect with the *desiderata* set out above.

Most distributors have a long box hopper mounted on two wheels, but distributors designed for attachment to the rear of a trailer have recently become popular. These may either be mounted on the vehicle or towed behind it. Power take-off driven machines, both tractor-mounted and trailed, are now available; but the lack of a drive which is independent of the gearbox introduces complications in setting.

Feed Mechanisms

Roller Feed. A delivery roller rotates at the bottom of the box with an action similar to that of an external force-feed drill. The roller may be smooth, corrugated or spiked, and generally has a cleaning mechanism to brush or scrape off the manure when it has been carried outside the hopper. The rate of sowing is controlled by the speed of the roller and the size of the slots through which the fertilizer is delivered. An agitator of some kind may be used in the hopper of this and of most types in which the feed mechanism is at the bottom of the hopper.

Worm Agitator Feed. In this type, a large auger slowly rotates inside the hopper, and its lower edges scrape fertilizer through a series of adjustable holes in the bottom. It is one of the simplest and best types for sowing heavy fertilizers.

Endless Chain Feed. In this type, an endless chain with finger projections is driven along the bottom of the hopper. Fertilizer is pushed by the fingers through slots, and the rate of sowing is controlled by the size of the slots and the speed of the chain. This type has disadvantages in the wearing, stretching and corrosion of the long chain, and the fact that lumps in the manure are drawn to one end of the hopper, thereby causing irregular distribution.

The Star Wheel Feed. The "star" wheel or finger wheel mechanism is commonly employed on combined grain and fertilizer drills of American manufacture. Finger wheels driven from a spindle by worm or bevel gearing rotate in the bottom of the hopper, and push the manure through slots of variable size to a position where it falls to the ground or into the seed tubes. The rate of sowing is controlled by the speed of the star wheels and adjustment of the shutters.

FIG. 116.—STAR-WHEEL FERTILIZER FEED FOR COMBINE DRILL, (MASSEY-HARRIS.)

Reciprocating Plate Feed. The bottom of the hopper of this type consists of three perforated plates. In the machine illustrated in Fig. 117 the top and bottom plates reciprocate and the middle plate is stationary; but in some distributors only the middle plate moves. Sowing rate is controlled mainly by the distance the plates travel at each stroke, and partly by using plates with different-sized slots. For very heavy sowing rates distance pieces may be fitted between the plates.

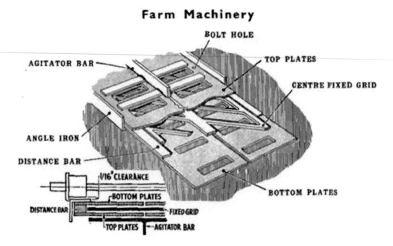

Farm Machinery

BOLT HOLE

AGITATOR BAR

TOP PLATES

CENTRE FIXED GRID

ANGLE IRON

DISTANCE BAR

1/16" CLEARANCE

BOTTOM PLATES

DISTANCE BAR

BOTTOM PLATES

FIXED GRID

TOP PLATES   AGITATOR BAR

FIG. 117.—RECIPROCATING PLATE FEED. (BAM-FORDS.)

Inset shows section of assembly with machine upside down ready for dismantling.

FIG. 118.—*TRAILED CENTRIFUGAL DISTRIBUTOR
WITH VARIABLE-SPEED FRICTION DRIVEN SPIN-
NER AND AUGER FEED. (GLOTRAC.)*

Centrifugal Delivery. This type of machine has a
hopper, generally of truncated cone shape, situated above
a single disc or a pair of horizontal discs, mounted side by
side. The discs rotate at high speed. The fertilizer is agitated
in the hopper and is fed in a stream, either by gravity or by
a finger wheel or conveyor mechanism, on to the rotating
discs, which distribute it by throwing it from their edges. In
good weather conditions and with careful operation these

machines give an even distribution, but may give poor results with mixed manures and in windy weather.

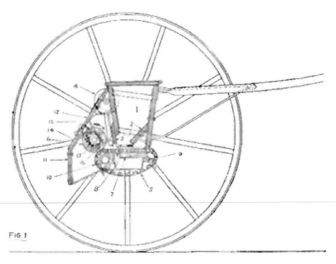

FIG. 119.—SECTION OF CONVEYOR-AND-BRUSH TYPE FERTILIZER DISTRIBUTOR. (COULTAS.)

1, hopper; 2, baffles; 3, adjustable slide; 5, conveyor; 6, brush.

A centrifugal distributor suitable for sowing lime can be constructed from the wheels and back axle of a car or lorry. The complete back axle assembly is used, and it is towed behind a trailer with the propeller shaft coupling uppermost. A single large disc is boiled on to the flexible disc that normally connects propeller shaft and differential, and as the wheels are towed the disc is driven at high speed by the normal final drive acting in reverse. All that remains to be done is to support a hopper of truncated cone shape above the disc, and to arrange for adjustment of the feed from hopper to disc. A light flexible shaft extending from the centre of the disc to the inside of the hopper agitates the contents of the hopper and ensures a regular flow.

Commercial spinning disc machines are now available. They are particularly useful for orchard work and lime spreading.

Rising Hopper Type. Machines which deliver the fertilizer from the top of a rising hopper give no trouble in sowing lumpy or damp manure, but they suffer from the disadvantage of a very irregular distribution immediately after filling up, owing to the fact that the fertilizer becomes compacted in the hopper as the machine is shaken in its passage over the soil.

Conveyor-And-Brush Feed. In the conveyor-and-brush type distributor (Fig. 119), the bottom of the box hopper is formed by a conveyor which carries an even layer of fertilizer-through an adjustable opening to a revolving cylindrical brush. The brush sweeps the conveyor clean and spreads the fertilizer evenly over the ground. The conveyor is designed to move slowly, so that a comparatively large opening is required even for small quantities of fertilizer; clogging of the opening is thereby largely eliminated, even with sticky or damp fertilizers.

Revolving Plate Feed. A distributing mechanism now adopted by several manufacturers has a series of saucer-shaped plates mounted below the hopper in such a way that when they rotate, fertilizer is carried out of the hopper through an adjustable gate. A spindle fitted with spikes rotates at high speed above the lips of the plates, and broadcasts the fertilizer. This is a simple type of distributor which is easy to clean and to operate.

Transport of Wide Distributors.

On large farms it is necessary at certain times to distribute fertilizer over several hundred acres within a few days. This necessitates using wide distributors, which require special arrangements for transportation from field to field. One wide distributor has a hopper which swivels about its centre on the narrow-based 2-wheeled chassis. Another has the outer sections of the hopper arranged to swivel forwards for transport, and a platform is built above the drawbar to provide for transporting the fertilizer.

Distributing Attachments for Lorries or Trailers.

FIG. 120.—REVOLVING PLATE FERTILIZER DIS-
TRIBUTOR INSET: SECTION OF HOPPER. (MASSEY-
HARRIS.)

FIG. 121.—FERTILIZER DISTRIBUTOR ATTACHED TO REAR OF TRACTOR TRALIER. (SALOPIAN.)

FIG. 122.—TRAILER-MOUNTED CENTRIFUGAL DISTRIBUTOR WITH ENGINE DRIVEN SPINNER AND LAND WHEEL DRIVEN FEED AUGER, (TASKERS.)

The drive for this type of distributor is normally by chain and sprocket from the lorry or trailer wheel. One machine (Fig. 121) has a hopper with two compartments and an independent feed from each, so that a mixture of fertilizers in the desired proportion can be distributed.

Another (Fig. 122) has the feed mechanism driven by the trailer's land wheel, and the centrifugal distributing disc driven by a small engine. This ensures that the disc is driven at constant speed and effects uniform distribution, even though the forward speed may vary considerably.

Combined Grain and Fertilizer Drills.

In recent years there has been a great increase in the practice of drilling seed corn and fertilizer together—a practice encouraged by the production of various types of granular compound fertilizers. Most large-scale corn growers now use combine drills for those corn crops requiring fertilizer dressings.

Combine drills normally sow seed and fertilizer down the same seed-tube and coulter, but a few can be set to sow the two separately. There is a risk of damage to the seed corn if too high a concentration of soluble fertilizer is sown in contact with it. Generally speaking, phosphatic fertilizers are harmless, but potassic salts, such as muriate, are liable to delay germination and burn the young roots. Nitrogenous fertilizers are intermediate between phosphatic and potassic ones in this respect, and root crops are more susceptible than cereals. Damage is most likely to occur in dry weather. It should be emphasized, however, that the small amounts of fertilizers sown with combine drills on many thousands of acres of corn crops throughout the country during recent years have almost invariably produced excellent results.

When properly cared for and used, the modern combine drill is a very efficient machine. The star wheel fertilizer distributing mechanism deals effectively with fertilizers that are in good condition, and is especially suitable for granular materials. The machine is, however, a complicated one; and unless it is well cared for, the whole drill may depreciate rapidly. It should be a rule, both with the combine drill and any other fertilizer distributor, to empty out the fertilizer every night; to cover the machine with a sheet; and to give the whole machine a thorough clean as soon as sowing is completed.

FIG. 123.—FIFTEEN-ROW COMBINED SEED AND FERTILIZER DRILL. (MASSEY-HARRIS.)

A useful first step in cleaning is to sow a hopper-full of sand or ashes. The machine should then be taken to pieces, washed in hot dilute washing soda solution ($\frac{1}{2}$ lb. to a gallon), dried, and covered with a rust-preventive. With combine drills the seed tubes should not be neglected, or they may need renewing after a single season.

A disadvantage of using combine drills is that the rate of seed drilling is inevitably slowed down a little. The practice of sowing seed and fertilizer together in an ordinary corn drill is not to be recommended, for great vigilance is necessary if sowing rates are to be accurate. With fertilizer at all damp, the cups on a cup-feed drill or the corrugations on a force-feed one become partially filled, and the sowing rate is cut down. The germination of grain may be impaired if it is mixed with damp fertilizer for only a few days.

Farmyard Manure Handling Machinery.

Application of farmyard manure to the land is an essential feature of the management of most British farms, and though new systems of grazing management and the use of such devices as portable milking bails may reduce the size of the problem, there seems no likelihood that the amount of manure that will need to be handled on most farms will diminish. When mechanization of other farming operations has reached a certain level, depending on the size and type of farm, the replacement of horses by tractors and the need to handle the manure with a limited labour force necessitates consideration of mechanizing this operation also. Many factors are involved in the choice and use of mechanized aids to manure handling. On some farms the daily task of getting manure out of cowsheds and pigsties is troublesome; on many others getting the manure out of the yards and into a field clamp causes difficulties: on yet other farms this latter job can be done at a slack time, and distribution of manure on the fields calls for more labour than is conveniently available. Limitation of the amount of capital invested in specialized manure handling equipment necessitates considering the possibility of using equipment, such as tipping trailers, which the farm may already possess, or could use with advantage for other work such as the carting of root crops.

Front-mounted Tractor Loaders.

Front-mounted tractor loaders are available in a variety of makes for fitting to almost any modern general-purpose tractor. These machines are normally operated by the tractor's built-in hydraulic lift pump, and their individual capabilities depend as much on the tractor as on the loader itself. The variables which need to be studied are manœuvrability of the tractor in difficult ground conditions and in obstructed enclosures; ease of control of the lift and tip; height of lift; capacity of bucket; and power of the tractor's lift system. Some loaders employ a hydraulic ram to push the manure off the bucket on to the load, but most have a bucket which tips on a hinge. Hydraulic loaders arc simple machines, and if carefully used they have a long life. The first cost is moderate, and the loader can be used for a variety of other jobs, especially if a range of fittings, e.g. root bucket, etc., is provided.

335

FIG. 124.—FRONT-MOUNTED HYDRAULIC LOADER WITH HYDRAULIC PUSH-OFF MECHANISM. (FERGUSON.)

It is usually best to work with the vehicle to be loaded at right angles to the face of the heap or exposed side of the yard. It is an advantage to have removable partition supports which fit into sockets in the yard bottom. The loader should work on the yard bottom, which should be kept as free from spilled muck as possible, to reduce wheel spin. Wheel spin, which results in damage to tyres and yard surface, and to loss of manœuvrability, is reduced by adding weight to the rear of the tractor. Added weight also improves stability.

Manure should be pulled off the top in layers, and the forkfuls built into a load on the spreader, not merely tipped in the middle of it. On rear-delivery spreaders, it is best to start building the load at the front end.

FIG. 125.—CONVEYOR-AND-FORK MANURE LOAD-ER. (WILD-THWAITES.)

Cable-fork and Conveyor Loader.

A type of loader which has proved its usefulness on many farms comprises a self-contained unit driven by an independent engine. The manure fork is fixed to the free end of a light steel cable, the other end of the cable being attached to a winch. One operator digs the fork into the manure and holds on to its handle. A second operator then engages the winch, which pulls the forkful of manure to the loader and drops it on to the foot of an elevator, which delivers it to the vehicle. Among the advantages of this type of loader are the fact that it can work in situations where there is very little head-room, while by standing the machine outside, it can be used for emptying small yards or boxes. Where a straight haul from the elevator is not practicable the cable may be passed round a snatch-block and the manure moved in two stages. Two men are needed to operate this loader, and the one who operates the fork has to work fairly hard. This system is suitable for use where several transport vehicles are employed, since the drivers of the transport vehicles can help with the loading. The machine has a rate of work roughly similar to that of a front-mounted tractor loader, and may be preferable to the latter where working conditions are difficult. The elevator is a versatile device which can be used for loading roots, sacks and bales.

Tractor-mounted and other Jib-Cranes.

Other types of mechanical loaders include jib-cranes, which may be tractor-operated and either mounted on the rear of the tractor or trailed. The cranes may be either mechanically or hydraulically operated. Some machines are transportable and driven by an independent engine. The manure is lifted by a grab with long tines which bite into the heap as the grab is raised. The loaded grab is slewed over the vehicle, either by means of the power unit or by hand, and the load is usually released by means of a trip-rope. Most of these devices require a considerable amount of head-room.

Other Types of Loader.

Self-propelled drag-line excavators equipped with a special manure bucket can work effectively at a high speed on open sites, and are sometimes employed on contract work.

Other devices include fixed installations for cleaning cowsheds, etc., designed to load the manure direct into trailers or spreaders. These may take the form of overhead gantries, with trollies which are filled by hand and can be used for feedingstuffs or manure; such devices have long been available but are not widely used in Britain. More recent developments include endless cowshed gutter cleaners which are usually operated by electric motors.

An interesting recent development is the use of front-mounted tractor scrapers for clearing central dung-passages. At the end of the run the manure may be pushed up an incline until sufficient height is obtained to enable it to be dropped on to the transport vehicle.

Trailer Spreaders.

The trailer manure spreader consists of a low wagon with a travelling bed which carries the load of manure slowly and uniformly towards a revolving beater or toothed cylinder at the rear of the machine. The beaters tear the manure into small fragments, and an additional cylinder scatters it over a strip wider than the machine.

FIG. 126.—2-WHEELED TRAILER SPREADER.
(MASSEY-HARRTS.)

The box is mounted low on the steel frame in order to reduce the labour of loading. Various types of conveyor are employed to carry the manure to the beaters. The endless apron and the drag chain type are commonly used, with either worm or ratchet drive. In the ratchet drive, the rate of spreading is easily varied by adjusting the throw of a rocker arm; from 3 to 20 tons per acre can generally be distributed.

One type of spreader has the spreading mechanism at the front and is convertible to a general-purpose trailer. Another conveying mechanism consists of a moveable hopper front, which is slowly pushed rearwards by a hydraulic ram mechanism operated from the tractor. Some machines are driven from the tractor P.T.O. and some from the land wheels.

FIG. 127.—P.T.O-DRIVEN TRAILER SPREADER WITH DISTRIBUTING MECHANISM AT THE FRONT. (BEN-TALL.)

FIG. 128.—MACHINE FOR SPREADING MANURE FROM SMALL HEAPS. (WILD-THWAITES.)

The beginning of one run should overlap the end of the previous one to maintain a uniform dressing. In order to avoid sharp turns and idle time on the headland the field should be worked in lands. Regular attention to cleaning of the machine and adequate lubrication of bearings and chain drives is essential.

Field Heap Spreaders.

The heap spreading machine is designed to spread small, closely-spaced heaps which have been set out by ordinary transport trailers. It is trailed and P.T.O. driven, and arranged so that the tractor straddles the rows of heaps. A revolving shredder tears up the heaps, which are then spread by a right and left hand auger mounted on a second shaft. With small heaps well laid out, spreading rates of 20-30 tons per hour are possible.

Rotating Disc Spreaders.

Hand spreading direct from the rear of a vehicle may be facilitated by the use of a spinning disc carried on and driven by a pair of land wheels. The manure is dropped on to the disc and is thrown from it by centrifugal force. Evenness of spreading is generally poor.

Manure Handling Systems,

The operation of getting manure from yards on to the land may be split up, either by choice or due to the necessities of husbandry, into a number of distinct stages. For example, the fact that yards often have to be cleared during slack periods in summer when the land is growing crops, or in winter, when fields are too wet for transport, often necessitates the making of field clamps, regardless of whether this fits in with a fancied manure handling system or not. Spreading directly from where the manure is made is usually advantageous when it is practicable, both on account of economy in labour and also in regard to avoiding unnecessary loss of nutrients. On the other hand it often does not fit in with the necessities of farm management, and the making of field clamps must continue to be a frequent requirement. Some systems of operation are clearly preferable for particular circumstances on account of economy in expenditure on specialized equipment, or the efficient use of labour, or for both reasons. Occasionally, however, the requirements of farm management make it necessary to employ large gangs which will get the job done quickly, and this may justify a high capital investment in equipment. There is, of course, no justification for a high investment and use of a large gang unless the operations are properly balanced to produce a high rate of work and efficient labour utilization.

Operating with a Small Gang.

One of the most efficient manure handling systems, both as regards minimum capital investment and labour economy, is the "one-man" system whereby the man, equipped with one tractor, a front-mounted loader and an easily hitched trailer spreader, does the whole job of loading, transport and spreading. With a transport distance of $\frac{1}{4}$ mile, about 3 tons per hour can be loaded and spread. There is nothing in this job to make the man tired, and if he works steadily he can deal with a great deal of manure in a few days. Similarly, a one-man outfit consisting of tractor, loader and hydraulic tipping trailer can clear yards and cart to a field clamp at a reasonable rate, doing any necessary tidying-up at the clamp with the mechanical loader.

For direct spreading, a 2-man gang consisting of one loader and one spreader tends to be uneconomic; but two men can work efficiently with two spreaders or at transporting manure from yards to clamp, using one loader and two trailers. Both men are required for operating the fork-and-conveyor type loader, and where a hydraulic loader is employed the second man can help in the building of good loads.

Larger Gangs and Long Hauls.

For high-speed loading and spreading on short hauls, e.g. when spreading from a field heap, a 3-man gang using one loader and two spreaders can be very efficient. This system is unbalanced if the transport distance is more than a few hundred yards. When carting to a field clamp the remedy for longer hauls may be to introduce more transport vehicles; but few farms can justify a fleet of mechanical spreaders, and where direct spreading on long hauls is needed, some other system of operation which does not call for continuous loading is necessary. This is quite easy to devise; for example, by cutting down the number of men to two, the spreaders can make their journey together and the two men can operate a fork-and-conveyor type loader at the yard. Alternatively, the two spreaders can work well apart and their drivers can operate a hydraulic loader mounted on a tractor which stays at the yard. For rapid direct spreading over long transport distances where good trailers are already available, there is much to be said in favour of making small field heaps and using the mechanical field heap spreader.

Sufficient has been written to indicate the importance of giving considerable thought to proposed handling systems before deciding what equipment to buy. Mechanical spreaders are far from ideal machines from a maintenance standpoint, and their rate of operation when actually spreading is so high that there can seldom be much justification for having a fleet of them for carting manure along the road. Farmers who are used to handling manure with a "running set" of horses and carts are apt to imagine that a minimum of three spreaders is essential for successful mechanization. In fact, however, as is apparent from what has been written, three spreaders are very seldom needed, and entirely new systems, such as the one-man method, which ensures that nobody wastes time because he is waiting for a machine, are likely to be the most attractive.

The Distribution of Liquid Manure.

A high proportion of the manurial residue of concentrated foods, especially the soluble nitrogen and potash, is found in the urine of livestock. When the animals are kept in straw yards or in well-littered boxes, the urine is absorbed and applied in the farmyard manure; but with modern cow-sheds and Danish type pig-houses, the conservation of the liquid excrement becomes a problem. The most satisfactory method is to construct a dual drainage system, so that the urine passes to an underground tank and the water used for washing the floors drains away elsewhere. Excessive dilution of the urine is thus prevented, and it may be stored until it is convenient to apply it to the land.

The urine may be conveniently pumped from the storage tank into the distributor by a centrifugal pump driven by a small engine or a $\frac{3}{4}$h.p. electric motor. The common type of distributor consists of a large barrel or tank mounted on a two-wheeled cart and fitted with a tap and delivery pipe at the bottom. On opening the tap, the liquid flows into a long trough and falls to the ground through holes or slots. Small holes or slots should be avoided, for they quickly become blocked by sediment or straw. A type of distributor that is popular in Denmark, partly because of its freedom from choking, is illustrated in Fig. 129. Another type of distributing mechanism which is quite satisfactory is a high-speed horizontal spinning disc, similar to those used for distributing artificial fertilizers. This can be a home-made device, constructed from the rear axle of an old car or truck; or the liquid can be run into the hopper of a commercial centrifugal type fertilizer distributor.

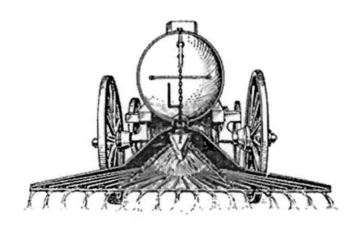

FIG. 129.—LIQUID MANURE DISTRIBUTOR EM-PLOYED IN DENMARK

On the Continent, devices which deliver the liquid into the soil behind cultivating points, between the rows of growing-plants, are now widely used in some areas.

In most of the Scandinavian and Central European countries conservation and distribution of the liquid manure is an important link in the farming systems. In this country, very few farmers have equipment for conserving the urine from the pig- and cowhouses, and on some farms where equipment is installed, it is considered too much trouble to empty the cisterns and distribute the manure. The low price of fertilizers has in the past been partly responsible for this situation, but another cause was the slow and laborious method of pumping the liquid from the cistern to the distributor, by a hand-operated chain-pump. The use of

349

light, portable power driven pumps would remove much of the prejudice against utilizing liquid manure.

CHAPTER TEN

HAY AND SILAGE MACHINERY

"As the work of haymaking lias to be done timeously and with speed, the value of a first-class haymaker not liable to get out of order cannot be over-rated."

Field Implements and Machines. John Scott, 1901.

The harvesting of the hay crop requires a heavy expenditure on hand labour unless it is carried out with the help of mechanical power and labour-saving machinery. Of the total of about 25 million acres of crops and grass in England and Wales, about 6 million acres are annually devoted to hay. Yields and quality are very variable, but the quality of the product depends to a great extent on "getting" the hay in good order. This is largely a question of the speed with which the hay can be made and stacked when the conditions are right. The methods practised in making and carting hay vary greatly. On small or rough patches of land where it is impracticable to use any machinery, the scythe, hand-rake and hand-fork are the only tools that can be used. In these conditions, over ten hours of man labour may be required for the cutting, making and stacking of a ton of hay. At the other extreme, where it is possible to use modern machinery in all the operations, it may be possible to harvest a ton of hay with the expenditure of little more than one hour of man labour. Modern labour-saving devices may enable haymaking to be done at a lower cost, but the extra speed at which the work can be completed is sometimes much more important than the cost of the operations.

Before the advent of haymaking machinery, the number of people required for haymaking was enormous. An early eighteenth century picture shows 122 people in one field. Most of these were at work on the hay crop, but the large number was partly made up of a troupe of Morris dancers, some musicians and people dispensing refreshments. Up till about 1850, hay was made almost entirely by hand. A "haymaker" which resembled the modern tedder was invented about 1814 and was gradually developed, but it was not till the middle of the century that haymakers and horse-rakes began to be at all generally used. A machine resembling in action the modern swath turner was introduced in 1896, and a convertible swath turner and side-delivery rake in 1903. In 1876, an American hay loader was exhibited at the Royal Show, and similar machines were built in this country a few years later; but neither loaders nor sweeps were much used until quite recent times. Great advances have been made in the mechanization of the hay harvest by the use of tractor mowers, hay loaders, tractor and motor sweeps, power stackers and pick-up balers. Haymaking is one of the outstanding examples of what may be achieved by mechanization in favourable circumstances.

Cutting the Hay.

As hay becomes mature, there is, after the flowering stage, a loss in the amount and digestibility of the protein and a marked increase in the amount of indigestible crude fibre. The maximum bulk of hay is obtained by delaying cutting until there is seed in the heads; but the quality of the product in such circumstances is often no better than oat straw. Hence, hay should normally be cut when most of the grasses are in the early flowering stage. This gives the highest yield of protein, a product of good digestibility, and a good aftermath.

The Mower.

The history of the development of the mower is closely bound up with that of the reaper and binder, and is dealt with in Chapter Eleven. The modern mower is a high-speed close-cutting machine which has almost entirely superseded the scythe for cutting grass and forage crops. Its use is not, however, confined to these crops, for it may be used in the harvesting of cereal crops that are badly laid and in other crops of various kinds where close cutting is necessary.

Construction of the Mower. Modern mowing machines are all very similar in essentials. The trailer type has a main frame, which is usually cast in one piece, mounted on two wheels, from which power is transmitted through a live axle to the cutting mechanism. It is a side-cut machine, with the cutting mechanism connected to the main frame by a double-jointed dragbar. This connection permits the position of the cutter bar to be controlled by means of levers.

FIG. 130.—TRACTOR TRAILER MOWER. (ALBION.)

The live axle is driven from the land wheels through a pawl and ratchet mechanism. This provides a differential drive which ensures that the cutting mechanism is driven when the machine is being turned in either direction, and allows the machine to be backed. The gears of modern mowers run in an oil bath and are on some machines made of hardened steel. In the usual type of gear box a large gear wheel, keyed to the main shaft, drives a small pinion which is mounted loose on a counter-shaft and may be connected to a second large gear wheel by means of a dog clutch. The drive from the second large gear wheel is to a compound pinion and bevel wheel mounted loose on the main shaft, and this, in turn, drives the bevel pinion that is screwed to the end of the crankshaft. These gears speed up the drive so that the crankshaft makes nearly 700 r.p.m. when the machine is drawn at two miles an hour.

The crankshaft, running at right angles to the main axle, carries at its forward end a heavy flywheel and the crank-pin to which the "pitman" or connecting rod is attached. The kinetic energy of the rotating flywheel assists the machine in running smoothly over the positions at which the knife changes direction, and through patches where there is a high resistance. The connecting rod may be of wood or of heat-treated high-carbon steel. It needs to be light but strong.

The cutting mechanism consists of a finger-bar containing a reciprocating steel knife, made up of triangular sections riveted to a steel bar. The knife sections can be easily removed and replaced when broken or worn, by punching out the two rivets which hold them, and riveting on new sections. The knife reciprocates in slots in fingers made of case-hardened steel or of malleable iron with hardened steel "ledger plates". It is essential that the ledger plates of all fingers should be in alignment, so that the knife sections slide over them smoothly. It should be understood that the action of a. mower is similar to that of a pair of scissors; cutting is performed by a shearing action, and there are TWO cutting edges, viz. the knife sections and the ledger plates. The ledger plates have square edges when they are new, and the hardening treatment assists in keeping these edges sharp. It is necessary to replace the plates when the edges have become rounded by long wear. When only slight wear has taken place the sides of the fingers or of the ledger plates can be ground away so that a sharp edge is again produced.

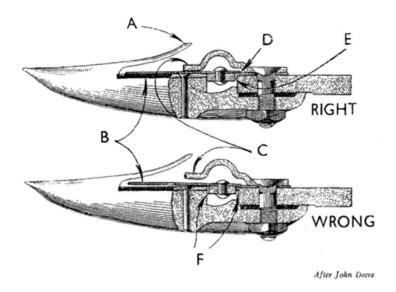

After John Deere

FIG. 131.—CONSTRUCTION AND SETTING OF THE MOWER FINGER BAR

A, lip of finger; **B,** knife section should fit flush on ledger plate; **C,** clips should press gently on knife; **D,** wearing plate; **E,** knife back; **F,** wearing plate needs adjustment to lake up play.

For clean cutting, the knife must be kept gently and firmly pressed on to the ledger plates. This is done by adjustment of the malleable "clips". The back of the knife generally runs on renewable hardened steel "wearing plates". The knife head is that part of the knife to which the connecting rod is fitted.

When the mower is at work, the cutter-bar should bear lightly on shoes situated at each end of it. The inner shoe carries bearings in which the knife head slides, and has attached to it the drag link and lifting mechanisms. The outer shoe, which may be provided with a small wheel to give easy adjustment of the height of cutting, carries a divider and a swath board. The divider separates the swath to be cut from the rest of the crop, and the swath board, with the grass stick attached, rolls the cut swath together so that there is room for horses or tractors and machines to pass between the swaths.

The pitch of the cutter-bar may be altered to suit varying conditions by means of a hand lever. This adjustment permits close cutting when the ground is smooth, and elevation of the fingers when stones or uneven ground render a close setting undesirable.

Adjustments and Overhauling. A cutter-bar that is out of alignment may cause poor cutting, heavy draught or breakages. The cutter-bar is in proper alignment when the crank-pin, the knife head and the outer end of the knife are in a line at right angles to the direction of travel, with the mower at work. When the machine is stationary, the pressure of the grass on the cutter-bar is released, and the outer end of the bar should be set slightly in advance of the inner edge. This "lead" of the cutter-bar should generally be about two degrees. The correct lead may be easily obtained by adjustment of the relative lengths of the members of the stays, or drag link, or by turning an eccentric sleeve fitted in the stirrup holding the cutter-bar.

When the knife reciprocates, the knife sections should "register" with the fingers; i.e. the point of each section should pass from the centre of one finger to the centre of the adjacent one. This may be adjusted by altering the length of the drag link. It is an important adjustment, for cutting may be ragged and difficult if registering is not accurate.

The lifting levers and lifting spring should be adjusted so that when at work the cutter-bar lies level on the ground, carrying just enough weight to hold it steady. The lifting spring should, at the same time, be sufficiently tight to facilitate raising of the cutter-bar; and the lifting linkage should be so adjusted that the inner and outer shoes commence to rise at the same time.

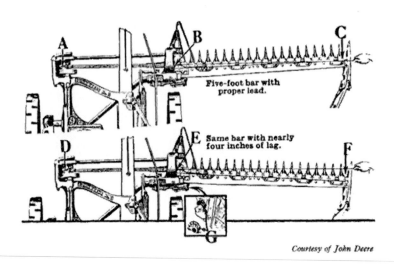

Courtesy of John Deere

FIG. 132.—*CORRECT AND INCORRECT LEAD FOR CUTTER-BAR*

Constant attention should be paid to lubrication, and bearings should occasionally be inspected. The main axle is usually carried on roller bearings, but replaceable brass or bronze bushings are generally employed for the crankshaft and connecting rod, and these should be replaced when they become worn. The pawls and pawl springs may also occasionally need renewal owing to wear or breakage. The care of the mower consists largely in attending to such points as these as they arise.

Power-Driven Mowers.

An early type of power-driven mower which is still useful for difficult ground conditions, such as sewage farms, is the one-horse mower which has the mechanism driven by a small internal combustion engine mounted on the frame. Early power-driven tractor mowers looked very much like the trailer machines already described, having a pair of large wheels, a heavy frame, and complicated gearing. Power drive, however, permits a complete departure from this design, and most modern machines are much lighter and simpler. Some, which may be described as semi-mounted, are more or less rigidly attached to the tractor drawbar and have either a pair of small castor wheels or a single one. Others dispense almost entirely with any sort of mower frame and have the cutter-bar and linkage bars carried directly on the tractor, either at the rear or at the side.

All forms of power-driven mowers have the advantage that knife speed and forward speed are, to a certain extent, independently variable by choice of gear. This is an advantage, both when dealing with heavy crops which require a low forward speed, and when topping pastures at high forward speeds.

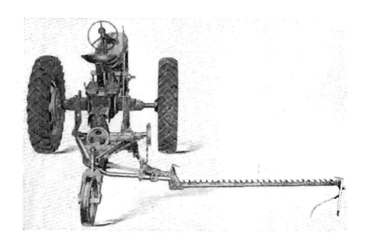

FIG. 133.—SEMI-MOUNTED POWER DRIVE MOWER, (INTERNATIONAL.)

Mounting the cutter-bar at the side is the standard practice in Germany, and is rapidly finding favour in Britain. The drive is usually very simple, consisting of a direct V-belt drive from a pulley mounted on the power take-off to a counter-shaft: the latter runs forward beneath and on the near side of the transmission housing, and has the crank which carries the pitman mounted directly on its forward end.

On all forms of power-driven mowers, safety devices must be incorporated. On most machines, the cutter-bar can swing back out of harm's way if it strikes a serious obstacle. Many direct mounted mowers have a safety device which slops the tractor automatically when the "break-back" mechanism operates. Such devices are particularly useful on mid-mounted mowers, where the cutter-bar cannot swing back very far before it fouls the rear wheel of the tractor. The modern method of employing V-belt drive eliminates the necessity of providing an independent slipping clutch in the drive to the cutter-bar.

FIG. 134.—P.T.O.-DRIVEN SEMI-MOUNTED MOWER.
(BAMFORDS.)

Some advantages of the mid-mounted mower over other types are readily apparent. The cutter-bar and control levers are in front of the operator; the close-coupled arrangement permits square cornering; the drawbar is left free for using any other equipment at the same time (e.g. a bruiser or a side rake); and the mower, with the cutter-bar having been put in the vertical position, may be left on the tractor if the latter goes to do some quite different work elsewhere. It therefore seems likely that the mid-mounted types will become much more widely used.

(N.I.A.E. Photo)

FIG. 135.—LIGHT TRACTOR (B.M.B.) WITH MID-MOUNTED MOWER IN TRANSPORT POSITION.

Rotary Grass Cutters.

There are many circumstances, e.g. in grass orchards and on pastures which are being grazed, when it is not necessary to collect the grass that is cut. The crops to be dealt with are usually but not always thin, and for one reason or another the reciprocating cutter-bar is somewhat unsuitable.

Grass cutters which employ high-speed horizontally rotating knives have been developed for use in such conditions. The cutting blades consist of renewable knife sections which are usually attached to the corners of rectangular steel plates, these plates being driven by V-belts and gearing from the tractor P.T.O. shaft. A shield covers the rotors to provide protection from injury by flying stones. These machines can keep going in conditions which are too rough for gang mowers.

Gang Mowers. Extension of the practice of grassing down orchards has resulted in the use by fruit-growers of gang mowers, of the type generally used for cutting sports fields. Such mowers can be quite satisfactory provided that sufficient care is taken to produce a level and uniform sward which is reasonably free from stones. Gang mowers are totally unsuited to the cutting of tussocky swards on rough land, and growers who have such conditions should make every effort to improve the turf by harrowing and rolling, or even if necessary by re-seeding, rather than expect the gang mower to do the impossible. Given a reasonably good start, gang mowers will steadily improve the sward, but care must always be taken not to cut it up unnecessarily in wet weather, during transport and spraying operations.

*FIG. 136.—REAR-MOUNTED ROTARY GRASS GUT-
TER. (HAYTER.)*

Repair bills on gang mowers are apt to be a heavy item, and keeping them within reasonable limits depends more on attention to the points mentioned above than on anything that can be done during the mowing operation itself.

Tractor-mounted gang mowers, which can be raised clear of the ground by the hydraulic lift, are now available; and these have obvious advantages in transport from field to field, and in turning at the ends.

Gang mowers which will collect and load the grass from aerodromes are now employed in conjunction with grass drying, and these are particularly suitable where chopped grass is needed for supplying high-temperature pneumatic and drum driers.

Making the Hay.

In haymaking, an average loss of 25 per cent, or more of the dry matter contained in the fresh material usually occurs, and this loss falls mainly on the most digestible portions, so that the true loss of feeding value is greater, generally ranging from 30 to 50 per cent, according to the conditions. For example, good hay on average suffers a loss of about two-fifths of the starch equivalent and one-third of the digestible crude protein originally present in the fresh crop. The losses occur through respiration, shattering of leaves, leaching by rain, bleaching in sunlight, heating in the stack, and so on.

Generally speaking, it may be stated that, provided partial over-drying and bleaching can be avoided, the best hay is that for which the interval between cutting and carrying is shortest. Haymaking machinery should be judiciously used to shorten this time by facilitating drying in every possible way. The exact machines to be used and the correct timing of their use is a matter on which little useful information can be given, owing to the wide variations in the condition of crops and the weather. In good weather, tedding immediately after mowing produces more rapid drying. The practice of haymaking varies a great deal from district to district. In the north of England, where heavy crops are common and the weather is often very wet in the haymaking season, use is commonly made of the system of cocking; and tripods, racks or wire fences are sometimes used to assist in reducing the losses which may occur in wet weather. In the south and cast, on the other hand, where the weather is finer, haymaking is often a comparatively simple operation.

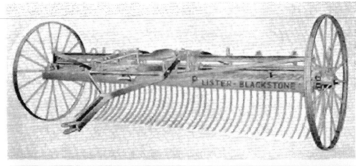

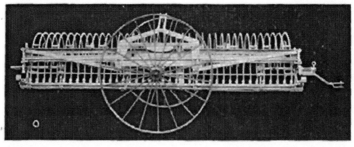

FIG. 137.—TRACTOR RAKE. (LISTER-BLACKSTONE.)
ABOVE, READY FOR WORK; BELOW, ADJUSTED FOR
TRANSPORT.

Horse Rakes.

The horse rake is widely used for dragging hay into windrows and for raking clean both hay and corn stubbles. A large number of curved teeth are carried on a bar mounted parallel to the axle, the teeth being made of hardened and tempered high-carbon steel. On some machines, the section of the teeth is round or oval, but I, T and H sections resist bending better. On the older types, the teeth may be raised by a mechanism operated by foot and hand levers, but a pawl and ratchet self-lift mechanism is provided on most modern machines. The pawls are tripped by a light pressure on a lever, and they then engage ratchets fixed to the wheels. When the mechanism has raised the teeth and released the hay, the pawls automatically move out of gear, and the teeth are allowed to fall again unless kept up by a retaining lever.

Tractor Rakes.

Wide tractor rakes, with self-lifts, have been introduced for speeding up the work and fitting in with the system on those farms where most of the other work is mechanized. These rakes are so constructed that travelling from one field to another is a simple operation. On one machine, for example, the wheels are removed from their ordinary axles and fitted to auxiliary axles provided at the front and back. The drawbar is folded back on top of the machine, another drawbar is clipped on to one of the free axles, and the rake is pulled from this end (Fig. 139). Other machines have removable sections and a telescopic axle.

FIG. 138.—*TRACTOR SIDE-RAKE, SWATH-TURNER AND TEDDER ARRANGED FOR SIDE-RAKING. (LIST-ER-BLACKSTONE.)*

Light tractor-mounted rakes, with a separately articulated dumping mechanism operated by the tractor's hydraulic lift, are now available.

Rake-bar Type Combined Side-delivery Rakes and Swath-Turners.

The side-delivery rake is used for rolling swaths sideways into windrows. The rake usually covers two swaths and can therefore roll five swaths together in one bout across the field. A common type of machine is one which can be used either as a side rake, as a swath-turner or tedder (Figs. 138, 139, 143). Three or four horizontal rake bars which carry a number of vertical teeth are mounted at each end on a rotating disc, and set obliquely across the frame of the machine. Rotation of the discs, which are driven from the land wheels, causes the hay to be swept forwards and sideways so that it is delivered in a light windrow at the side of the machine. Such a machine may be used as a swath-turner if the teeth in the centres of the rake bars are removed. It used to be necessary for the operator of one of these machines to carry a knife for the purpose of cutting free the hay which soon became entangled with the mechanism, but the best modern machines are free from this trouble.

FIG. 139.—REAR-ACTION COMBINED SIDE-DELIV-
ERY RAKE AND SWATH-TURNER, SET FOR SWATH-
TURNING. (NICHOLSON.)

Rotating-Head Type Combined Haymaking Machines.

The swath-turner is used for turning the swaths left by the mower so that further drying of the under side may take place. It rolls the swath over in a ribbon on to the dry ground between the swaths. The action is a gentle one and, with good machines, few leaves are broken off in the operation. The swaths cannot, however, be turned in a clean ribbon unless the butts are turned over the heads. The machine therefore usually works round and round the field, following the mower, though some machines can turn the swaths either to the right or to the left, and are able to work along the swaths in either direction. One type of machine has two sets of forks which have a rotating action similar to that of a potato spinner. Both sets of forks may usually be driven independently in either direction (Fig. 140).

FIG. 140.—TRACTOR-MOUNTED P.T.O.-DRIVEN ROTATING-HEAD TYPE SWATH-TURNER AND COL-LECTOR.

FIG. 141.—SIDE-RAKE WITH INDEPENDENT GROUND-DRIVEN INCLINED WHEELS. (BLANCH.)

FIG. 142.—HAY TEDDER. (BAMFORDS.)

Light tractor-mounted P.T.O. driven rotating-head type combined haymaking machines arc now available.

A new type of side-rake recently introduced in U.S.A. and now manufactured in Britain has a number of independently-mounted large-diameter wheels with spring teeth attached radially. These are set obliquely in the frame and obtain their drive by contact with the ground. These machines arc designed to work at fairly high speed, and it is claimed that the free-floating action makes them suitable for use on uneven land. By removing the centre wheels the machine can act as a swath-turner (Fig. 141).

Tedders.

The tedder is sometimes used to facilitate drying when the hay lies in thick, tightly packed swaths. One type is a rotary forking machine, with teeth revolving about a horizontal shaft mounted parallel to the axle. Tedders which have a "feathered" action produced by the use of a linkage mechanism are sometimes preferred to those with a direct action, since the latter type tends to damage the herbage by handling it roughly.

Many modern haymaking machines whose main use is side-raking can also be used as tedders.

Power-driven Haymaking Machines.

Side-rakes, swath-turners and tedders of various types are now available with P.T.O. drive. This leads to a more positive action, and the use of enclosed gears which give little trouble. Some of the lighter types are tractor-mounted and power-lifted, while the heavier machines are enabled to run on pneumatic-tyred castor wheels, a fact that is made use of in providing for transport of the machine (Fig. 143). In a German machine advantage is taken of the use of castor wheels to provide for swivelling the whole machine in relation to the direction of travel, thereby producing a variety of effects on the crop.

(A) WORKING POSITION.

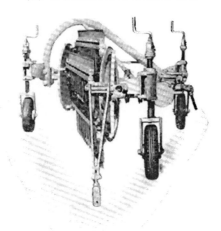

(B) READY FOR TRANSPORT.

FIG. 143.—P.T.O.-DRIVEN COMBINED SWATH-TURNER, SIDE-RAKE AND TEDDER. (NICHOLSON.)

Grass Crushers and Bruisers.

During the 1930's experiments in Mississipi, U.S.A., with a prototype machine, which crushed the swath between a pair of rollers soon after cutting, showed that in the dry atmospheric conditions normally prevailing there, grass could usually be dried more uniformly and quickly after the crushing process. Crushing hastened the drying of the tough nodes of certain types of crops, with the result that the hay could be made more quickly. The effect was most marked in good drying weather.

Since that time work has continued in other parts of U.S.A. and in Canada, and a commercial machine which picks up the swath and crushes it between rollers has been developed. The results confirm that it is in really good drying weather, when the leaves of uncrushed grass rapidly become brittle, that real advantage is achieved by crushing. Hay crushers are still only rarely used commercially in North America. Preliminary experiments in Britain with roller type machines indicate that very heavy pressures are needed to achieve any worth-while result with our heavy crops and thick swaths.

In Britain, the commercial development of machines and their practical application has preceded fundamental research by public institutions. One commercial machine employs a bruising mechanism which is not unlike the drum and concave of a threshing machine. The machine is mounted on a rectangular chassis with large-diameter wheels at the rear and small castor wheels at the front. It is P.T.O. driven, the drive passing from the gear-box by chain drive to the rotor, which revolves at 1,500 r.p.m. and carries a number of swinging beaters. The distance of the concave from the rotor is adjustable, and the height of the rotor above the ground is regulated by altering the distance between the chassis and the rear wheel axles.

Farmers' experiences show that in good weather- the crop should be bruised soon after mowing, and that more rapid drying, and hay of improved quality, can result. A powerful tractor is needed to drive the machine, and the rate of work in good crops is not high. In view of the generally unpredictable British climate and the rather high capital and running-costs involved in carrying out the bruising process it is possible that its most important application in Britain will be as an aid to artificial drying processes, rather than to natural drying. Grass which has been treated by the bruiser can be elevated directly on to a vehicle if it is desired to cart it immediately for drying or for silage.

Hay Loaders.

The hay loader is a transportable forking and elevating mechanism which, when hitched behind a wagon, automatically picks up the hay from the swath or windrow and delivers it on to the wagon. It saves time and man-labour in carting hay that is to be stacked out of the field, and is successfully employed on large farms and small; many occupiers of family farms consider the loader one of their indispensable machines.

FIG. 144.—RECIPROCATING FORK TYPE HAY LOAD-ER. (MASSEY-HARRIS.)

The loader has a light frame of angle steel, carried on two large drive wheels at the rear and a swivelling fore-carriage with small wheels at the front. Two main types of elevating mechanism are in common use, viz. the oscillating fork type and the continuous apron type with a cylindrical forking device to pick up the hay.

In the apron or conveyor type, a gathering cylinder, which rotates in the same direction as the land wheels, picks up the hay from the ground and delivers it to the foot of an endless conveyor of ropes and slats. The conveyor, which generally has teeth in the slats, may travel either above or below the hay. A second smaller gleaning cylinder, which rotates against the direction of travel, may follow behind the gathering cylinder. This picks up any hay left and delivers it into the tines of the gathering cylinder. The mechanism is driven by gearing through a pawl and ratchet drive on the land wheels, and a dog clutch is provided for putting the machine in and out of gear.

In the fork type, both the collection and elevation of the hay may be carried out by oscillating forks driven by cranks; but in some machines the actual picking up of the hay is performed by a cylinder. In the fork type, the platform extends right down to the ground, and the hay is lifted on to it and pushed up it by the forks, which oscillate with an action not unlike that of the shakers of a threshing machine.

The rake bars are driven by a crank geared to the travelling wheels.

Improvements incorporated in some new fork-type machines are four-throw balanced cranks, with two rake bars coming into operation every quarter revolution, so that dead-centre points are eliminated and an even flow of hay up the chute is ensured. One British machine has a chute which is tapered from 5 ft. 6 in. wide at the bottom end to 3 ft. 6 in. at the top. This machine delivers the hay on to the centre of the load, and is particularly advantageous in windy weather.

The fork type is more generally employed in this country. It is more simple and less expensive than the conveyor type, and continues to force the hay on to the load even when the loader does not keep the upper end clear. The conveyor type, on the other hand, tends to drag the hay back with it unless there is a free delivery. An advantage of the conveyor type is that it handles the hay rather more gently than the fork type.

Hay Sweeps.

When hay is to be stacked in the field, the sweep is a useful labour-saving device. It may also be employed when the stack is made in an adjacent field if a stacker capable of transporting the hay over the fence or hedge is available. The sweep consists of a frame with a set of long wooden teeth about a foot apart, projecting forwards. The teeth should be made of ash and fitted with metal shoes at the tips. When the sweep is at work, the points of the teeth run along the ground and the hay is pushed up to the back, where it is prevented from falling off by wooden back and side rails.

Horse sweeps, tractor sweeps and car sweeps of various types and sizes are used. Horse sweeps generally have a wheel at each side and a swivelling wheel at the rear, with a pole hitch for a horse at each side. The sweep is 10-12 ft. wide, and when loaded, holds about half a ton of hay. When a full load is obtained, the teeth may be raised clear of the ground and the load is then taken to the stack, where it is deposited by lowering the teeth and backing.

{N.I.A.E. Photo}

FIG. 145.—SIMPLE TRACTOR SWEEP AND LOW-HOPPER ELEVATOR.

One type of tractor sweep is very similar to the horse machine, except that it is arranged to be pushed by the tractor. The more usual type is a simpler and cheaper device which is attached direct to the front of the tractor and has no wheels of its own. On some general purpose tractors, it may be fitted with a power lift, but many simple tractor sweeps are successfully used without any lifting mechanism.

In U.S.A. sweeps are often home-made, and equipped with a power lift made partly from the back axle of a car or lorry. The complete back axle is mounted on the rear of the tractor and driven from the power take-off. It will be readily-understood that if one wheel is replaced by a suitable winding drum, and the brakes are independently operated, a fairly satisfactory winch is obtained. Many farmers have built their sweeps on the rear of the tractor, the advantages of this being that there is more cooling air for the radiator, the sweep is easier to manœuvre, there is no loss of traction when the load is lifted, and there is good visibility to permit driving in top gear to the stack or barn.

The general use of hydraulic lifts on modern all-purpose tractors makes the lifting of either front-mounted or rear-mounted sweeps a simple matter. Power-lifted sweeps cost little more than those without lifting arrangements, and the slight extra cost is usually well worth while.

Hay Elevators.

The elevator may be used for stacking hay, corn or straw. It consists of a long, rectangular wooden or steel trough, mounted on a transport truck; endless chains with cross slats and upstanding teeth travel along the bottom of the trough and carry the hay from a hopper at the bottom, up to the top of the stack. The conveyor is usually driven, by a $1\frac{1}{2}$ to 2 h.p. engine, which may be independently portable or may be carried on the elevator frame. The elevator trough is adjustable to the height of the stack and is generally in two hinged sections so that it may be folded for transport. The design of the hopper at the foot of the trough may be varied according to the main use for which the machine is required. For use with hay sweeps, it is a great advantage to have the hopper extending down to the ground to facilitate feeding.

Hay and straw elevators are now frequently required to deal with baled hay or straw, and most can do this satisfactorily with a little modification. One method is to have a special set of elevator chains and slats for this work, while, on some makes adapters can be fitted to the standard, chains, and the latter need not be removed. There is usually a limit to the number of heavy bales that can be put on the elevator at any one time, but in practice it is seldom, necessary to have more than a couple of bales on at a time to keep the man stacking fully occupied.

Horse Forks.

The horse fork is a simple crane carrying a special hay fork in the form of a harpoon or grapple. These machines have been used on some farms for many years. The crane may consist of a derrick or a swinging boom attached to a vertical mast. A guyed mast or tripod is erected beside the stack, and the mechanism is operated by a horse which raises a load of hay by pulling on a rope as it is led away from the stack. A power-driven friction hoist may be used instead of the horse, and this gives a more rapid hoisting, with easier control.

The common type of grapple fork has hinged arms which grip the hay until it is in position over the stack, when a jerk on a release-cord deposits the load. The harpoon fork is plunged into the hay to be elevated, and on being hoisted the prongs of the harpoon open out. To deposit the load, the operator pulls a cord which causes the harpoon prongs to turn downwards and release the hay. Slings (i.e. nets made of rope and wooden slats) may be placed on the cart before loading so that a whole load or half-load may be lifted at once by attaching the ends of the slings to the hook of a swinging stacker.

FIG. 146.—LOW-HOPPER ELEVATOR. (MASSEY-HAR-RIS.)

Some farmers who have both horse fork and elevator prefer the grapple fork for hay loaded with a hay-loader, and prefer the elevator for hay swept or loaded by hand.

FIG. 147.—WHEELED TRACTOR HAY SWEEP AND OVERSHOT STACKER AT WORK.

Overshot Stackers.

The increased speed of bringing hay to the stack brought about, by the use of tractor and car sweeps has made it desirable to employ stackers which will transfer a sweep-load to the stack in one operation, without any hand-forking. The overshot stacker comprises a frame which rests on the ground and has hinged to it two telescopic arms, carrying a cradle similar to a sweep. The sweep pushes its load of hay right on to the stacker cradle and then backs away, leaving the load on the cradle. The arms of the stacker are then raised and, extending as they rise, they lift up the hay and throw it over backwards on to the rick. Means are provided for varying the height of the lift. The stacker may be driven by a horse, a tractor, a car, or an engine with a friction hoist. It is also possible to build an overshot stacker around a tractor so that the outfit may be used as a combination sweep and stacker.

The "Push-Off" Tractor Stacker is a device which, after being developed in the United States, is now available in Britain. It consists of a small sweep attached to a hydraulic front loader, and having a back which can be pushed forwards to deliver the load. The lift and push-off device can be operated independently, being actuated by separate hydraulic rams.

The stacker may be used exactly like a sweep until the load is collected, and it can then deposit its load either on to a trailer or directly on to a small stack. The tractor is naturally somewhat unstable when carrying a heavy load in the raised position, and it is advisable to keep the load only slightly raised on journeys to the stack, and to avoid making sharp turns.

Push-off stackers clearly have great potentialities, especially for small farms. There may be considerable advantages in carrying sweep-loads of hay or silage at the front and rear of the tractor simultaneously, as either used alone tends to cause difficulties in steering and general balance of the tractor.

Hydraulically operated push-off tractor stackers seem likely to supersede other types on account of their basic simplicity, and the fact that they utilize for the most part the hydraulic front loader which is likely to become, on account of its many uses, an item of equipment that will be found on almost all farms.

Field Baling.

Experiments on baling hay direct from the field were begun in England several years ago, and the method has been found quite suitable for use in our climate, providing that it is properly carried out.

Hay is now generally considered to be fit to bale only a very little earlier than it is fit to stack, and though very good green hay of high feeding value can sometimes be made by early baling, the risks are great and the temptation to start before the crop is reasonably fit should be avoided. Hay that is fit to stack usually has a moisture content of about 20-25 Per cent. With pick-up baling, it is sometimes possible to get good results with material having a moisture content of up to about 30 per cent, provided the windrows arc uniformly dried and that fairly loose bales are made so that drying can continue in the bale. But with very damp green hay there is always a possibility that serious moulding may result, and the bales must be left out for a time or stacked so that air can pass through them to avoid heating. The tougher the hay when baled, the looser the bales should be, a suitable density for very damp hay being 8-10 lb. per cubic foot. Damp patches in the windrows will cause trouble with mould. Hay should never be baled when there is surface moisture on it, since this is almost certain to result in a mouldy product. On the other hand, the crop should not be allowed to get over-dry, since this will result in the loss of a high proportion of the leaf through shattering, and will inevitably produce a fodder' of low feeding value.

As this method of haymaking makes progress, farmers will be able to judge the correct moisture content and density from experience.

Types of Balers.

Stationary Ram Balers are usually large, heavy machines, arranged for belt drive. Compression is produced by a gear-driven horizontal piston, and about 15 b.h.p. is required to drive the machine satisfactorily. Two men are needed for wiring, and the standard size of bale is 3 ft. 6 in. long and 18 in. by 23 in. in section. The capacity of such a machine, with three men feeding, two wiring and one sweeping the hay, is about 2 tons per hour.

The pressure put on the bale in these machines depends on the action of two screw clamps which control the distance apart of the ends of the baling chamber. With greenish hay the density can be varied from about 10 to 22 lb. per cubic foot.

A light type of ram baler needs only about 5 b.h.p. and is sometimes used with small grass driers. Tying is by twine.

Hand-Wired Pick-Up Balers. This type of pick-up baler comprises the pick-up unit, consisting of spiked tines attached to a flexible chain elevator; a horizontal cross-conveyor, which feeds the hay to the baling chamber; and a ram-type baler.

Balers of this type must now be regarded as obsolete, though they are still useful machines, particularly where the hay is to be sold, and transported long distances. With these machines, quite apart from the extra cost for labour, there is the disadvantage of a low rate of work. The maximum rate that experienced operators can maintain with such machines is about one bale a minute, and output is usually not more than 2 tons per hour with the smaller machines which have a baling chamber 14 in. by 18 in. in section, or 3 tons per hour with the large machines which have a baling chamber 22 in. by 18 in. Some of these machines can be converted to string-tying by hand methods, but the advantages of such a conversion are doubtful. The conversion, where efficient, can save one operator, but the speed of work is still restricted, while the cost of the heavy twine needed is a little more than that of wire. Nevertheless, many farmers greatly dislike baling wire, owing to the harm that can be done by cattle swallowing short pieces, and some prefer twine if only for this reason.

FIG. 148.—HEAVY RAM-TYPE STATIONARY BALER. (RANSOMES.)

Pick-up balers may be driven either from the power take-off or by an engine mounted on the machine. There are some advantages in having a separate engine, but with a fairly powerful tractor the power take-off drive is satisfactory. The pick-up unit is generally driven from the ground, at the same rate as the forward travel of the machine.

Automatic Tying Ram Balers. This popular type of machine makes a compact bale with a section usually about

18 in. by 16 in., and length variable from about 30 to 40 in. The bales are automatically tied lengthwise with medium weight twine having a "runnage" of about 225 feet to the pound. Bale density can be varied easily, but in general the bales are not quite as tight and "square" as those from a hand-wired machine. The plunger which pushes hay into the baling chamber carries a knife which slices the hay, making the layers easy to separate for feeding. About. 4-5 lb. of twine is used per ton of hay.

FIG. 149.—*ENGINE DRIVEN SELF-TYING PICK-UP BALER. (INTERNATIONAL.)*

With such a machine the labour gang for haymaking can be entirely dispensed with, for two men can pick up the hay at a very satisfactory speed. Some machines employ automatic wire-tying devices.

A self-propelled machine with a very high rate of work has been developed for big farms and contract work (Fig. 150).

FIG. 150.—SELF-PROPELLED AUTOMATIC RAM TYPE PICK-UP BALER. (JONES.)

FIG. 151.—SELF-TYING BALING PRESS. (STANHAY.)

Press Balers differ considerably in design from ram balers. The bales are pressed along their long side instead of from one end. The section of the bales is generally 40 to 50 in. by 12 to 16 in., and the length adjustable from 16 to 26 in. Pressure is controlled by a screw clamp, and the density with green hay cannot exceed about 11 lb. per cubic foot. With hay fit to bale, therefore, the bales cannot be made too dense with a press type machine.

The press type machine has automatic tying. With a stationary press baler and a team of five men (two sweeping and three feeding), up to $2\frac{1}{2}$ tons per hour can be baled.

Pick-up press type balers are available, and these, equipped with a wooden ramp at the rear, can easily deliver the bales directly on to a 4-wheeled trailer. This arrangement makes it simple on level land for two men to pick up from the windrow, bale and load in one operation.

The bales are apt to be rather loose and untidy, and will not stand much rough handling; but they have the advantage that handling with a fork is easy, the weight usually being only 20-40 lb. Another advantage is that ordinary binder twine is used for tying, and only about 2-3 lb. per ton of hay is needed.

Roll-Type Roto-Baler. The roll-type pick-up baler picks up hay from the windrow and rolls it into cylinders 36 in. long. The sloping conveyor delivers the windrow beneath a press roll into the space between two sets of endless rubber belts running in opposite directions. These belts roll up the windrow much in the same way as a carpet is rolled. When the bale reaches the desired size the conveyor automatically stops feeding more material and a length of binder twine is wrapped spirally round the bale without being tied. The twine is then automatically cut and the bale discharged from the rear of the machine.

The bale diameter can be regulated from 14 to 22 in. by adjusting the length of a trip rod, while the density can be varied by tightening or loosening an external brake. About 30 to 36 ft. of twine is generally used per bale, though shorter lengths may be employed by adjustment if desired. With bales weighing about 60 lb. and twine having a runnage of 400 ft. per lb. about 3 lb. of twine is used per ton of hay baled.

When once the mechanisms of the roto-baler are understood, the setting and adjustment are easy, and there is little to go wrong. The bales have the advantage of being fairly weatherproof if left lying as they fall.

Automatic pick-up balers can have a high rate of work as well as a low labour requirement.

A disadvantage of the pick-up is that the bales may be scattered over the field; but this need not be so if a little ingenuity is employed in devising a suitable trailer hitch, or a bale sledge.

(N.I.A.F., Photo.)

FIG. 152.—P.T.O.-DRIVEN ROLE-TYPE PICK-UP BALER. (ALLIS-CHALMERS.)

Collecting the Bales.

Whether it is better to cart the bales immediately, or to leave them in the field to dry out, depends both on the condition of the hay when baled, and on the type of baler used, as well as on weather conditions. If the hay is baled in perfect condition it is foolish to leave it out in the field to get bleached on the outside, to get wet if it rains, and to require moving every few days to avoid mould at the bottom or spoiling the grass on which the bale stands. On the other hand, there is no doubt that in average weather appreciable drying of the hay takes place if damp baled hay is left out in the field. Moreover, both rolled bales lying flat and ram-type bales stood on end are resistant to damage by showers, though they do suffer appreciably in continuous rain. When ram-type bales are stood on end, the end which is formed first should always be put at the top. Bales produced by press-type machines are particularly vulnerable and should never be left out in unsettled weather.

Many farmers find collection of the bales something of a problem, since bale-loaders are not yet fully satisfactory, while the extra cost of a specialized machine can often not be justified. Several farmers have tackled this problem by attaching a bale sledge to the rear of the baler in such a manner that the bales can be dropped in windrows. Fig. 153 shows a typical homemade sledge designed for carrying about nine bales. It is made very simply out of a pair of heavy gauge corrugated iron sheets. The bales are collected and carried on the sledge until the windrow is reached, when they are pushed off without stopping. If desired they may be shipped off by sticking a bar into the ground through a slot designed for this purpose. Subsequently, it is a simple matter to come along the thick windrows with the transport, and collect the bales without much running about. A "universal" sack, bale and root-loader can be efficiently used when the bales are grouped in this manner.

(N.I.A.E. Drawing)

FIG. 153.—HOME-MADE BALE SLEDGE MADE FROM TWO SHEETS OF CORRUGATED IRON.

Stacking.

There can be no doubt that the best place to stack baled hay is in a barn, and that the earlier fit hay is there, the better. A great deal of baled hay is spoiled by bad stacking, and even if great care is taken with the stacking, the bales often move so much on settling, that what appeared to be a good stack becomes an untidy heap which is easily damaged by rain. Wire-tied bales should be stacked on their narrow sides with gaps of 3-4 in. between the rows if they are taken in at once; but if the bales are really dry, or with string-tied bales, there is usually no need to make any special provision for ventilation. The stacking of roll-type bales in rectangular stacks presents some difficulty. On an open site the easiest method is to stack in a pyramid, as with drainpipes; a base 12 bales wide often gives a suitable size. All that is necessary to prevent slipping is to pass baling wires round the 4 or 5 outer bales at both ends of the bottom layer. Some farmers favour rectangular stacks, and build with all bale ends to the outside, except at alternate corners. Such stacks, however, need great care in building, unless they are in a barn and can receive a little support from the walls.

Other Methods of Handling Hay.

During recent years many studies of methops of handling hay have been made in the United States. Hay is generally stored in a loft above the cowshed over there, and one new method consists of chopping it into short lengths with a pick-up chopper, and blowing it into the loft. More hay can be stored in a given space, and the chopped material is easier to feed. The chopping method does not, however, reduce weather risks at haymaking time, and needs more equipment. There does not seem to be any possibility of this method proving of value in Britain.

Another method consists of completing the drying of partially made hay by employing forced ventilation. This method is discussed in Chapter Nineteen.

Economic Haymaking Systems.

Successful mechanization of haymaking usually depends on getting a high rate of work from a limited labour force. The equipment chosen must be related to the amount of work to be done, and small farms must often make do with inexpensive machines. Quality of the product is more important than a low labour requirement or a high rate of work. Some systems—e.g. using an automatic pick-up baler—have the advantage that the vital baling operation can be done very rapidly by only one man, the bales being collected later. Other systems—e.g. sweeping and stacking—require all operations to proceed simultaneously and necessitate a larger gang.

The following table gives a general indication of results that can be achieved by employing various haymaking systems, using the size of gang required to carry on all operations simultaneously.

The field chopper (forage harvester) method is popular in U.S.A. mainly because it takes all the hand work out of the loading operation. With this method, overall outputs of ·66 tons collected and stored per man hour can be obtained, and the whole operation can conveniently be carried on as a one-man job.

TABLE III.—HAYMAKING SYSTEMS—TYPICAL RE-
SULTS.

System	Size of Gang*	Rate of Work Tons per Hour	Tons per Man-hour
Hand Work (Horses and Waggons) ..	6	1	·17
Tractors, Hayloader and Elevator ..	6	$1\frac{1}{4}$	·21
Tractor Sweep and Stacker	4	$1\frac{1}{2}$	·37
Two Tractor Sweeps and Stacker ..	6	$2\frac{1}{2}$	·42
Automatic Pick-up Baler, Bale Loader and Elevator	6	$3\frac{1}{2}$	·58

* For complete operation, including collection of bales.

Making Silage.

Making silage enables green crops to be conserved in succulent condition for feeding to stock during winter. Silage was mentioned in agricultural literature in 1843, but at the time it attracted little attention. In 1885 Mr. George Fry published a short account of his experience with silage, and the practice was at that time considered important enough to justify the attention of a Royal Commission. After that the process made little headway in this country until very recent times, but it became so important in North America that about a million silos are in use on farms there. The spread in the use of silage in the United States has been facilitated by the fact that conditions there are very favourable to the growth of maize, one of the best silage crops.

During recent years silage has been gradually achieving success in other countries than America, partly owing to advances in knowledge concerning the correct methods of ensiling such crops as grass, lucerne and other legumes. New methods of ensilage have been developed, and some, such as the A.I.V. process, introduced by Professor A. I. Virtanen in 1928, have entirely changed the position with regard to the practicability of making good silage from such crops. For the production of good silage from high-protein materials like young grass or legumes, conditions in the silo must be kept such that butyric acid fermentation is prevented. This may be done by adding a mineral acid to the silage, as in the A.I.V. method, sufficient being added to reduce the pH value of the material to below 4·2. There are obvious disadvantages in the use of acids such as sulphuric and hydrochloric acids; but phosphoric acid can be utilized to a certain extent by animals, and serves as a fertilizer when excreted.

A second method, viz. that of adding molasses or sugar to the material, achieves the same object and does not suffer from the disadvantages associated with the use of acids. The principle is to add sugar to the green material, and this is readily fermented by bacteria to form lactic acid. Sufficient sugar must be added to produce so much lactic acid that butyric fermentation cannot occur. The amount of sugar added is usually from 1 to 2 per cent, of the weight of the green material, and one of the great advantages of the method is that any excess of sugar is not harmful, but forms useful food for the stock.

The molasses method of making grass and similar crops into silage had just been perfected in 1939 when war conditions brought about a shortage of protein feeding stuffs and gave a great fillip to silage making in Britain.

To-day the position is that most British farmers are well aware that grass, clover seeds, mixtures of oats and vetches, or beans, oats and vetches, maize, kale, etc., can be made into good silage successfully.

Silage can be made with most certainty of success if the material is chopped and blown into a tower silo. Chopping makes it possible to pack the material well, and if molasses or acid is to be added the most convenient method of adding it is usually to feed it on to the herbage as it passes through the silage cutter. The tower silo also has advantages over other types in utilization of the silage; for with a high and narrow tower a layer can be removed from the whole of the surface each day, and this prevents any spoiling. Moreover, little labour is required for the removal of the silage from the tower, compared with that required with some types of silo. There is, however, much to be said for pit and clamp silos, in that the capital outlay is small and that for most materials no cutter is required. Portable silos, too, have advantages, in that they can be used in fields situated a great distance from the homestead.

Pit silos have many advantages which are now generally appreciated by farmers. They are suitable for any type of soil, providing that they can be drained. Suitable dimensions for a large pit to hold 200 tons of silage are 15 ft. across the top, tapering to 12 ft. at the bottom; 4 ft. deep; and 40 yd. long. Such pits may be constructed with a ramp at each end.

On small farms, where it is necessary to provide for only 60 tons of silage, the section of the pit may remain about the same, and the length will be cut down to about 12-15 yd. Construction of the large pits is best carried out with the help of a contractor's industrial-type earth mover. For the small ones, however, a very satisfactory job can be done with a small tractor pulling a scoop that moves about 5 cwt. of earth at a time. One man and a tractor can easily excavate a small pit in a day.

Pit silos are easy to fill; for the tractor and trailer can be driven right through, and various mechanical devices can be fitted up to pull off the load and distribute it roughly over the surface.

Getting the silage out without wastage often presents some difficulty. After a vertical cut has been made, a front loader with a manure bucket or a special grab (Fig. 154), or a cable-fork-and-conveyor loader, may be used.

FIG. 154.—HYDRAULICALLY OPERATED SILAGE GRAB. (BAKER-SELIGMAN.)

FIG. 155.—PATERSON BUCKRAKE ON LIGHT TRAC-TOR.

Machines for Collecting Silage.

Buckrakes or Sweeps. One of the most valuable types of machine for collecting green crops for silage is the cheap and simple buckrake, which may be mounted either at the rear of the tractor on the hydraulically operated 3-link system, or at the front on a hydraulically operated front loader.

Rear mounting on the 3-point hitch is usually advantageous where the silage is to be transported to a pit or clamp in the same field, or only a short distance away. The weight of the load on the rear wheels of the tractor improves adhesion, and makes it possible for the tractor to run right over the silage in the pit or clamp, thus aiding consolidation. On the other hand, there may be advantages, especially in wet conditions, in avoiding running over the silage with a tractor which has dirty wheels; and in these conditions mounting the buckrake on the front loader, which facilitates unloading from the side of the clamp, may be advantageous. There is much to be said for a fitting which can be used either at front or rear, and there are also obvious possibilities, where a mechanical catch for the rear-mounted buckrake is fitted, in loading the tractor at both front and rear.

The tines for collecting short grass need to be fairly closely spaced, and are usually much shorter than those used for hay sweeps. The buckrake for collecting green crops is also usually narrower than a typical hay sweep. Pointed metal tines are employed, and it is important that they should be smooth and perfectly straight. Chief limitations of the buckrake are failure to pick up cleanly crops that are very short, and difficulties where the ground is uneven or very soft.

The loads carried are usually small, some 3-6 cwt., and this fact, together with the difficulty of avoiding losing a little of the load on the way to the silo, makes buckrakes unsuitable for carrying to silos situated a long way from where the crop is grown. Nevertheless, with fairly bulky crops in good condition, the pick-up is perfectly clean and the amount dropped in transit usually negligible. One man can collect, transport and ensile the crop, in a pit or clamp; and if he takes care in depositing loads in the silo only a very small amount of hand spreading is required when he gets off the tractor to sprinkle molasses.

The buckrake is designed to collect two undisturbed 5 ft. swaths at a time, travelling in the same direction as the mower. With very light crops a cleaner pick-up is obtained if the direction of travel is across the swaths, or if a side rake is used to put several swaths together. If a 6 ft. mower is used to cut the crop, double swath boards should be fitted, otherwise the width of the two swaths may be wider than the buckrake, and some of the crop will be left. With good crops it is a mistake to side-rake into windrows, as the crop is then usually more difficult to collect.

The buckrake or sweep must be set so that the points of the tines will follow the ground closely. This is done by making sure that all tines are in line, particularly at the points. Bent tines should be straightened or replaced. Rusty or dirty tines should be polished, since young grass will not push back unless tines are clean. The height and angle are adjusted by lengthening or shortening the top link of the hitch. The beam should be set just clear of the ground. Too slow a speed may reduce the size of load, since grass will not drive back on to the tines. "Double loading" will increase the size of load that can be transported, and is worth while when working a long way from the silo.

If the beam is set too high, trouble may develop through the tines digging into the ground. This trouble may also be caused by the ground being too soft, and in this case, broad flat points should be tried. Another cause of digging in is bent tines. If for this or any other reason the tines have dug into the ground, the outfit should first be driven forward away from the load.

A method of working which may be worth consideration when the field is a long distance from the silo is that of collecting by buckrake mounted on a front loader, and loading into trailers. Experience will soon show how far the buckrake has to travel to collect a load, and a good method of operation is to place the trailer in position across the swaths ready to receive the sweep-load, so that the tractor with the buckrake does not have to manœuvre with a load.

GREEN CROP LOADERS.

Where loading green crops on to trailers is a regular need, some form of automatic trailer-type pick-up loader is almost essential. Many types of loaders are available to suit the varying needs of picking up crops for silage and of collecting crops for various types of grass driers. There are machines which cut and load, those which cut, chop and load, and those which pick up and load from the swath or windrow. Several types of green crop loader which pick up grass or forage crops from the swath or windrow are now available. One common type (e.g. Fig. 156) resembles in some respects a reciprocating-fork type hay-loader, but is stronger and has at the foot of the trough a special pick-up reel designed for handling fairly short grass. The elevator trough is fitted with triangular steps to prevent short grass rolling back between the upward strokes of the rake-bars. An important advantage of this general type of loader for most farms is its versatility. It will handle grass as short as necessary for the general farmer, and will also deal with bulky crops of arable silage or with hay (Fig. 157).

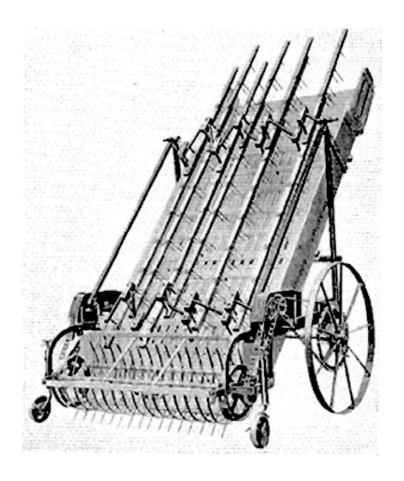

FIG. 156.—GREEN-CROP LOADER. (BAMFORDS.)

The tines of the pick-up cylinder revolve in the opposite direction to the land wheels of the loader, and raise the crop just sufficiently to enable the curved teeth at the bottom of the oscillating forks to pick it up.

Another type of green crop loader employs a rotary pick-up cylinder in conjunction with an endless conveyor.

The small-diameter retractable-tine pick-up cylinder raises the swath or windrow above itself, and deposits it at the foot of an elevating conveyor which may consist of endless chains and slats, or, in one make, two rubber belts. In the Wilder-Steed loader, where the crop is elevated between rubber belts, it is completely protected from the wind during elevation.

FIG. 157.—COLLECTING LONG SILAGE WITH CYL-INDER AND RAKE-BAR TYPE GREEN-CROP LOADER.

A third type of loader (e.g. Blanch-Snook) employs endless chains with cross-bars and spring tines to perform both the pick-up and conveying operations. The conveyor runs above the crop in the loader bed, and is adjusted for height at the foot so that the tines just sweep the ground. Independent drive of the conveyor mechanism by means of an internal-combustion engine permits variation of forward speed to suit the crop conditions.

FIG. 158.—GREEN-CROP LOADER WITH ELEVATION
BY RUBBER BELTS. (WILDER-STEED.)

(*Power Farmer photograph*)

FIG. 159.—ENGINE DRIVEN GREEN-CROP LOADER
PICKING UP SUGAR BEET TOPS. (BLANCH-SNOOK.)

Operation of Green Crop Loaders. Green crop loaders vary greatly in design, and only a few general observations on efficient operation can be given. Most machines require a man on the load, and these cannot usually work safely at speeds of more than about $2\frac{1}{2}$ m.p.h. In light crops it is worth while to side-rake several swaths into a windrow before using the loader. Trailers with high sides invariably make loading easier, and it is worth while to have sides specially made where there is much work to be done. It pays to load systematically, beginning layers at the front, as this facilitates unloading.

Where crops are picked up direct from the mower swath it is usually best for the loader to work in the same direction as the mower. It is important to set mower swath boards carefully, so that each swath lies separately, otherwise the loader may "poach" on the next swath. Poaching, if it continues unchecked, may cause serious blockages, with considerable loss of time and the possibility of damage to the mechanism. To obviate troubles occurring on corners it is advisable first to mow and clear about 10 swaths round and round the field and then to cut the rest in lands parallel to any furrows. This gives the crop loader a clear run and greatly facilitates the work.

Most crop loaders have the mechanism driven from the land wheels, so good adhesion is important. Where steel wheels slip, a set of spuds may be fitted with advantage. Many modern machines are equipped with land-grip pneumatic tyres.

"One-Man Loaders." One of the most important recent developments is the use of "one-man" machines which eliminate the need for a man on the load. This has many advantages which can result in a great saving of time and labour. If the machine is suitably designed and constructed it permits forward speeds of up to 5-6 m.p.h., giving loading rates two or three times as high as those obtainable with simpler machines. Elimination of treading of the load also facilitates unloading.

To achieve successful one-man operation it is necessary to have a machine which can deliver the crop to the front and middle, as well as on to the rear of the transport vehicle. This necessitates a long reach over the vehicle, and a telescopic drawbar. High trailer sides are, of course, an essential for one-man loaders.

CUTTER-LOADERS.

The Wilder "Cutlift" was the pioneer British machine for cutting and picking up short grass in one operation. It consists of a mower unit, fitted with twin fingers for close cutting; an elevator which takes up almost every blade of grass cut; and a trailer which is towed behind and is easily detached. The machine is driven from the tractor power take-off. There is a flexible connection between the cutter-bar and elevator, so that the cutter-bar has freedom of movement in following uneven ground. The elevator on modern machines can be adjusted to deal with kale by reversing direction of travel of the elevator and raising the cutter-bar.

FIG. 160.—THE "CUTLIFT", USED FOR CUTTING AND COLLECTING SHORT GRASS. (WILDER.)

The Taylor-Doe "Silage Combine" consists of the chassis of a Massey-Harris self-propelled 8 ft. 6 in. combine harvester with a cutter-bar and conveying mechanism suitable for cutting and delivering silage crops at very high speed to vehicles pulled behind the machine.

416

Many other types of cutter-loaders have been tried, and among those which have created interest are machines which either blow, or suck and blow the fodder into a trailer after it has been cut by an ordinary cutter-bar.

FIG. 161.—ENGINE DRIVEN FORAGE HARVESTER WITH PICK-UP ATTACHMENT COLLECTING GRASS FOR DRYING. (FOX.)

Note truck with high box sides.

FORAGE HARVESTERS.

In the United States, where silage is normally chopped into short lengths before being filled into the container, and where the chopping of hay is of practical interest, combined cutter-chopper-loaders generally called forage harvesters are widely used. A few typical American machines have been imported and used for silage-making and grass drying in Britain. Examples of these include the Fox-Rivers, a sturdy machine which is now normally engine-driven and can be fitted either with a cutter-bar or a pick-up attachment; the Allis-Chalmers, a light, P.T.O.-driven machine which employs a cutter-bar and reel, with rubberized canvases to convey the crop to a lawn-mower type chopping cylinder; and the John Deere, which may be fitted either with a conventional cutter-bar, a special maize gatherer or a pick-up reel, and has a heavy knife similar to that used on a chaff-cutter or silage cutter-blower. Most of the American machines blow the crop into the transport vehicle.

Among the many British machines either on the market or being developed are the A.R.M., an engine-driven machine which is somewhat intermediate in action between a cutter-loader and forage harvester. It cuts the crop by means of a cylinder similar to that of a gang mower and elevates the cut material to a trailer at the rear; the Wild-Thwaites, an engine-driven machine designed for picking up from the windrow, chopping in a cylinder fitted with curved blades, and blowing the crop to the transport vehicle; and the Fisher-Humphries, a light P.T.O.-driven machine not unlike the Allis-Chalmers in general layout.

*FIG. 162.—LIGHT P.T.O.-DRIVEN FORAGE HARVEST-
ER. (FISHER-HUMPHRIES.)*

Note.—Trailer being used is unsuitable.

Forage harvesters are unfortunately too expensive for use on most small farms.

Self-emptying trailers may be well worth consideration for the collection of green crops. They are described in Chapter Twenty-one.

Silage Choppers and Shredders.

The great extension of silage making which has taken place in Britain in recent years has been achieved, in the main, without the use of machines for chopping or shredding the silage. It has been found that with young grass and leguminous crops, satisfactory silage can be made without the extra operation of chopping. With coarse crops, however,

419

such as kale, maize, and sugar beet tops, there is much more certainty of success if the crop is chopped, though there has been no thorough research into the problem of how small the material should be chopped or of the results achieved by various kinds of bruising processes. An extensive survey in Denmark has shown that the average quality of chopped silage is generally superior to that of unchopped, especially with the coarser types of fodder which are difficult to consolidate effectively without chopping. A further reason for chopping is that the chopped material can easily be blown into elevated silos, and for this reason alone chopping is justified where tower silos are employed.

FIG. 163.—SILAGE CUTTER-BLOWER. (MASSEY-HARRIS.)

The common type of silage cutter and blower is a development of the chaff cutter, with an added arrangement for blowing away the cut fodder. The general layout of the conveyor, feed rollers, shear plate and knives is the same as that of the power-driven chaff cutter. The only important difference is that after the silage has been cut, a powerful fan, rotating in the flywheel casing, blows the cut silage along a pipe leading to the top of the silo. On some machines the fan and flywheel have independent drives, so that the fan rotates at a higher speed than the knives. The pipe along which the silage is blown must be erected as nearly as possible in a vertical position, and with the projecting ends of the pipe sections which occur inside the pipe always pointing upwards, to prevent blocking of the material at the junctions. Modern machines have one control lever to start, stop or reverse, and a gear shift lever permits easy change of cut from about $\frac{1}{4}$ in. to $\frac{3}{4}$ in. In the selection of a silage cutter the strength and capacity of the machine are important considerations. The speed of filling the silo is usually governed by the capacity of the cutter, and since a tractor with plenty of power is generally employed to drive the machine, the capacity should not be too small.

FIG. 164.—TRACTOR-MOUNTED P.T.O.-DRIVEN CROP SHREDDER. (MITCHELL-COLMAN.)

A type of silage chopper developed in Britain has a high-speed cylinder with a number of swinging steel hammers which pass between slots in a steel grating. This type is suitable for dealing with crops such as sugar beet tops, and has a reasonable through-put provided that an even feed is arranged and adequate power is available.

Machines designed to bruise rather than chop the silage have also been developed, and claims are made that such a process results in the production of better silage.

There is need for a comprehensive research programme to determine how small various types of crop should be chopped for best results, whether bruising achieves worth-while improvement, and what types of mechanisms are most efficient at producing the desired results. There is clearly an economic limit to the power that can be expended on this job, and also to capital expenditure on special-purpose machines.

For a detailed account of methods of making silage, reference should be made to books on the subject.[19,20]

REFERENCES

(19) *Ensilage.* Ministry of Agriculture Bulletin No. 37.

(20) Moore, H. I. *Silos and Silage. Farmer and Stockbreeder.* London.

CHAPTER ELEVEN

BINDERS

"It was curious to see on the soil of a Cleveland farm two implements of agriculture lying side by side in rivalry, respectively marked 'McCormick, Inventor, Chicago, Illinois'; 'Hussey, Inventor, Baltimore, Maryland'—American competing with American on English ground."

Rudimentary Treatise on Agricultural Engineering. G. H. Andrews, 1852.

The sickle and scythe are hand tools which have been gradually developed during many thousands of years. There is evidence that bronze reaping hooks were used in France round about 2,000 B.C., and these followed flint knives fixed to wooden handles. The scythe developed from the sickle, the blade being lengthened and the handle being adapted so that both hands could be used. The addition of the cradle enabled the corn to be deposited in a bunch at the end of the stroke. The sickle and scythe are still the only implements used for cutting grass and corn in some remote districts, but in general their use is now confined to opening up fields for harvesting machines and cutting small areas which cannot be satisfactorily dealt with by machinery.

Pliny, writing in the first century, described a reaping machine used in Gaul. It consisted of a large comb attached to the front of a two-wheeled box which was pushed along by an ox. The comb was run through the corn just below the heads, and a man walking beside the machine swept the heads backwards so that they were broken off and gathered into the box.

Little progress was made in the development of harvesting machines until the end of the eighteenth century. In 1799, W. Walker, in his *Familiar Philosophy in Twelve Lectures*, described a machine which consisted of a comb and two horizontal rotating knives, driven through bevel gearing and pulleys from a live axle connecting the two wheels on which it ran. This machine was clearly a derivation from the Gallic reaper, and the principle involved, viz. the comb and horizontal rotating knife, was employed in many machines invented during the period 1799 to 1843.

In 1812, John Common, of Alnwick, Northumberland, submitted a model of a reaper to the Royal Society of Arts. This machine had an angular reciprocating knife in a finger-bar, and was driven from a live axle through bevel gearing, an eccentric and a connecting-rod. It possessed the essentials of the modern mower, but never became a commercial success. In 1822, Henry Ogle, one of Common's neighbours, produced a machine which had a reel for gathering the corn on to a smooth reciprocating knife. This machine also had a hinged sheafing platform behind the knife, but it achieved no more success than Common's.

The first successful reaping machine was produced in 1826 by the Rev. Patrick Bell of Carmyllie, Forfarshire; the original machine may be seen in the Science Museum, London. Bell's reaper was fairly widely tested in Forfarshire and other parts of Scotland, and it continued to be used in some parts for several years. But it was not until the middle of the century, when the great 1851 Exhibition drew attention to two American reapers, that any great interest was taken in Bell's machine by people south of the Tweed. By that time, the American machines had progressed to such an extent that the story of the later stages of evolution is largely one of developments in America.

Cyrus H. McCormick patented his first reaping machine in the United States in 1834, and by the middle of the century, machines which differed little in essentials from the trailer mower and side-delivery reaper were being sold. There seems little doubt, however, that although McCormick's machine received the premier award at the 1851 Exhibition, the machine invented by Obed Hussey was at that time better in many details. Hussey's machine had a knife and finger-bar which were very similar to those incorporated in modern reapers.

From 1850 onwards, various patents were taken out in America for canvas conveyors, sheafing and tying devices, etc. In some cases, men stood on a platform on the machine and tied the sheaves by hand, but in 1869 J. F. Appleby patented a wire binder. Wire binders were built and used in 1874, but by that time Appleby had begun to experiment with the twine knotter, which was taken up by the Deering Company in 1878. Since that time, improvements have been in details of workmanship rather than in essentials. The most important fundamental development of recent years has been the direct driving of the cutting and binding mechanism through the tractor power take-off.

The Reaper.

The "side-delivery" or "sail" reaper is used to a limited extent for harvesting certain crops which either shatter in binding or must be left in a loose condition to permit drying. Buckwheat, peas, white mustard seed and clover seed are some of the crops harvested with the reaper for these reasons. For corn, the reaper has been generally superseded by the self-binder or combine harvester, and many of the crops which cannot be combined direct are now windrowed with a tractor-mounted swather and later threshed by a combine fitted with a pick-up attachment. (See Chapter Thirteen.)

The cutting mechanism of the reaper does not differ in essentials from that of the mower. The cutter-bar is attached to the front of a quarter-circle platform. The cut grain falls on to this platform and is swept off and delivered at the side in sheaves by the "rakes" or "sails". The sheaves have to be tied by hand when tying is necessary. There are commonly four sails, two of which may be "dummies" if the size of sheaf is too small with four rakes. The drive is from a large "travelling" or "bull" wheel which carries most of the weight of the machine. The outer end of the platform carries the "grain" wheel and a long divider. The driving and cutting mechanism are very similar to those on the binder described below.

The Binder.

The modern binder is one of the most complicated of farm machines. It cuts the standing crop and ties it into neat and uniform sheaves. A land-wheel-driven machine comprises the following main parts: (1) a main FRAME on which the various mechanisms are assembled; (2) a large BULL WHEEL with a cleated rim, which furnishes power through sprockets, chains and gearing to the cutting and binding mechanisms; (3) the GUTTING MECHANISM, consisting of a knife reciprocating between a row of fingers and driven by a connecting-rod and crank; (4) a REEL to gather the crop on to the knife and cause it to fall uniformly on to the platform when it is cut; (5) an elevating system of endless CANVASES with slats of wood attached, to carry the cut material over the bull wheel to the binding mechanism; (6) the BINDING MECHANISM which forms the cut corn into sheaves and ties them with a band of twine; and (7) ADJUSTMENTS to nearly all these parts to permit dealing with a wide variety of conditions.

The ERAME is made of steel throughout and is braced to give rigidity. The BULL WHEEL is slung in a rectangular opening in the frame and is supported by roller bearings with thrust washers. It carries most of the weight of the machine and must, therefore, be strongly constructed. It is set in a toothed quadrant and provided with a mechanism for raising and lowering it.

FIG. 165.—TRACTOR BINDER. (ALBION.)

The MAIN DRIVE is taken from the bull wheel by sprockets and a pintle chain to a countershaft on which are mounted a dog-clutch and gearing. The countershaft transmits power through bevel gearing to the crankshaft and through chains and gears to the reel, canvases and binding mechanism. The chains, gears and bearings of the best modern machines illustrate the great advances that have been made in the details of agricultural machinery in recent times. Malleable hook-link chains on cast-iron sprockets have been superseded by steel roller chains on machined steel sprockets; cast-iron gears have given way to enclosed hardened steel ones; and plain bearings with oil-can lubrication to ball and roller bearings with grease-gun lubrication.

The GUTTING MECHANISM is similar to that of the mower. The fingers are spaced 3 in. apart, but the throw of the crank is 6 in. Each knife section therefore passes over two spaces between fingers instead of over one as in the mower, but the speed of the crankshaft is much lower. The cutter-bar is attached to a rectangular steel platform which is strengthened by cross-bars and riveted to the main frame. The height of cutting is controlled by raising and lowering the main wheel and the grain wheel in the frame. The whole machine may also be tilted backward or forward by a lever adjustment, thus raising or lowering the cutter-bar, and altering the slope of the platform. For cutting lodged grain, long lifting fingers which extend 18 in. or more in advance of the ordinary fingers are very effective.

FIG. 166.—EIGHT-FOOT CUT POWER-DRIVE TRAC-TOR BINDER. (MASSEY-HARRIS.)

The REEL has laths which bend the straw over slightly towards the knife, so that it falls evenly on the platform as it is cut. It must have a wide range of adjustment in both a vertical and a fore and aft direction in order that it may be adjusted to suit the height of the crop and the direction in which it is leaning. Neat cutting and collecting of the grain depend much on skilful use of the reel.

The DIVIDER is an important part of the binder, for it is here that blocking often occurs in tangled crops. Modern machines usually have a large sheet-iron outside divider ("torpedo" divider). In Germany a revolving divider has been developed, and is there generally considered a great success. This divider consists of a large worm that is caused to rotate inwards, and it has the effect of pushing the corn which lies over it upwards and inwards, towards the platform.

The Canvases. When the corn falls on to the platform, the platform canvas carries it along to the foot of the two elevating canvases. It then passes between the elevating canvases up to the binder deck. The upper canvas "floats" so that it can accommodate varying quantities of corn. The tensions of the canvases are all adjustable, and should be kept just sufficiently tight to prevent slipping at the driving rollers. Special looseners, enabling the canvases to be made slack in a moment when the machine is being left overnight, are provided. Rubber-impregnated canvases do not shrink and stretch as much as plain canvas; but although the rubber-impregnated material has many advantages over the plain, the cost of its manufacture has so far limited its commercial use. In order that the canvases may run evenly over the rollers, the frame supporting the rollers must be adjustable so that its two ends can be kept parallel. Adjustment is effected by varying the lengths of the diagonal braces. Canvases should always be put on with the buckles running first.

With heavy crops, the ears tend to go up the elevator canvas first. On some machines a strip of metal resting on the platform canvas helps to prevent this.

The Binding Mechanism. When the corn is delivered to the binder deck, the packers and buttor pack it into sheaves and deliver it to the knotting mechanism. The packers are oscillating fingers which act from below through slots in the binder deck. The buttor is a reciprocating board placed at the forward end of the deck and driven by a crank. It works the base of the sheaf into a level butt.

The twine, as it comes from the ball, passes first through tension rollers, and then through the needle and over the knotter bills to the retainer, where the end is held. While the corn is accumulating on the binder deck, the needle is below the deck, and a single strand of twine lies below the corn, with the free end held in the retainer. When the sheaf reaches its full size, it presses down the trip lever, which releases the shipping mechanism and engages a clutch operating the tying mechanism. The needle rises from below the deck, carrying the twine over and around the sheaf, and over the knotter bills to the retainer. Thus, the two ends of the piece of twine which is to tie the sheaf are placed in the retainer. The retainer is driven a fraction of a turn, and this holds the two ends of the twine fast. The needle then moves back and the tying of the knot is performed by the knotter bills. The process is completed by the cutting of the twine between the knot and the retainer, and the throwing out of the sheaf by the ejector arms. (See Figs. 167-169.)

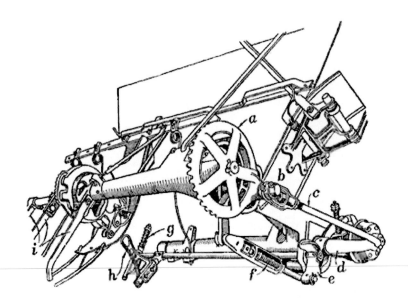

FIG. 167.—BINDING MECHANISM

a, cam gear; **b,** butt adjuster; **c,** needle pitman; **d,** needle shaft; **e, f,** tripping mechanism; **g,** adjustment for trip spring; **h,** compressor arm; **i,** ejector arms.

FIG. 168.—ACTION OF THE NEEDLE

Above, Needle lies below binder deck; one end of twine is held in retainer. *Below,* Needle moves over and places other end of twine in retainer.

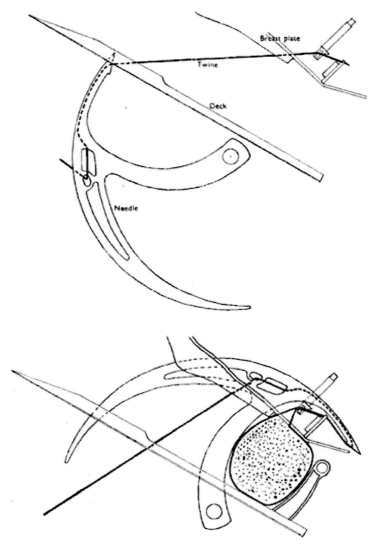

Adapted from Harrison, McGregor & Co.

The Knotting Meghanism. The knotting mechanism is rather complicated. In tying a knot, the knotter bills, which are driven by the knotter cam gear, make one revolution, winding the two strands of twine around them, while the ends are held fast in the retainer. As the revolution is completed, the bills open and grasp the twine between the loop and the retainer. Then, as the twine is cut free from the retainer and the sheaf is ejected, the loop around the knotter bills is stripped off over the two ends grasped between the bills, thus forming the knot.

Courtesy of John Deere

FIG. 169.—TYING OF THE KNOT

The knotting mechanism should not be altered while it continues to work well; but if tying of the sheaves is erratic, the action of the knotter should be examined to discover the cause of failure. The following are some of the causes which may be responsible for failure of the tying mechanism: (1) weak twine which may break as the sheaf is being tied. Good twine withstands a tension of 90 lb. (2) Wrong adjustment of the spring on the tension rollers through which the twine passes as it leaves the twine can. The spring should be adjusted so that a tension of 6-8 lb. is required to pull the twine through the rollers. (3) Wrong adjustment of the retainer clip. The retainer generally consists of a notched disc with a clip pressed over one edge by means of a spring. The twine is held between the disc and the clip after the disc has been partially rotated by the knotter cam gear. If the clip is too loose, the twine is pulled out as the knot is being tied; and if too tight, the twine may be crushed and broken between the clip and the disc. The spring should generally be adjusted so that a tension of about 35 lb. is required to pull out the twine. (4) Faulty adjustment of the knotter bill clip. When the knotter bills grasp the two ends of the twine after forming the loop, the pressure should be just sufficient to ensure that the twine is not pulled out before the knot is tied, but should not be so great that the sheaf cannot be ejected because the twine cannot be pulled off the knotter bills. The knotter bill clip, which presses the bills together, has a spring adjustment; and this should be set so that a pull of about 12 lb. is required to free the ends of the twine.

Adjustments of the knotter bill and retainer tensions should not be attempted unless it is quite clear that the trouble is not due to something else. (5) A dull knife, a worn or bent needle, or rust on any of the parts of the tying mechanism may be responsible for the "missing" of sheaves. Rust prevents many binders from working properly until harvest is well advanced. Attention to the vital parts before putting the machine away for the winter may prevent trouble from this source.

FIG. 170.—ACTION OF THE KNOTTER BILLS

Above left.—End of twine held in retainer.

Above right.—Other end of twine in retainer, and knotter bills start to rotate.

Below left.—After rotating, knotter bills open and grasp twine between knotter and retainer.

Below right.—Knotter bills close, and knot is tied as sheaf is stripped off.

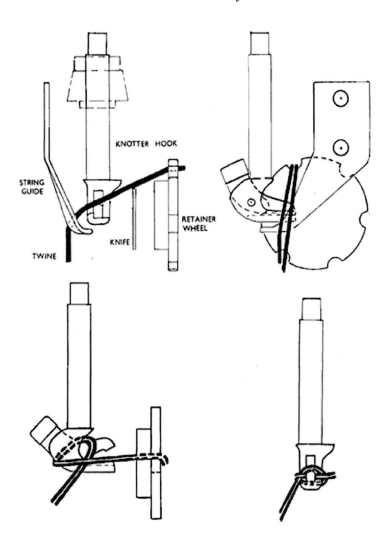

KNOTTER HOOK

STRING
GUIDE

RETAINER
WHEEL

KNIFE

TWINE

The cause of failure of the knotting mechanism may often be determined by examining the untied or broken band. In Fig. 171 **A** shows a correctly formed knot and **B, C** and **D** three of the common types of failure. A loop knot similar to **B** may be caused by a blunt twine knife, or by the retainer clip or knotter bill clip not being tight enough.

If there is a slip-knot round the sheaf and the free end of the twine runs back to the needle, the needle has failed to place the free end in the retainer. The cause may be a bent needle, rubbish blocking the retainer, or the timing of the needle being too late, so that the retainer has moved on before the twine was there to be gripped.

In failure **C**, if the free end of the twine is cut off square, the retainer clip is too loose or the twine tension is too tight, and twine is pulled from the retainer instead of from the twine can as the needle advances. If the free end of the twine is ragged and crushed, it is an indication that the retainer clip is too tight. Where both ends of the twine are "chewed" and the twine is found with the sheaf, the retainer clip is too tight.

In the failure **D**, the ends of the twine show signs of having formed a knot, but the knot has not been pulled tight and has come undone. This may be due to worn knotter bills or to the knotter bill clip not being tight enough.

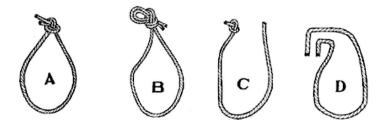

FIG. 171.—THE BINDER KNOT AND COMMON
TYPES OF FAILURE

A, correct knot; **B**, loop knot; **C**, knot in only one end of twine; **D**, knot formed but not completed.

The Knotter Shaft Drive. If the binding mechanism is taken to pieces, care must be taken to reassemble it so that the "timing" of the drive is not upset. The devices concerned in the driving of the binder mechanism from the packer shaft are the shipping pawl (or trip dog), the shipping pinion (trip-dog pinion), and a drive from these to the cam wheel on the knotter shaft. If the mechanism is not assembled or "timed" correctly the binder shaft may operate continuously, producing "baby" sheaves, or, on the other hand, it may not work at all. The correct timing is usually facilitated by punch marks on gear teeth which should engage with one another.

Incorrect functioning of the knotter shaft drive may be caused by a worn trip dog and stop, a broken dog spring, or worn rollers on the drive from the packer shaft. It is sometimes possible to correct the trouble by grinding the trip dog into shape, but it is generally necessary to replace the worn or broken part.

ADJUSTMENTS OF THE BINDER.

The following adjustments may be made from the operator's seat: (1) the position of the reel; (2) the pitch of the platform; (3) the position of the tying mechanism, and (4) the position of the buttor. The making of neat, well-formed sheaves depends upon these adjustments and also upon two others which cannot be made by means of levers, viz. *(a)* adjustments of the positions of the compressor arm and of the trip lever to control the size of the sheaves; and *(b)* adjustment of the trip spring to control the weight of the sheaves. Independent adjustments of the size and weight of the sheaves control their tightness.

FIG. 172.—BINDER CONTROLS. (MASSEY-HARRIS.)

A, tilts platform; **B**, reel adjustment; **C**, position of binding mechanism; **D**, sheaf carrier; **E**, reel; **F**, height of platform; **G**, twine can; **H**, buttor.

Where the trip lever and compressor arm are in one (as on International binders), it is necessary to adjust the tension of the trip spring as well as to adjust the position of the compressor arm, in order to vary sheaf size. For small sheaves the arm is set in and the tension decreased; for large ones the arm is set out and the tension increased. On other machines, the tension should not be altered unless it is desired to vary the tightness.

Accessories. A sheaf-carrier permits from 4 to 6 sheaves to be collected and dropped at will by the operation of a foot-controlled trip lever. This is a useful device for preventing obstruction at the corners and for dropping the sheaves in windrows as an assistance to stooking when the crop is thin.

FIG. 173A.—BINDING MECHANISM ADJUSTED FOR LONG STRAW

FIG. 173B.—BINDING MECHANISM ADJUSTED FOR SHORT STRAW

For transport the binder is drawn endways. The draught-pole is placed beneath the platform opposite the grain wheel, and two special transport wheels are set under the frame, the bull wheel being drawn up clear of the ground.

Draught of Binders.

As with the mower, the horse-drawn binder may be used for tractor work by substituting a stub-pole for the long horse-pole. Such an outfit gives satisfactory results, provided that the tractor is not driven too fast. Binders built for horse work will not for long withstand the battering they receive if drawn at 4 to 5 m.p.h. Work on the binder is very tiring to horses, partly because the draught of the machine fluctuates so much. At the moment when a sheaf is being ejected the draught may be twice as much as it is when the tying mechanism is not in operation. The speed of horses at

the beginning of the day is about $2\frac{1}{2}$ to $2\frac{3}{4}$ m.p.h., but it falls to about 2 m.p.h. when they tire. They are then apt to stop on occasions when the draught is above average. Each binder requires two 3-horse teams, working 5 or 6 hours each, and these 6 horses cannot do as much work in a day as a single small tractor. Horses have, therefore, been largely superseded by tractors for binder work.

FIG. 174.—SHEAF GARRIER (ALBION.)

As with mowers, the mechanism of horse-drawn machines may be driven by a small internal-combustion engine placed on the machine. If the binder is also mounted on pneumatic tyres, two horses can easily pull it; but though machines of this type were developed, they found little favour in Britain, owing to widespread use of tractors.

Power-Driven Binders.

Tractor binders are generally strongly constructed machines which are driven from the tractor power-take-off.

Safety clutches are provided in the main drive, and also on the packer shaft and the crankshaft.

Power-drive shafts should be fitted to run as straight as possible, and must always have a guard fitted. With some tractors it is necessary to fit steering stops to the front axle so that very sharp turns cannot be made.

The main wheel, since it is not concerned in the drive, may be made smaller than on horse machines. Small pneumatic-tyred wheels are a good fitting; for in addition to giving a reduced draught they also eliminate many of the shocks. Smaller wheels permit elevator canvases to be shorter and the binder deck to be lower, and these modifications result in improved performance.

Owing partly to the rapid growth of interest in combine harvesters, improvements in binders have been mainly in details, and it is only very recently that the logical step of making the tractor binder into a semi-mounted machine has been taken. This enables the main weight of the machine to be carried partly on the tractor and partly on a pair of small pneumatic-tyred wheels at the rear. The main wheel is entirely eliminated, and as a result the elevator canvases are shorter and the whole machine becomes much simpler.

Care and Maintenance of Binders.

A modern binder has over 100 lubrication points; and while ball bearings and covered chain drives on the latest machines do not need attention as frequently as similar parts on older machines, it is advisable to grease all parts every morning, and to attend to fast-moving parts three times a day. Exposed chains should not be lubricated in dusty soils of a cutting nature.

Canvases should be tightened evenly, and always slackened off and covered at night. When moving from field to field the platform canvas should be removed if equipment is to be carried on the platform; and even then, heavy loads should be avoided.

Before storing the binder at the end of the season, canvases should be removed and stored in the dry, away from rats and mice. All chains should be taken off, washed in paraffin, and then stored in a tin of waste oil. All bright parts, especially the knotter and knives, should be smeared with a rust preventive. Binders fitted with pneumatic tyres should be blocked up so that the tyres are clear of the ground.

Flax Pulling Machines.

A great increase in the flax acreage during the war years was facilitated by the development of efficient flax pulling machines. On one successful machine, the pulling mechanism consists of a system of belts and pulleys driven from the tractor power take-off. Two pulling drums rotate in opposite directions and the machine depends for its pulling power on the contact between rubber belts ($2\frac{1}{2}$ in. wide by $\frac{1}{2}$ in. thick) and rubber facings on the drums.

The pulling width is 3 ft. and the swath is separated into two equal parts by three long dividers, which bend the flax over and lead it to the two points, at the lower parts of the drums, at which the pulling takes place. The lower pulleys must be adjusted so that the belts are squeezed tightly between them and the pulling drums. The flax is gripped between the drums and belts at a distance of about 12 inches from the roots; and with its length horizontal and head foremost it is carried upwards round the outsides of the drums. At the top the belts leave the pulling drums and the two swaths of flax intermingle and pass between the front pulling belt and the top canvas belt to the binding apparatus, which is exactly like that of an ordinary binder. The operator riding the machine must see that when the flax reaches this point it is fed evenly to the packers.

For good results it is necessary to run the machine exactly at its full pulling width, otherwise the straw of the outside half of the swaths tends to bend over, and uneven butting of the sheaves results.

The pulling mechanism has the advantage over such a mechanism as a binder cutter-bar of having no reciprocating parts. The machine operates at a speed of about $3\text{-}3\frac{1}{2}$ m.p.h.

Although flax should only be grown on clean land, the machine achieves a measure of selective pulling in weedy crops. Deep-rooted weeds such as thistle are usually left growing, while weeds and flax less than about 12 inches high do not reach the pulling points and are also left.

CHAPTER TWELVE

THRESHING MACHINES

"Some useth lo winnow, some useth to fan,
Some useth to cast it as clean as they can.
For seed go and cast it; for malting not so,
But get out the cockle and then let it go."

Five Hundred Pointes of Good Husbandrie. Thomas Tusser, 1573.

Threshing originally meant the simple operation of knocking grain free from the ears of corn, but the term now embraces the processes of separating the grain or seed from its straw and chaff, freeing it from impurities and grading it ready for use. The many operations required to achieve this are now all carried out in one machine. From very early times corn has been threshed by means of the flail, an instrument consisting of a wooden handle with a wooden club attached to it by a leather strap. The unthreshed corn was placed on the threshing floor, and was struck with sharp blows from the flail until the grain was separated from the ears. The straw was then cleared away, and the chaff separated from the grain by throwing the mixture of chaff and grain into the air and allowing the wind or an artificially created blast of air to blow away the light chaff. This operation is called "winnowing". The flail was generally used in this country until the eighteenth century, and is still widely employed in some parts of the world. Another primitive method of threshing, employed in Biblical times and still the accepted method in some Eastern countries, is the treading out of the grain by oxen, sometimes pulling a "Norag" or weighted sledge over the corn on the threshing floor.

In 1636, Sir John Van Berg patented a thresher which consisted of several flails operated by cranks. Later, in 1732, Michael Menzies invented a machine which, according to his advertisement, "in a minute gives 1,320 strokes, as many as 33 men threshing briskly. But as men rest sometimes and the machine never stops, it will give more strokes in a day than 40 men and with as much strength. It goes by a water wheel". It was later recorded that Menzies' machine had enabled one man to do six men's work, and a consequence was that labourers attempted to break up the machine. In 1753, Michael Stirling invented the first machine which employed a drum rotating within a cylinder, but it was not until 1786 that the first really useful thresher was made by Andrew Meikle. This consisted essentially of a drum, fitted with oak pegs, revolving in a concave frame, also fitted with pegs. This was the forerunner of many similar machines, and the same principle is employed in the modern peg drum. In this country, the peg drum was later displaced by the beater-bar type.

FIG. 175.—MEDIUM FINISHING THRESHING MA-
CHINE. (RANSOMES.)

At the beginning of the nineteenth, century threshing by machinery was adopted with avidity. The process was gradually made more and more efficient by the addition of extra winnowing and riddling devices, until by the middle of the century practically all the parts of the modern thresher had been invented. Since that time, improvements have been more concerned with details of design than with essentials.

The modern thresher not only threshes the grain from the straw but also winnows the chaff from the grain, riddles and grades the grain, and delivers from different parts of the machine straw, cavings, chaff, small weed seeds and various grades of corn. The grain delivered from the modern finishing thresher should be ready for the market without any further winnowing or dressing. The thresher is one of the most expensive machines used on farms, and it has been the custom in most districts for the work to be done by contractors who hire out travelling sets driven by steam-engines or tractors. But the general use of tractors on large farms in recent years has resulted in a great increase in the number of farms with their own threshing drums. A standard 54-in. thresher can be driven by a tractor developing 20-25 b.h.p.

The Sequence of Operations in a Thresher.

The corn is fed between the drum and concave at the top of the machine, either by hand or by a self-feeder. If no self-feeder is used, the corn must be carefully fed into the machine in a regular stream. When it passes between the drum and concave the grain is rubbed out of the ears, some of the grain and chaff falls through the bars of the concave and the rest passes with the straw on to the shakers (Fig. 176). There, the grain, chaff and cavings (small pieces of straw) are shaken out and the straw is delivered by the shakers out of the forward end of the machine. The grain, chaff and cavings pass by way of an oscillating grain-board on to the cavings riddle. Oscillation of the cavings riddle carries the cavings forward and delivers them out of the end of the machine, below the straw. The grain, chaff, etc., pass through the cavings riddle and *via* another collecting board on to the first dressing shoe, where a strong blast of air removes the chaff and delivers it either on to the ground or, more usually, to a bagging apparatus. The first dressing shoe sifts off chobs (pieces of unthreshed ears), large weed seeds, stones, etc., and allows small weed seeds to fall to the ground. The grain then passes down an oscillating spout to the foot of an elevator, which transfers it to the top of the machine. From the elevator the grain may either be delivered direct to the corn spouts or it may pass through one or more of the following devices: (1) the awner and chobber, which removes awns and rubs the grain; (2) the second dressing shoe, which performs a further cleaning; and (3) the rotary screen, which grades the corn.

Setting the Machine. Two spirit levels are fixed to the machine to facilitate levelling, one being situated at the side and one at the back so that the machine may be set level in both directions. If the machine is not level it does not work efficiently, partly because the grain is not spread uniformly over the riddles and will not pass freely down the collecting boards. Setting is rendered easy if each wheel is provided with a good set of lifting chocks.

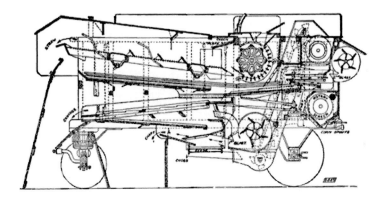

FIG. 176.—SECTION OF THRESHER FROM SIDE.
(RANSOMES.)

The Self-Feeder. The type of self-feeder most generally used consists of an endless canvas conveyor fitted with slats, running beneath a series of oscillating forks which have a tedding action on the sheaves and even out the flow of corn to the drum. The man who feeds the drum cuts the bands and, with a sweep of his arm, distributes the corn over the conveyor.

An automatic band cutter is sometimes fitted on the self-feeder, but this is uncommon on British machines. The self-feeder economizes labour and generally increases the speed and efficiency of threshing. Experiments have shown that both the efficiency of threshing and the output are very largely dependent upon an even feed of corn to the drum.

Construction and Operation.

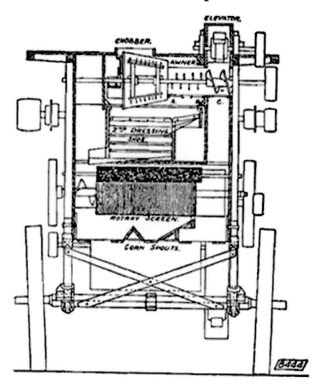

FIG. 177.—SECTION OF THRESHER FROM REAR.
(RANSOMES.)

An extension feeder with a long hopper running down to the ground or on to the stack may be employed to achieve a further saving in the labour of feeding the corn into the machine. It is difficult to understand why a simple self-feeder is not always fitted.

FIG. 178.—CANVAS SELF-FEEDER FOR THRESHING
BOX. (FOSTER.)

The Drum is situated near the top of the machine. It is carried on a shaft of high-grade steel which runs right through the box and carries the main driving pulley at one end. A number of steel "chairs" or "beater beds" are keyed to the spindle, and the corrugated steel beaters are mounted, parallel to the spindle, on the periphery of these chairs. The drum of a "standard"-size machine is 54 in. long and 22 in. in diameter. It is heavy and acts as a flywheel, supplying energy in case of a sudden demand. It normally revolves at 1,000 to 1,100 r.p.m., and must be very carefully balanced to prevent vibration.

The Concave consists of a concave grating made up of wrought-iron bars and wires, in the form of a quarter of a cylinder. It is situated round one side and the lower part of the drum, and is usually made in two reversible parts, hinged together. The distance between the drum and concave can be regulated by the adjustment of screws which attach the concave to the frame of the machine. The revolving drum beats the grain out of the ears by striking them against the concave, and it is of the utmost importance that the concave be correctly adjusted to suit the crop being threshed. If the setting is too wide, grain passes out of the machine with the straw, while too close a setting results in bruised and broken grain. When threshing cereals, the usual distance between the drum and concave is about 1 in. where the corn enters

at the top, $\frac{1}{2}$-in. at the hinge, and $\frac{1}{4}$-in. at the bottom. After long use, both the drum and concave become worn more in the middle than at the edges, and satisfactory threshing cannot be secured without renewing the beaters and planing or renewing the concave bars.

FIG. 179.—BEATER-TYPE DRUM AND CONGAVE

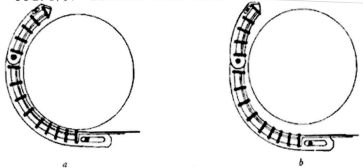

a   *b*

FIG. 180.—SETTING OF THE CONCAVE

a, correct; **b,** incorrect.

Peg-Tooth Drums and Concaves. The drums of most American threshers differ from the common European type in having heavy steel teeth or pegs bolted to the periphery of the drum and to the inside of the concave. The rows of pegs on the drum pass between those on the concave. These drums break up the straw much more than the beater type. They are used in this country on American threshers and a few combine harvesters of American origin, but it is doubtful whether they are as good as the beater type, even for the difficult threshing sometimes encountered by the combine.

FIG. 181.—PEG-TYPE DRUM AND CONCAVE

The Shakers. Straw, chaff and grain pass from the drum and concave on to the straw shakers. These are long wooden racks which are given a continuous oscillating motion by a crankshaft at each end. The shakers toss the straw upwards and forwards at each stroke, and deliver it out of the end of the machine. Grain, chaff and cavings pass through the shakers on to the grain-board below.

Adjustable check-plates are suspended over the shakers. These prevent grain being thrown out of the end of the machine by the drum, and also prevent the straw from going through so quickly that the grain is not completely separated from it.

The Cavings Riddle is a long reciprocating screen with large perforations that allow everything but the cavings to fall through. It is supported on ash hangers and is driven from a crankshaft (the jog crank) by ash rods (the jog connecting-rods). The riddle slopes down towards the front of the box, where the cavings are delivered. Separation of the grain and chaff from the cavings is assisted by the "divided" or "third" blast, which plays beneath the riddle and prevents any tendency to choke. (The numerical nomenclature of the various blasts expresses the order in which their use was generally adopted.)

The Lower (First) Dressing Shoe contains a set of sieves. The "first" blast is directed upwards through the top sieve, and the light chaff is blown away. The blast must be adjusted to suit light or heavy grain by altering the wind shutters; for if the blast is too strong, grain may be found among the chaff. The sieves of the dressing shoe must be selected so that the chobs are sifted off by the top sieve, and small weed seeds pass through the bottom one. The grain is then delivered, partially cleaned, to the elevator.

The Corn Elevator consists of a canvas belt with buckets of steel plate riveted to it, passing round pulleys at the top and bottom of the machine. The upper pulley is usually supported by bearings which may be adjusted to tighten the belt.

The Awner and Chobber assembly consists of a shaft carrying a number of knives and bars, rotating at about 700 r.p.m. inside a stationary iron drum or wire cage. A large auger carries the grain along from the elevator to the barrel of the awner. The knives, which are mounted spirally around the spindle, cut off the awns from the barley and at the same time move it along to the chobber, where it receives a final rubbing. The chobber may be adjusted to give severe or light treatment by varying the number and type of the bars on it. The chobber casing is conical in shape, and the severity of the treatment in both awner and chobber may be controlled by adjustment of the size of the outlet from this casing.

After the corn leaves the elevator it may be discharged by means of slides (I) direct to the corn spouts; (2) to the second dressing shoe without passing through the awner or chobber; (3) to the second dressing shoe after passing through the awner but not the chobber; or (4) to the second dressing shoe after passing through both awner and chobber. The treatment to which the corn is subjected must depend upon the nature of the grain being threshed.

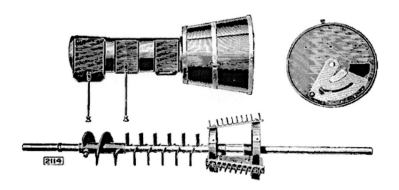

FIG. 182.—AWNER AND CHOBBER. (RANSOMES.)

The Second Dressing Shoe usually has two sieves, placed one above the other. The "second" blast plays beneath and through these sieves and removes the awns, chaff and dust loosened in the awner and chobber. Any grain that has been insufficiently treated in the awner and chobber rolls off the sieves and is conveyed back to the first dressing shoe, to be passed a second time through the cleaning devices.

The Rotary Screen receives the grain from the second dressing shoe and separates thin grains and weed seeds from the head corn. It is made up of steel wires, forming a cylinder of nearly the width of the machine. The distance between the wires is small at the end where the grain enters, and gradually increases towards the other end. Thin grains fall through the wires, and the largest grains pass out through the far end of the screen. The corn is usually delivered to the corn spouts separated into "tailings", "seconds" and "head" corn, unless fewer grades are required. The screen can be adjusted to suit different samples of grain by turning a handle which opens out or closes the wires (Fig.183).

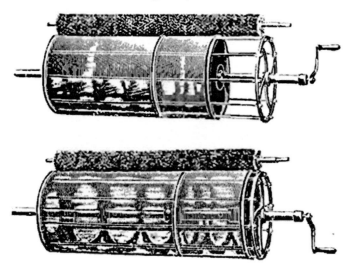

FIG. 183.—*ADJUSTMENT OF THE ROTARY SCREEN*

A revolving brush is fitted to keep the wires clear, but when the screen is to be closed it should always first be opened wider in order to liberate any grains trapped between the wires.

Lubrication, Bearings and Belting. The many rotating spindles and cranks on the thresher necessitate careful attention to bearings and lubrication. Modern machines, with self-aligning ball-bearings and grease-gun lubrication, are greatly improved in this respect compared with older ones. The belts also require regular attention *(vide* Appendix Two), and it is important that the engine or tractor be set square with the thresher, so that the main driving belt shall not be out of line.

Starting Work. When a travelling thresher is employed, it is advisable to clean the riddles and screens and run the machine at full speed with the air blasts at maximum strength before starting to thresh, in order to remove weed seeds, which may be present in large quantity. Feeding of the drum should not commence until the machine has attained full speed. Since feeding largely controls the speed and quality of the threshing, it should be in the hands of a capable and active man.

Threshing the Common Crops.

When threshing wheat, $\frac{7}{32}$ in., $\frac{1}{4}$ in. or $\frac{5}{16}$ in, riddles are generally used in the first dressing shoe, and $\frac{5}{16}$ in. mesh in the upper riddle and $\frac{1}{4}$-in. in the lower one of the second shoe. Bunted wheat should not be passed through the awner and chobber, but normal grain should be passed through the whole of the cleaning mechanism.

The riddles for barley should generally be of $\frac{5}{16}$ to $\frac{3}{8}$ -in. mesh. It is most important for a malting sample that no grains should be broken and that the awns should not be cut off too short.

Oats require larger riddles than wheat and barley (about $\frac{1}{2}$ in. in the first shoe and $\frac{5}{8}$ and $\frac{1}{4}$ in. in the second), The corn is often sacked off direct from the elevator.

For beans and peas the drum should revolve at a lower speed, or should have some of the beaters removed if it is important to thresh an unbroken sample. The concave should be moved away from the drum so that it is wide open at the top, and not less than $\frac{5}{8}$ in. from the drum at the lower part. A breast-plate may be fitted to cover the lower part of the concave. The corn should be sacked off direct from the elevator.

With proper adjustment the modern thresher may be used to thresh out such other crops as mangold, sugar beet, mustard and grass seed.

Threshing Grass Seeds.

All common grass seeds except Timothy can be threshed with the ordinary machine, if care is taken with the adjustments and a few simple alterations are made. Drum speed should be a little below normal, and the back of the upper third of the concave should be covered by a piece of sheet iron, or packed behind with straw. This prevents the heads slipping through the bars before the seed is threshed out. The concave should generally be set about $\frac{1}{2}$ in. from the drum at the top, $\frac{3}{4}$ in. at the centre and $\frac{3}{8}$ in. at the bottom.

A special cavings riddle is almost essential for good work. It should preferably be of wood, the first 2 ft. being blank, and this being followed by 2-3 ft. with $\frac{3}{8}$-in. to $\frac{1}{2}$-in. holes. The holes should be tapered out so that the aperture is slightly larger underneath, as this prevents clogging.

The first blast must be reduced by fitting a larger pulley or by partially closing the shutters, and special riddles must be fitted in the dressers. In the first dresser the top sieve should be of approximately $\frac{3}{16}$-in. mesh, while the bottom weed sieve should be replaced by a plain metal sheet. In the second dresser the top sieve should be of approximately $\frac{1}{8}$ -in. mesh and the bottom of $\frac{3}{32}$-in., the blast again being reduced.

It may be necessary to use the awner and chobbcr to break up seed spikelcts, but too drastic rubbing may sometimes do harm by skinning the seed.

The rotary screen must be closed right up, and then the best grass seed falls through the tail-corn spouts, while screenings which pass on to the best-corn spouts should be put through the machine again.

Timothy seed generally needs to be threshed in a clover huller for satisfactory results. The threshed seed should be of a silvery white colour. A high proportion of dark brown seed indicates that it has been skinned or hulled; and since such seed tends to lose its germination on storage, adjustments should be made to reduce the rubbing, even at the expense of leaving some seed in the heads.

Threshing Clover Seed. Clover seed cannot be threshed without a very severe rubbing, and special threshers or *hullers* are usually employed for this purpose, though it is possible to thresh clover with an ordinary machine by fitting a suitable hulling apparatus. One type of apparatus consists of a special concave of sheet iron which is attached to the ordinary concave and is used together with clover riddles. The clover is first put through the ordinary machine to knock off the seed heads, which fall to the ground from the front of the first dressing shoe. The hulling apparatus is then fixed to the concave and the heads are threshed in the usual way.

An improved hulling apparatus, for use with a standard drum, permits threshing and hulling to be carried out in one operation (Fig.184). A plain back-plate is fitted to the concave; and a small aspirator picks up the clover heads from the first dressing shoe and delivers them to the huller, which is mounted on top of the thresher. The huller itself consists of a cylindrical cage of steel wires, inside which a drum rotates at high speed. The seed is rubbed through the wire mesh, and the heads and chaff pass through an opening at the bottom. Thence the mixture passes to the dressing shoes.

For detailed advice on the harvesting and threshing of various kinds of grass, root and vegetable seed crops reference should, be made to a Ministry of Agriculture Bulletin on the subject.21

Straw Stackers.

When the straw is to be stacked near the thresher, an elevator or "straw jack", driven from a conveniently situated pulley on the thresher, is employed. With a good elevator, two men can put up a large and well-built stack. On American machines a *wind stacker* is now almost always used. This consists of a fan which blows the straw along a telescopic pipe. The pipe may be set to swing (with a power drive) in the are of a circle. The pipe is easily adapted to varying conditions, and its use makes possible rough field stacking without any labour on the stack. The broken straw that comes from American peg drums is, of course, more easily handled by a wind stacker than the long straw from English drums could be.

Chaff and Cavings Blowers.

Chaff and cavings blowers capable of blowing the material to a heap 60 ft. from the drum are available for some British threshers. Such devices save at least one man—often two. It is surprising that more use is not made of these blowers, and it may be expected that their use will steadily increase as their value becomes more widely appreciated.

The chaff and cavings are usually blown away with the straw when a wind stacker is used. On the Continent many machines that are fitted with a trussing or baling device for the straw have a blower which will deliver the chaff and cavings to a heap clear of the machine.

Straw Trussers.

When the straw is to be stored in lofts or barns, or is to be transported a great distance, it may be trussed or baled. The trusser is an attachment very similar to the tying mechanism of a binder. Two bands are placed around each cylindrical truss by employing two knotting mechanisms which operate simultaneously. The trusser does not compress the straw appreciably. While this device is hardly ever seen in the Eastern Counties, it is widely used in the West.

For easy storage and transport a baling press that makes compact rectangular bales must be used. A description of baling machines is given in Chapter Ten. Power balers driven from a pulley on the thresher are commonly employed in Denmark and Germany, where threshing is frequently done in barns, and the straw is stored in the buildings. With a suitable layout of the buildings and equipment it is there possible for threshing to be done at a good speed by only four men—one pitching sheaves to the thresher; one feeding the drum; one sacking off corn and occasionally attending to the chaff and cavings blowers; and the fourth stacking away the bales of straw as they are pushed by the baler up an incline to the loft.

Cost of Labour.

The overall importance of the labour-saving devices mentioned above is emphasized by a report of the Farm Economics Branch of the Cambridge University Department of Agriculture22 on the saving in labour and cost of threshing when American threshers were compared with typical British ones. It was found that an average of

8.7 men were needed with the British machines, compared with 5.1 for the American, while the comparison in cost per quarter of wheat threshed was 2S. 8d. per quarter with the American and 4s. IId. per quarter with the British. It should be emphasized, however, that these figures do not tell the whole story; for the difference in disposal of straw, chaff and cavings has already been outlined, while the sample of grain from the American machines often leaves something to be desired. The figures do, however, underline the necessity for more attention to the use of labour-saving devices in normal British threshing practice.

REFERENCES

(21) "Threshing and Conditioning of Seed Crops." Bulletin No. 130 of Ministry of Agriculture. H.M.S.O., 1951.

(22) "Threshing by British and American Machines." Cambridge Univ, Dept. Agric, Farmers' Bulletin No. 10.

CHAPTER THIRTEEN

COMBINE HARVESTERS

"I had the pleasure of seeing this wonderful machine at work in California in 1887. It was propelled by 16 mules, harnessed behind, so as not to be in the way; but steam-power is now used."

The Wonderful Century. Alfred Russel Wallace, 1898.

The idea of harvesting in one operation is now more than a hundred years old. Two separate combined harvesting and threshing machines were patented in the United States in 1836, but neither achieved any practical success. One of the earliest practical machines was a stripper type combine invented in Australia in 1845. Little advance was made until about 1860, when machines began to be rapidly developed in California, where conditions are especially favourable to the combined operation. Improvements in both reapers and threshers were incorporated in the combined machines, and by 1890 combines were not at all uncommon in California.

It was not until the period just after the 1914-18 war that the combine began to spread appreciably to the east of the Rocky Mountains into the semi-arid sections of the United States and the Canadian Prairies, and it was only after about 1930 that small machines began to be generally used in the mixed farming areas of the Middle West and the Eastern States. The combine is essentially a labour-saving machine, and the main reason for its spread throughout North America has been the economic necessity of reducing production costs by reducing man-labour.

The combine was introduced into this country in 1928. In 1930, four privately owned machines were at work here; in 1931 the number had risen to ten, and the next year it was doubled. There were just over fifty machines at work in England during the 1937 harvest. Shortage of labour during the war years greatly accelerated adoption of combining. About 500 machines were in use in 1941 and nearly 1,000 in 1942. By 1947, when about 5,000 machines were in use, the practice of harvesting by combine had spread to all important grain-producing areas in Britain, and since that time the numbers used have increased substantially, about 20,000 being on British farms by 1952.

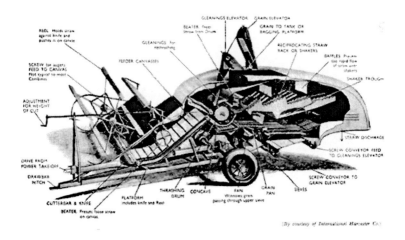

FIG. 185.—*SMALL TRAILED P. T.O.-DRIVEN COM-BINE HARVESTER IN SECTION*

Methods of Harvesting by Combine.

Though combine harvesters are mostly used to cut and thresh in one operation, they may also often be used to advantage to pick up and thresh crops which have previously been cut and left in the swath, or as stationary or portable threshers for dealing with crops which have been stooked or stacked.

Windrowing. The technique of "windrowing" or "swathing" has attracted considerable interest on account of the possibility that it might assist the combining of grain

476

of low moisture content and eliminate the need for drying. In fact, however, substantial drying of corn in the windrow does not take place except in weather when standing crops dry out well, and it is for other reasons that windrowing of corn is worth consideration. It provides a rapid means of saving crops from shedding if too many fields ripen at the same time. It is also a useful technique when weeds are present in the crops, for mixed corn or any crop which ripens unevenly, or occasionally for crops in which clover has grown too high. Peas differ from corn in that the crop often needs to be windrowed in order to hasten drying.

When windrowing is practised the corn may be cut either by a special windrower or swather, or by a binder from which the compressor arm has been removed. Figs. 186 and 187 show two types of windrowing machines. The trailer type is designed primarily for corn, and is similar to the machines now widely used throughout the Canadian Prairies. The 10 ft. cut makes a good swath, and allows plenty of room for delivering the swath on to stubble which has not been run over by the wheels of either tractor or windrower. The height of cut is adjustable from about 3 in. to 18 in. A pick-up reel can be fitted. A clutch is provided to stop the conveyor at corners. It is a one-man machine, and can windrow a standing crop at a high speed.

The tractor-mounted type of swather was originally designed for harvesting crops such as peas, so has different characteristics. This machine is designed for close cutting and for dealing with crops which require lifting. It is usually equipped with a pick-up reel. Both front-mounted and rear-mounted machines are available.

FIG. L86.—10-FT.-CUT TRAILED P.T.O.-DRIVEN WINDROWER. (HARRISON MCGREGOR AND GUEST.)

FIG. L87.—FRONT-MOUNTED SWATHER UNDERGOING OUT-OF-SEASON TESTING. (LEVERTON.)

478

In general, the technique of windrowing can only be recommended where the corn is standing, and has a reasonably sturdy stubble which will support the crop well clear of the ground. As a general rule a stubble 6-8 in. high is needed, and laid crops are unsuitable for windrowing. On the other hand, provided that the combine follows fairly close behind the swather, a type of tractor-mounted machine which can almost shave the ground has been proved invaluable for cutting and windrowing crops which are laid so flat, that no combine harvester cutter-bar can get under the straw. Tractor-mounted swathers are now widely used for harvesting peas, both green for delivery to green pea threshers and also ripe peas for seed and packeting.

When a binder is used for windrowing, great care must be taken to ensure that the swath does not fall into a wheel-mark, and it may be necessary to offset the tractor.

When seed crops are windrowed and heavy rain beats them down, there are few machines suitable for raising the windrows so that they can dry out again. Most of the swath turners used for haymaking are quite unsuitable. They knock out seed and disturb the swath so that it does not feed evenly into the combine when it is picked up later.

FIG. L88.—WINDROW AERATOR. (TASKERS.)

The WINDROW AERATOR is a machine specially designed for lifting such crops. It consists of a pick-up attachment, usually of the draper type, which gently raises the windrow and delivers it over the rear of the machine. Some farmers who practise windrowing consider such a machine almost indispensable.

A combine harvester which is used for picking up from the windrow must, of course, be equipped with a pick-up attachment in place of the cutter-bar.

Main Types of Combine.

Practically all combines consist essentially of the cutting mechanism of a binder attached to a travelling thresher. An exception is the McConnell-Wild Harvester, which threshes the standing crop and leaves the cutting and collection of the straw to be dealt with as a separate operation. Excluding such machines, there are several quite distinct types of combine. The simplest type is the small machine of up to 6 ft. cut, mounted on two fixed wheels, and with cutter bar, elevator, threshing drum and shakers arranged so that the straw travels straight back through the machine, A modification of this is a simple type in which the corn goes straight back from cutter-bar to drum, and then on to shakers disposed at right angles to the direction of travel.

Larger machines generally have the cutter-bar off-set to the right, and carried either by a wheel on a sliding axle which can be adjusted according to whether the cutter-bar is attached or folded up; or by a separate detachable wheel. With this type the corn has to be carried to the left, as on a binder platform, and then passes back to the drum. Some combines employ canvases similar to those on the binder, but others have a large Archimedean screw or auger, which works well in most conditions, especially with short straw. With long straw, rubberized canvases may be better.

The other main type of combine is the self-propelled machine, with four wheels, cutter-bar at the front, and canvases or auger running towards the centre to deliver the corn to a centrally disposed drum. Some features of the construction of the. various types are briefly described below. The great advantage of the self-propelled machine, apart from the fact that a tractor is saved in a busy time, is that it is able to go straight into the crop without any opening out, and can cut it in the most convenient direction.

It also has a large number of gears, and this enables it to deal effectively with a wide range of crops. The driver has a full view of the cutter-bar, and is able to leave laid or otherwise difficult patches if the weather is damp, or to get on with the difficult patches and leave the other when the crop is really dry. Some combines are constructed with a view to bulk handling of the grain, while others are equipped for bagging.

Power Required.

The cutting and threshing mechanism is usually driven by an auxiliary engine mounted on the combine, but on small machines up to 5-7 ft. cut it may be driven from the tractor power take-off.

FIG. 189.—TWELVE-FOOT CUT SELF-PROPELLED COMBINE WITH GRAIN TANK, HARVESTING WHEAT. (MASSEY-HARRIS.)

The tractor for driving a P.T.O. machine requires adequate power and a good selection of gears at low speeds if it is to deal efficiently with a range of heavy crops. The governor must be efficient, so that difficult conditions do not cause speed to drop unduly. Combining is one of the operations for which a P.T.O. independent of the main drive to the tractor wheels is a great advantage. A tractor of not less than about 30 b.h.p. is needed to provide adequate power for a P.T.O. driven 6-ft. combine in average conditions, but a small tractor may be adequate on flat firm land for one which has an independent engine drive.

Capacity of Combine Harvesters.

Experience has shown the difficulty of giving a figure for combine harvester capacity which is of general use in Britain's widely varying conditions. In the Southern and Eastern Counties, if a drier is available when necessary, a combine can usually be relied on to deal with 25 acres for each foot of cutter-bar width. Such a figure may, however, be much too high for the wetter conditions often prevalent in the North and West, where a figure of 15-20 acres per foot is a more realistic estimate. On the other hand, there are many farms in the more favoured Southern and Eastern districts where 30 acres per foot is a normal figure, and where 40 acres has often been exceeded.

The average amount cut per machine in Britain is steadily declining as combines are used on smaller farms. A decline in average annual utilization is not necessarily altogether wasteful. It. provides a reserve of combining capacity for use in unfavourable seasons, and enables the combine to be used only when conditions are really suitable, and the grain needs little drying. Little-used combines, if well maintained and carefully stored when not in use, should have a long life.

The Cutting Mechanism.

The platform can be adjusted to leave various lengths of stubble, the range frequently being from 3 in. to over 2 ft. The position of the reel may be easily adjusted when the machine is at rest, but is not usually moved during work, as it is with the binder. A knife with serrated teeth is employed on some machines. Such knives will work for long periods without sharpening, and are seen at their best when cutting dry, upstanding corn; but they compare unfavourably with knives fitted with plain sections when cutting corn containing a high proportion of weeds or with undersown "seeds". The cutter-bar and platform are usually balanced so as to allow easy adjustment. Large machines often have a power-operated adjustment for length of stubble.

Pick-Up Reel. For use in laid crops a pick-up reel may be fitted in place of the standard type. The hanging tines are maintained at an adjustable angle by an eccentric mechanism. The reel works best when the crop is laid across the direction of travel.

The Threshing Mechanism

does not differ in most essentials from that of the stationary thresher. The feed to the drum is often assisted by a rapidly revolving "beater" but some machines have an upper feeder canvas for this purpose. There is usually a revolving beater just behind the drum, and its purpose is to free straw from the drum and open it out on its way to the shakers. There are three main types of drum and concave, viz. (1) the beater bar type as used on British threshing machines (Fig. 179); (2) the peg type, which chops up the straw and absorbs more power (Fig. 181); and (3) the flail type, a light variant of the beater type, which knocks the corn out rather than rubbing it out (Fig. 191). The flail type may be rubber-faced. It runs at high speed and is used only on small machines.

FIG. 190.—SELF-PROPELLED COMBINE HARVESTER FITTED WITH PICK-UP REEL FOR LAID CORN, AND BAGGING ATTACHMENT. (MASSEY-HARRIS.)

Some of the larger machines have shakers similar to those on a thresher, but small machines often have a single wide shaker or straw-rack.

Grain from the shakers and from the concave passes to two or three sieves which reciprocate together. The upper sieve is generally of the louvred type, with serrated adjustable louvres. An air blast winnows out the chaff, and the grain passes through to a second sieve which may be either louvred or with round holes. Grain passing through this sieve goes to the grain elevator. The material off the top of the sieve passes to the tailings or gleaning elevator, which delivers unthreshed ears or pieces of ear back to the drum again.

FIG. 191.—FLAIL TYPE DRUM AND CONCAVE

The passage of the corn through a modern self-propelled combine may be followed by reference to Fig. 192. The corn is carried by the auger to the centre of the table, where it is picked up by a chain and slat elevator and raised to the drum. The feeding beater helps it on, and threshing takes place between drum and concave. Grain, chaff and straw pass on to the shakers, and grain return chutes bring back chaff and grain to join that which passed straight through the concave on to the reciprocating grain pan. The mixture of chobs, chaff and grain passes on to the sieves, where an adjustable blast blows the chaff out of the back of the machine. A return pan brings the grain to the grain auger and it passes up the grain elevator to the grain tank or bagging device. Pieces of unthreshed heads which are not blown out with the chaff and will not pass through the sieves are returned by an auger to the tailings elevator, which delivers them again to the drum.

Combines do not normally grade the grain. Weed cleaners which take out small seeds for separate bagging are often fitted, but light weed seeds are blown out with the chaff.

The modern combine will work efficiently on fairly steep slopes. In U.S.A., for situations where there are very steep slopes, a special type of machine, called the "hillside" combine, can be employed.

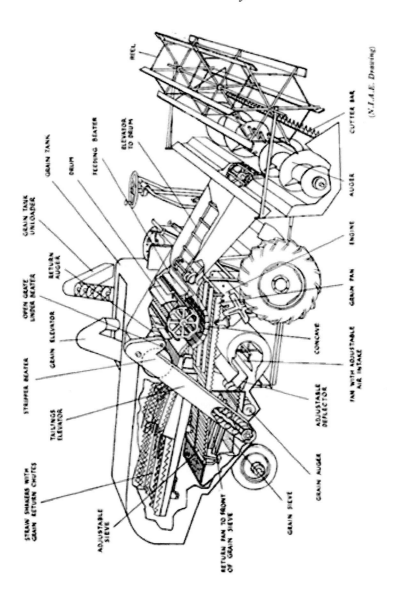

(N.I.A.E. Drawing)

REEL

CUTTER BAR

GRAIN TANK

DRUM

FEEDING BEATER

ELEVATOR TO DRUM

AUGER

GRAIN TANK UNLOADER

ENGINE

RETURN AUGER

OPEN GRATE UNDER BEATER

GRAIN FAN

GRAIN ELEVATOR

CONCAVE

STRIPPER BEATER

FAN WITH ADJUSTABLE AIR INTAKE

TAILINGS ELEVATOR

ADJUSTABLE DEFLECTOR

STRAW SHAKERS WITH GRAIN RETURN CHUTES

GRAIN AUGER

ADJUSTABLE SIEVE

GRAIN SIEVE

RETURN FAN TO FRONT OF GRAIN SIEVE

FIG. 192.—SECTION OF A SELF-PROPELLED COM-
BINE HARVESTER.

489

This contains adjustments which enable the threshing mechanism to be kept level while the machine is working on side gradients which would be impossible to cultivate without special equipment.

Grain Handling. The grain, may be delivered either into bags or into a grain storage tank mounted on the machine. The latter method is generally more satisfactory if central buildings, well equipped for bulk handling, are available. In ideal conditions the grain is delivered from the grain tank into a tipping motor lorry or trailer which can deliver its load into an elevator hopper at the buildings without the necessity of employing any hand labour. With fields up to two miles from the drier a 2-ton lorry can serve two 12-ft. combines.

Where a tractor and trailer are used for bulk grain transport it is usually convenient to have two trailers, one of which can be left at a corner, handy for the combine.

Bagging the grain is a simple method but it needs an extra man on the combine and may lead to a good deal of work in picking up bags of grain afterwards. Advantages are that no specialized equipment is needed, the picking up can be done when the combine is not at work, and bags known to contain damp or green corn can be kept separate. When a bagging combine is used it is generally a convenience to arrange either to dump the bags in windrows or to transfer them straight on to a trailer. Damp grain in small hessian bags can remain in the open overnight without much risk of damage by heating, but in warm weather, damp grain put into big heaps and covered by a tarpaulin may heat in a few hours.

Operation and Adjustments.

It is not usually possible to obtain a perfect sample of grain with the combine, nor is it desirable to overload the cleaning mechanism by attempting to do so. It is important that the drum and shakers should be large enough, in proportion to the size of the cutter-bar, to make possible efficient separation of the grain from the straw. But the threshed grain almost inevitably contains pieces of herbage that cannot be effectively removed while wet, and it is better to leave most of the cleaning to be done by barn machinery rather than to hinder the progress of the harvest.

Where both combine and binder are to be used, it is advisable to use the combine on the lighter and cleaner crops, and on those which are badly laid. Wheat and barley should be combined in preference to oats. It should be remembered that the rate of combining is governed by the volume of straw handled, rather than by the weight of grain. Weeds, damp weather and damp straw all reduce output. Thistles and other juicy weeds clog the screens and shakers, and it may be necessary to stop and clean them frequently when operating conditions are bad.

FIG. 193.—TWELVE-FOOT SELF-PROPELLED COM-
BINE HARVESTER USING WINDROW PICK-UP AT-
TACHMENT. (MASSEY-HARRIS.)

All grain should be dead ripe before being combined. Wheat, for example, generally needs ten days to a fortnight after it is fit for the binder. This longer growing period results in a higher yield and less tail corn. There is, naturally, a risk of oats blowing out and of barley necking when the crops are left to ripen. Clover may be directly combined on a fine, dry day, with the drum set close and a light draught on the sieves. Peas, linseed, mustard and other crops that ripen irregularly are generally better windrowed. Table IV shows the generally preferred methods for handling various types of seed crops by combine. Further information may be obtained from *Bulletin No. 130* of the Ministry of Agriculture.

Adjustments. Leaving a long stubble eases the work of the combine in many ways. On some combines it is shaker capacity that limits threshing capacity, and it is losses from the shakers which become serious if an attempt is made to deal with too much straw. With upstanding crops it is generally advisable to keep well above any weeds or undersown clover seeds.

On many combines adjustment of the drum is effected by raising or lowering the drum, and not by adjusting the concave as with the thresher; but the principles are exactly the same. The speed of the drum must be reduced for peas and beans, and increased for linseed and clover.

At the beginning of combining the drum speed and concave clearance should be set according to the maker's recommendations. If the combine does not thresh cleanly, the first adjustment should be to increase drum speed. More speed is needed for damp and tough crops. If increasing the drum speed is not effective, the concave clearance must be reduced, care being taken to adjust equally on both sides of the machine.

If threshing is clean and grain is being cracked, the first step should be to widen the concave clearance. After this has been done, if damage to the grain continues, drum speed should be reduced.

Adjustment of the fan needs careful attention. The speed is usually kept constant, and the draught regulated by means of a shutter. Too much draught blows the grain out of the back, while too little fails to keep the screens clear. It is advisable to keep the last quarter of the screen clear on most machines.

The fan blast must be varied in conjunction with adjustment of the sieve opening, and the general principle to be observed is to employ a strong blast and a large sieve opening rather than a weak blast and small sieve opening. Too close a setting of the grain sieve will cause an excessive amount of material to be returned to the drum, and this may result in the grain being cracked.

Farm Machinery

TABLE IV.— HARVESTING SEED CROPS BY COMBINE

Crop	Generally preferred method of harvesting by combine	Notes
Cocksfoot and Fescues	Generally preferred method of harvesting by combine	Medium drum speed and concave clearance. Very light wind. If windrowed, difficult to pick up. If direct-combind, risk of serious loss by shedding, and trouble due to green seeds.
Ryegrass	Windrow and pick-up, or use binder and stook.	High drum speed, small concave clearance, and light wind. Keep windrows small and turn gently. Light, standing crops can be direct combined.
White Clover	Windrow and pick-up.	Filler plates needed. High drum speed, very small concave clearance and light wind. Clover sieves needed. Theresh when perfectly dry. Watch for splitting of seed. Can occasionally be combined direct.

Red Clover	Direct combine. Adjustments as for white clover. unless heavy, leafy window if a heavy, tangled crop. crop.	
Sainfoin	Windrow and pick-up. Sometimes direct combined.	"Seed" is enclosed in a husk which has to be milled later.
Trefoil	Windrow and pick-up	Not generally desirable to attempt to thresh the seed from the pod.
Timothy	Binder and stook. Thresh from stook after several weeks.	Adjustments as for clovers. Threshing is difficult, and re-threshing is sometimes worth while.
Mustard, Rape, Swede, Turip	Windrow and pick-up.	Medium speed and concave clearance. Light wind. Avoid splitting seed.
Linseed	Direct combine if clean.	Medium drum speed, small concave setting, medium wind. Leave till dead ripe. Watch for leaks and stop up with palasticine. If much green rubbish use binder, and stook or windrow.
Sugar beet and Mangold	Binder and stook. Thresh form stook.	Low drum speed, wide concave clearance, medium wind. Very stout and heavy crops much better cut by hand.

Losses from combines may be serious unless a constant watch is kept for signs of unsatisfactory threshing. Conditions change from hour to hour, and it is often necessary to vary adjustments several times a day.

The moisture content of grain threshed with the combine varies greatly, according to the state of maturity of the crop and the conditions in which it is harvested. It cannot be stored in bulk with safety if its moisture content exceeds about 14 per cent., and it is often necessary to remove some of the moisture before storage. The construction and operation of grain driers are dealt with in Chapter Nineteen.

Handling of the Straw.

There is great diversity of practice in the handling of straw left by the combine. Where it is to be ploughed in the combine may be fitted with a straw spreader. Even distribution of the straw over the stubble greatly facilitates subsequent tillage operations.

On most farms the straw is needed for feeding or for litter and must be collected. Sweeps, hayloaders or balers may be employed, but by far the most popular method is to use a pick-up baler. Some combines may be fitted with a straw deflector which puts two windrows into one, and in light crops this reduces the time spent on baling.

One imported German combine (the Claas) is equipped with a low-density straw trusser which trusses the straw as it comes from the shakers. This fitting is particularly useful where long straw is needed for such purposes as potato clamping.

Much of the saving of time and cost brought about by using a combine rather than a binder may be nullified by inefficient straw handling methods. Some farmers who are concerned to get the straw collected in good condition find it most satisfactory to use a binder, and cart direct to a stationary thresher.

Choice of Crop Varieties.

Farmers who use combine harvesters need to take careful account of harvesting characteristics in their choice of variety. The advantage of growing short-strawed crops which are resistant to lodging is now becoming widely recognized, and it is fortunate that most of the short stiff-strawed wheat varieties are also early in ripening. The factor of earliness in ripening is often given insufficient weight in selection of varieties. Earliness is a good feature in itself, since there is more chance of getting long drying days in July and early August than in late September. Earliness is also useful in enabling the harvest season to be spread by the earlier start. Autumn-sown Pioneer barley and Picton oats are examples of very early-maturing varieties, but reference should be made to a series of farmers' leaflets, published by the National Institute of Agricultural Botany and to *Cereal Varieties in Great Britain,* by R. A. Peachey (Crosby Lockwood) for details of suitable crop varieties for various soil and climatic conditions.

There is little to be said for growing crops for combining which are known to mature very late in the year.

Combine Harvester or Binder.

Combining has many advantages over use of a binder, especially on large farms. Its chief advantages are reduction of the labour needed for the harvest, elimination of the tedious and costly winter job of threshing, and the fact that with efficient management and a drier available when required, there is less risk of loss due to unfavourable weather. Combining also enables more corn to be gathered from laid crops if devices such as a pick-up reel are used. Nevertheless, there is still a place for the binder on many farms where long straw is valued, and as has been shown already, there are now devices such as hydraulic push-off stackers which may completely alter the appearance of the sheaf-handling problem. Moreover, there is no real reason why stationary threshing should be so slow and costly in labour as it usually is. The binder method is more likely to give a sample of corn that will store safely without drying, and stationary threshing gets most of the weed seeds out of the sample without scattering them over the fields. There are, therefore, many reasons why the occupiers of small farms should interest themselves in possibilities of improving sheaf-handling systems as well as in combine harvesters. The combine, in solving one problem, often introduces two—grain storage and straw handling—which are just as troublesome. It seems likely that any further great extension of the combine method on to the small farms of Britain will require either an increase in contract work, or co-operative use of machinery, or some new cheap combine which either bundles the straw or facilitates its collection by machines which have several other uses.

REFERENCES

(23) "Harvesting by Combine and Binder." Cambridge Univ. Dept. Agric., Farmers' Bulletin No. 9.

(24) Hutt, A. C. *Combine Harvesting and Grain Drying.* London & Counties Coke Association. London.

CHAPTER FOURTEEN

ROOT HARVESTING MACHINERY

"All over the world, in England, America, France,. Russia, Germany and elsewhere, eager search is again being made to-day for a sugar-beet harvester."

Implement and Machinery Review, lxiii, 1937, 752, p. 756.

Potato Lifting.

Early potatoes are still often raised with the help of hand forks, but this can only be economic with high-priced crops, and only possible on a limited acreage, on account of the heavy labour requirement. Most commercially grown crops are now lifted by some form of potato digger. Potato raising ploughs are fairly widely used for all kinds of crops, while complete harvesters, which deliver the potatoes into bags or into trailers, are being used on an increasing scale for second-early and main-crops.

The conditions in which potato harvesting implements and machines have to work vary widely. Sometimes the soil is friable and stone-free, and separation of the potatoes from the soil and from their tops is a fairly simple matter; sometimes the soil is dry and stony or cloddy, in which case it is fairly easy to get a mixture of clean potatoes and stones or clods, but difficult to separate the potatoes from the contaminant; and sometimes the soil is wet and sticky, and any machine which attempts to separate potatoes from the soil merely succeeds in sticking a layer of soil over them, and often in covering its own mechanisms with so much wet soil that they completely cease to function.

Choice of potato harvesting equipment must therefore depend primarily on working conditions and secondarily on the acreage to be dealt with, capital that can be invested, etc. Thus, the simple potato plough will sometimes be found in use on quite big farms, not because it is an ideal implement, but because it is a cheap stand-by which can be relied on to do some sort of job when more complicated machines fail.

Avoidance of damage to the potatoes by bruising or cutting them is an essential requirement for British conditions, whether the crop is required for immediate sale or for storage. This is a factor which needs attention when choosing harvesting equipment, and also when using it.

FIG. 194.—POTATO PLOUGH, (RANSOMES.)

The POTATO-RAISING PLOUGH consists of a beam of the usual horse plough type, with a pair of equal-sized wheels at the head and a special frame which carries the share and raising prongs. The share is renewable and is generally made of chilled cast-iron. It is drawn beneath the row of potatoes and is followed by either cast-iron or wrought-iron prongs which sift the tubers from the soil. The plough does not usually expose potatoes for picking as well as the spinner, but they are raised with less bruising of the skins. Many growers use a scuffler or a special harrow to follow half a round behind the plough in order to expose more tubers immediately after the first picking. The potato plough may often be converted into a ridging plough by removing the raisers and putting on suitable ridging breasts, share and slade.

Some farmers have found that they can do good work with potato-raising bodies fitted to a tractor tool-bar frame. The bodies are generally set to raise alternate rows, two at a time; and a little extra speed is sometimes found to improve the work.

The conventional type of POTATO SPINNER has a strong wide share that passes beneath the potato rows and completely loosens the soil and tubers; it is followed by revolving digging forks, which throw the soil and tubers sideways, generally leaving the latter well exposed. In "direct action" diggers, the forks are attached directly to the revolving shaft. Many diggers, however, employ a linkage mechanism that keeps the revolving forks always in a vertical or nearly vertical position and produces a "feathered" action. The exact angle of the forks may often be varied by adjusting a socket controlling the link gear.

FIG. 195.—SELF-LIFT TRACTOR DRAWN POTATO SPINNER. (BLACKSTONE.)

*FIG. 196.—P.T.O.-DRIVEN DIRECT-MOUNTED POTA-
TO SPINNER WITH HANGING TINES. (RANSOMES.)*

A type of spinner which has become popular has the main spinner rotating on a nearly vertical axis, and an auxiliary spinner which helps to separate the tubers from haulm and soil (Figs. 195, 197 and 198). This type may have an all-steel frame and a drive by chain and bevel gearing from the main axle. Potato diggers are now usually operated by tractors, and trailed machines have self-lift devices similar to those used on ploughs.

The modern trend is towards P.T.O.-driven tractor-mounted machines with cither a single depth wheel or no wheels at all. Such machines have the advantage of simplicity, ruggedness and a positive drive. The mounted potato spinner does, however, in some instances reveal one of the weaknesses of many modern tractors, which have the hydraulic lift pump driven from the tractor P.T.O, shaft. Lifting of a mounted spinner necessitates the running of the

P.T.O., and as the implement is raised the angles at which the universal joints arc required to operate, often become extremely large. It would, in such cases, simplify implement design if the P.T.O. drive could be disengaged before lifting the implement. In spite of this small difficulty, direct-coupled spinners seem likely to displace trailed types.

Operation of Spinners. The machine must always be hitched so that the point of the share is exactly in line with the centre of the ridge to be lifted. With tractor-mounted machines this may necessitate altering the tractor wheel-track setting before starting work. The depth of work must be just sufficient to ensure that all the potatoes are lifted without being cut: deeper setting will result in more potatoes being buried. Failure to penetrate deep enough may be due to insufficient pitch or a badly worn share. On some machines, re-setting the share may also necessitate altering the working depth of the digger tines, which should normally run with their points level with the rear edge of the share. A screen is often fitted to facilitate gathering, by preventing the potatoes from being widely scattered.

The speed of the tines must not be excessive, otherwise serious bruising may occur. Where it is possible to vary speed, the rule should be to use the lowest tine speed consistent with clean digging. In dry, friable soils it may be necessary to fit rubber hose over the tines to reduce bruising. Bruising may cause serious losses in clamps, especially in seasons when late "blight" attacks are prevalent.

FIG. 197.—UNIT-PRINCIPLE POWER DRIVE POTATO DIGGER. (DAVID BROWN.)

FIG. 198.—P.T.O.-DRIVEN DIRECT-MOUNTED PO-TATO SPINNER WITH INCLINED ROTOR. (FERGU-SON.)

Elevator Diggers. The elevator digger consists essentially of a broad share, behind which rises a metal elevator chain composed of parallel links, running on agitator sprockets. The soil is shaken through the links of the chain web, and the potatoes are delivered at the rear of the machine in a narrow row. The agitator sprockets can be changed to give more or less shaking, according to conditions. Modern machines are all P.T.O.-driven, and this permits some variation of speed of the chain in relation to forward travel. Some American machines are fitted with a 3-speed gearbox in the P.T.O. drive. Two-row chain elevator diggers are commonly used in U.S.A. and are available in Britain. They put the potatoes from both ridges into a single row, thereby further facilitating picking where conditions are good.

Tractor-mounted elevator diggers are now available (Fig. 200).

Compared with spinners, elevator diggers save picking labour in conditions where they operate satisfactorily, because they expose more potatoes and leave them in a narrower row. Their great disadvantages are that they will not operate in really sticky conditions, and that heavy maintenance costs are almost inevitable. In sandy soils the chain web, agitators and sprockets may need complete replacement after doing only 20-30 acres, and on stony soils breakages may be frequent. On average land, where the life of a chain web is likely to be about 50 acres, use of the elevator digger in favourable conditions is usually well worth while, and it is advisable to keep a spinner in reserve in case of breakdowns.

Operation. Elevator diggers need a fairly wide headland, and it is usually best to plant the land adjacent to the boundary (i.e. the normal headland) with about 12 rows running parallel to it, and move the headland itself into the held. The headland rows are dug first, and almost all hand digging is eliminated. Continuous working with the digger is possible if a deflector is fitted at the rear (Fig. 200).

FIG. 199.—ELEVATOR-TYPE POTATO DIGGER WITH POWER DRIVE AND 3-SPEED GEAR. (INTERNATION-AL.)

FIG. 200.—TRACTOR-MOUNTED P.T.O.-DRIVEN EL-EVATOR POTATO DIGGER FITTED WITH DEFLEC-TOR FOR CONTINUOUS WORKING. (JOHNSON.)

In using elevator digger's it is advisable, where possible, to set both tractor and digger wheels to straddle two rows. An extension rim is needed on one digger wheel to keep the elevator level. The point of the share must run accurately in the centre of the row, and should work just beneath the potatoes. Elevator agitation should always be the minimum necessary to free the potatoes from most of the soil. Excessive agitation bruises the potatoes and results in more wear and tear of the machine.

With some machines it is possible to run the elevator chain in either one or two sections. In good conditions it is best to have the web in one piece, but in wet and sticky conditions better separation may sometimes be achieved by splitting it into a long front part and a short rear section.

The chain web should be run fairly tight, and excessive slackness caused by wearing of the links should be remedied by removing links. The web, rollers and agitators should not be lubricated, but all rollers should be inspected occasionally to ensure that they are rotating.

If the soil is very easily separated and the potatoes are being bruised, the front agitators should be replaced by plain rollers, and the front roller should be set as low as possible. On heavier soils in wet conditions, where separation is difficult, the front roller should be set fairly high, so as to break up the ridge immediately it comes to the elevator.

Shaker Diggers. The shaker digger has oscillating grids instead of a chain web, and this makes possible some reduction in maintenance costs compared with the usual type of elevator digger. The oscillating grids are usually mounted on ash hangers and oscillated by an eccentric. Two shaker grids oscillating in opposite directions are employed in order to reduce harmful vibration in the machine. Setting consists mainly of adjusting the depth of the share and the pitch of the grids. More pitch on the grid tends to keep the potatoes on it longer and so gives more time for soil separation.

Shaker diggers are employed in some areas for digging early potatoes, and it is claimed that the machine soon pays for itself because it exposes a higher proportion of the crop than the ploughs and spinners that it replaces. Bruising need not be serious if the machine is set so that soil separation is only just complete as the potatoes leave the grid.

With all types of diggers and harvesters which achieve soil separation by riddling it through grids or chain webs, very small potatoes inevitably pass through with the soil.

Complete Potato Harvesters.

When potatoes have been riddled free from soil in an elevator digger it is illogical to deliver them on to the ground again and pick them by hand. On many British soils such as the Black Fens, the silts and some sandstone soils, where there are few or no stones, there is no reason why the potatoes should not be delivered directly into a cart. Several inventors have designed machines to do this, and a few of these machines are proving quite successful.

The Johnson (Fig. 201) and Whitsed harvesters are typical of machines which incorporate a primary separating mechanism similar to that of a chain web elevator digger. The potatoes, clods, weeds and haulm pass from the primary elevator to a short double chain conveyor, consisting of an outer chain of wide pitch and an inner cross conveyor of short pitch.

(*N.I.A.E. Photo*)

FIG. 201.—*POTATO HARVESTER WITH SOIL SEPARA-TION BY CHAIN WEB. NOTE HAULM REMOVAL BY CHAIN PASSING ABOVE REAR CROSS CONVEYOR. (JOHNSON.)*

The outer chain carries the tops and rubbish straight over the back of the machine, while the potatoes, clods and stones are delivered by the inner chain to an elevator which returns them forward to the final delivery conveyor. There is accommodation on the machine for up to 6 pickers, who can remove stones, clods or other rubbish before the potatoes are delivered to the bags or trailer.

When the proportion of clods or stones is very high, arrangements can be made on some machines for the potatoes to be picked out and put into a separate compartment of the final conveyor, while the stones and clods pass back to the ground. This method of operation always necessitates a big gang of pickers on the machine, but their output is higher than it would be if they picked potatoes off the ground. Substantial economy of labour is only possible, however, where the machine separates out most of the stones and clods, and 2-4 pickers riding on it do the rest. In such conditions a total gang of 6-8 men can clear 2 to 3 acres a day in an average crop. Several other makes of harvester employ somewhat similar mechanisms. Most of the harvesters are trailed and P.T.O. driven, but some machines are engine driven and one (the Wota-Crawford) is semi-mounted on the tractor and has the digging share and primary separation mechanism at the side of the tractor. Some machines (e.g. the Byfleet) employ a combination of endless chain mechanisms and rotary riddles.

(Farmer & Stockbreeder Photograph)

FIG. 202.—ENGINE-DRIVEN POTATO HARVESTER WITH SOIL SEPARATION BY ROTARY RIDDLES. (PACKMAN.)

The Packman harvester (Fig. 202) relies entirely on rotary riddles and conveyors. Rotating riddles can have a fairly long life even in abrasive soils, due to the fact that there are no joints working in the soil. A rather flat share lifts the crop, and the mixture of potatoes and soil passes into a large inclined rotating riddle which conveys the potatoes upward while the soil falls through. Rubber fingers assist in keeping the riddle clean. The potatoes pass from the primary cleaning riddle to a horizontal circular riddle, from which operators can pick out rubbish. The crop finally passes to an endless belt type of delivery elevator, and here again operators can remove stones, clods or other rubbish. The whole machine has a gentle action and is particularly suitable for light soils and early lifting. The drawbar is mounted on the tractor's hydraulic lift linkage, and this gives greater flexibility for turning on headlands, while the extra weight on the tractor wheels is an advantage in wet conditions.

FIG. 203.—P.T.O.-DRIVEN POTATO HARVESTER
WITH SOIL SEPARATION BY INCLINED ROTORS.
(GLOBE.)

The "Globe "harvester (Fig. 203) also employs rotary mechanisms for soil separation. The ridge is lifted by an open share on to a pair of inclined rotors which are fitted with radial fingers. These fingers allow the soil to pass through, and after the crop has passed in turn over the two rotors it is delivered to a cleaning and elevating drum which is mounted transversely at the rear of the machine. From there the potatoes pass forward along a picking table. Operators stationed on either side of this table can pick either potatoes or rubbish according to conditions. The action of the machine is gentle, and it is particularly suited to light soils.

Separation of Stones and Clods. The problem of how to separate stones and clods from potatoes by means of automatic mechanisms, without damaging the crop, is one that still baffles the many inventors who have tackled it. It is unfortunately a fact that up to the present no mechanism which looks very promising has been demonstrated to the farming public.

A mechanism employed on one machine (the Johnson) consists of a steep elevator which picks up the angular clods and allows the potatoes to roll back. This and some other devices which have been tried are partially successful, and work fairly well in some conditions.

The whole future of potato harvester development depends on solving this problem, since only a small proportion of the potato-growing lands of Britain are normally free from either stones or clods at harvest time. Nevertheless, there are substantial potato growing areas which are comparatively stone-free, and some of the machines at present available can be effectively used on these soils.

On some soils the amount of clod is to a certain extent within the farmer's control. With the old methods of lifting, farmers on stiff soils could be reasonably well satisfied with a cloddy seed-bed which was just deep enough to allow the ridgers to run. There can be little doubt that on such soils the amount of clod that comes up with the potatoes at harvest time is largely determined by the condition of the land when the potatoes are planted, and it is possible that the regular use of complete harvesters will demand much more attention to the production of fine seed-beds. It is conceivable, for example, that in years when the potato land is cloddy in spring the use of rotary cultivators or other power-driven cultivating tools will be justified if it results in the soil being reasonably free from clods at harvest time.

Operation of Complete Potato Harvesters. The principles of operation of complete harvesters are similar to those already outlined for spinners and diggers. A minimum amount of soil must be lifted by the share, and both the share itself and any auxiliary attachments such as side discs must be set to pare away as much as possible of the sides of the ridge, without cutting the tubers. Accurate driving, and setting of the share, are of the utmost importance, and it is useless to expect a complete harvester to operate satisfactorily on a field where careless work in spring has resulted in very uneven ridges. Soil separating mechanisms must be set so that the loose soil is got rid of exactly at the right time, since too rapid separation on some machines can result in serious bruising. Headlands must be wide, and tipping trailers with reasonably deep sides are essential. Two 3-ton hydraulic tippers can nicely keep the harvester going when the clamp is in the same field. It is usually advisable to have one or two men gleaning good potatoes which are spilled from the share, from the clod-separating mechanism where this is fitted, or from accidents due to bad driving of the trailers. If the potatoes are not gleaned immediately, most of them are run over by tractors and trailers and spoiled for human consumption. When gleaning is carried out in this way the amount of potatoes left in the land should be so small as to make harrowing not worth while. Most of the very small potatoes which are riddled through with the soil are near the surface, and are killed by the first frost.

Haulm Strippers and Destroyers. Haulm often causes trouble at lifting time, both to potato diggers and complete harvesters, and several methods of minimizing the difficulties have been evolved. One common practice which assists in controlling potato blight damage, as well as in reducing mechanical troubles, is spraying the haulms with sulphuric acid or other haulm destroyer about 10 days before lifting.

This method is not applicable to early crops, since every day's growth is important.

Power-driven haulm cutters consisting of high-speed horizontally rotating knives have been employed for several years, and though they do not deal with haulm which straggles between the ridges they are useful in clearing the way for digging and harvesting where haulm growth is heavy. One machine has a pair of belt-driven cutters mounted on the front of the tractor, while some harvesters have smaller strippers working just in front of the share.

With the object of pulverizing the haulm completely, some manufacturers employ beaters attached to a horizontal P.T.O.-driven miller shaft, and operating much like the tines of a rotary hoe. These rotating beaters may consist of hoe blades, steel hammers or rubber hammers. They act mainly by virtue of their speed, and the exact shape and nature of the beater itself is probably important mainly in regard to wearing properties. To do a thorough job the revolving beaters need to be contoured to match the potato ridges. Haulm pulverizers require a considerable amount of power where the tops are long and tough, a medium-powered tractor being capable of doing only 2-3 rows at a fairly slow speed. The degree of pulverization needed to render innocuous haulm that is heavily infested with blight is not yet known, and trials to determine such points are proceeding. Destruction of haulm either by spraying or by mechanical methods is a suitable job for contract work, and many-farmers find contract work more economic than buying another special-purpose machine.

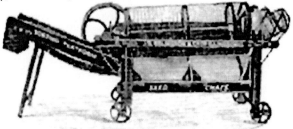

FIG. 204.—ROTARY POTATO SORTER.(EDLINGTON.)

Potato Sorters. The potato marketing regulations have made it necessary for all potato growers to produce properly cleaned and graded samples. Many growers rely entirely on hand riddles for this work, but on most of the larger farms machines are employed to assist in the sorting operations. Two main types of sorter are employed. One of these consists of a rotary screen with meshes designed to allow "chats" and "seed" to pass through into separate hoppers, while "ware" passes to an elevator at the lower end of the screen. As the ware passes along the elevator to a bagging device, "blights" and foreign material may be picked out. The other type of sorter has flat reciprocating riddles arranged above one another, and the ware passes from the upper riddle to an elevator where hand sorting may be carried out, while seed and chats are retained on the lower riddles. On some machines the sorting elevator may be fitted with a roller conveyor which slowly turns the potatoes as they advance. This is a useful fitting when the crop contains a number of blighted tubers. Potato sorters may be either hand- or power-operated.

FIG. 205.—ENGINE-DRIVEN RECIPROCATING POTATO SORTER, WITH LOW-LOADING HOPPER, (COOCH.)

A $1\frac{1}{2}$ hp. engine has ample power to drive the usual type of sorter, and is preferable to man-labour where the machine is fully employed. When an engine is used, the work is speeded up if a low-loading elevator is fitted.

Sugar-beet Lifters.

When sugar-beet was first grown on a large scale in this country many types of lifters were employed, but only a few of these have remained in general use. A common type is a simple two-horse implement with a strong vertical steel knife attached at the top to a plough beam, and carrying at the bottom a small chilled cast-iron share which runs beneath the row and loosens the roots so that they may easily be pulled. The share acts as a wedge which simply raises the beet a little and loosens the soil around them. On some implements two short, curved, wrought-iron prongs run behind the share and help to push the roots upwards out of the soil. In the hands of a good man the simple lifter can do good work at a rate of about 2 acres a day.

FIG. 206.—SUGAR-BEET PLOUGH. (RANSOMES.)

Tractor-hauled lifters which raise from one to six row sat a time are manufactured, and are now widely used. Their chief advantages lie in a reduction of the man-labour requirement of the lifting operation, and in adaptation of the work to tractor operation on those farms where few or no horses are kept. The tractor needs to have suitable wheels to fit between the rows of the crop. Pneumatic tyres do not damage the roots, but unless the wheels are fitted with guards to push the tops clear, serious slipping on the wet tops may occur. The difficulty may often be solved by fitting the tractor with a pair of" open "type steel wheels.

Where a tractor is used, four rows at a time may be lifted with the single-arm type, and it is not necessary to run over any ground that has already been loosened. When the double-arm type of lifter is used only two rows are normally raised at a time, and the lifters are spaced two row-widths apart, so that alternate rows are worked. The double-arm lifter does good work, especially on medium and light soils.

Other types of lifter employing two wheels to raise the roots are available, but they are generally suitable for use only on light soils.

Lifting should be done with as little disturbance of the soil as possible, so that the work of carting off is not made more difficult than necessary.

Sugar-beet Harvesters.

Work on the development of a complete sugar-beet harvester which tops, lifts and cleans the beet commenced in the early nineteen-thirties, and though a fairly successful prototype was soon produced it was not until the late nineteen-forties that large-scale production and use of harvesters began. Ninety per cent, of Britain's 41,000 beet growers each grow 20 acres or less, and a high proportion of these wish to preserve the tops in a clean condition for feeding to livestock. It was, therefore, a great advance when cheap and simple two-stage harvesters, suitable for use on such farms, were added to the more expensive complete harvesters which had held the field alone in Britain until 1948. 5,300 acres were harvested mechanically in 1947, and by 1951 the figures had risen to around 80,000 acres or 20 per cent, of the crop.

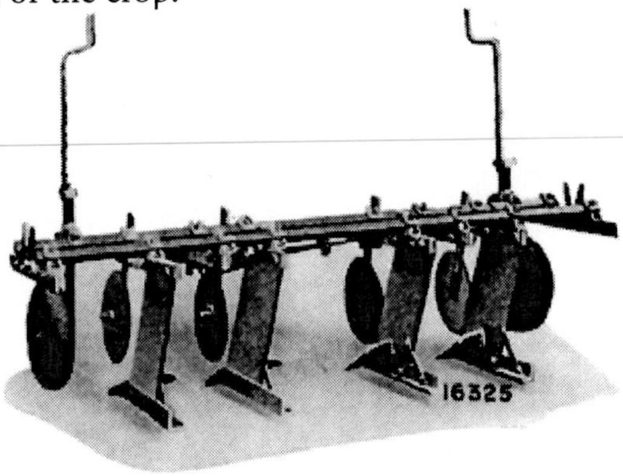

FIG. 207.—FOUR-ROW SUGAR-BEET LIFTER ON REAR-ATTACHED TOOL_BAR. (RANSOMES.)

Whatever kind of mechanical harvester is used, planning for use of the machine should start right at the beginning of the farming year. One-way or square ploughing is the first step to level seed-beds, and careful work on seed-bed preparation and drilling must follow if uniform crops that are really suitable for mechanical harvesting are to be obtained.

FIG. 208.—TWO-ROW DOUBLE-ARM (BOW_TYPE) SUGAR-BEET LIFTER. (FORSON.)

Row widths for convenient mechanical harvesting with tractors need to be about 20 inches to accommodate 11-inch-wide rear wheels, and the field should be drilled with a wide headland which is lifted first when harvesting begins.

The range of types of sugar-beet toppers, lifters and complete harvesters is now so wide that it is impossible in these pages to mention all types. In addition to looseners used in conjunction with hand topping, the following main kinds of machines are available:

Simple Toppers which top the beet and push the tops to one side. The tops must be cleared by hand or moved by side-rake before lifting the roots (e.g. Roerslev, Mern, Hudson, etc.). These machines may be trailed or tractor-mounted, and may do one, two or several rows at a time. Some machines are automatically steered by sledges. Others are independently steered by a man walking behind or riding on the topper.

Topper-Windrowers. With these, devices are added to move the tops over out of the way of the lifter. Some machines employ a short elevator to put them in a windrow parallel to the rows while others collect the tops in a hopper and dump them in windrows at right angles. (The latter type of collector is also used as part of a complete harvester, e.g. Salopian.) Where the tops are not needed for feeding, a rotary flail may be used to sweep them out of the way (e.g. Whitsed).

Topper-Loaders. These machines top the beet and deliver the tops directly into a trailer running alongside (e.g. Catchpole Top-saver, Standen Top-saver).

The general principles of operation of all the above types of machines are the same, whatever the nature of the topping mechanism employed. All rely on a feeler wheel, track or sledge of some kind, which is designed to run over the crowns of the beet and determine the setting of the knife or knives which cut off the top. The feeler mechanism must run true along the centre of the row of beet, and there must be just enough pressure on it to ensure that it gets down to the crown. The feeler wheel is slightly over-driven to hold the beet against the topping knife, and the knife must be set, according to the average size of the beet, so that the cut begins when the feeler wheel is well on top of the crown of the beet. Where a simple knife blade is used for topping it must be kept sharp. Some topping mechanisms used on complete harvesters have a power driven saucer-shaped knife (e.g. International and Robot).

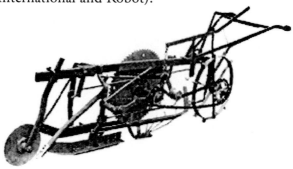

FIG. 209.—SINGLE-ROW BEET TOPPER. (ROERSLEV.)

'The Catchpole, which has a pair of plain horizontal discs for topping, has a power-driven tracklaying device to regulate their height.

When separate toppers are used the beet must not be topped too far ahead of the lifter, otherwise serious loss of sugar from the roots is likely. If beets are left in the ground only a few days after topping some small loss may be found. If they are left for more than a fortnight: the loss of sugar is likely to be appreciable.

(*Farmer & Stockbreeder*)

FIG. 210.—TOPPING MECHANISM OF THE CATCH-POLE SUGAR-BEET HARVESTER

Lifting and Cleaning Machines. Lifting and cleaning machines may deal with one or two rows at a time, may be trailed or tractor-mounted, ground driven or P.T.O. driven, and may deliver the beet in windrows parallel to the work or at right angles to it. In general, the type with a ground-driven cleaning drum (e.g. Roerslev, Mern, etc.) is mainly suitable for use on light soils, and power-driven machines are needed where cleaning is difficult.

With all types of lifter the following general principles of operation apply: (i) the machine must be accurately guided either by skids or by being steered, so that the line of beet is centrally between the lifting points or shares; (2) the lifting prongs or shares should be set to "pull" the beet, so far as this is possible, without breaking off too much of the tap root. The less earth that is lifted with the beet the easier the cleaning and carting off; (3) cleaning devices should be adjusted to give adequate cleaning without doing serious damage to the roots. Where necessary, forward speed must be reduced to give more cleaning.

FIG. 211.—TWO-ROW P.T.O.-DRIVEN HARVESTER
LIFTING PREVIOUSLY TOPPED SUGAR BEET.
(FORDSON.)

With all the machines that leave beet in windrows, care must be taken not to have a big acreage of beet exposed in frosty weather. A frost that will do no harm to beet stored in a big clamp will render useless for processing beet that lies uncovered in a thin windrow.

Lifter-Loaders. Some power-driven lifter-cleaners can be fitted with an elevator which will deliver the beet to a trailer running alongside (e.g. the Rota), and there is no reason why delivery to a trailer pulled behind the harvester should not be developed for this type of machine. In order to keep a side-delivery harvester going it is necessary to provide two tractors and trailers for transport, even if the clamp is only at the headland; and many small farms have neither the tractors nor the men needed to drive them. It is, therefore, somewhat illogical to provide for only side loading with an inexpensive harvester.

(*Farmer & Stockbreeder*)

FIG. 212.—COMPLETE SUGAR-BEET HARVESTER. (CATCHPOLE.)

Complete Harvesters. Most of the complete harvesters incorporate two or more of the types of mechanisms already briefly outlined. The type of complete machine most common in the past has been a two-wheeled power-take-off-driven trailed machine with a topping mechanism at the front, immediately followed by the lifting shares and cleaning mechanism, and arrangements for either dumping the beet in windrows or delivering them to a trailer drawn alongside or towed behind (e.g. Catchpole, Salmon, Robot, Minns, Ruhlmann). Some machines, however, have the topping mechanism either in front of the tractor, underneath it or alongside it (e.g. Standen, Johnson, International, Catchpole Minor). One harvester dumps beet and tops in separate windrows (Salopian) and another delivers the beet to a trailer pulled behind and the tops to a trailer running alongside (Catchpole Major).

The cleaning mechanisms employed in complete harvesters include a rotating spinner in a stationary cage, various kinds of rotating cages, elevator chains running on agitators, and revolving knockers which shake and rub the roots.

One of the most successful American harvesters, the Scott-Urschel, works on a principle different from that of any of the machines already mentioned, in that it tops the beet *after* lifting them.

FIG. 213.—COMPLETE SUGAR-BEET HARVESTER WITH SIDE-LOADING TOP-SAVER AND RFAR DE-LIVERY OF THE ROOTS. (CATCHPOLE.)

In this trailed power-driven machine, a pair of gathering points straddles the row and lifts the leaves. Just behind these points a pair of rubber-faced elevator chains grasps the beet tops at the same time as the roots are loosened by a small share running beneath the row. The elevator chains run backwards and upwards, and are geared to travel backwards at approximately the same speed as the machine travels forward. This ensures that the beet are pulled vertically out of the ground by means of their tops, which are firmly gripped between the elevator chains. The pulling of the beet is one of the machine's strong points, for much of the soil is stripped off in the process.

At the top of the elevator the beet, still carried by their tops, are delivered to the topping knives, which consist of two rotating discs with serrated edges. The topped beet drop on to a cross conveyor which carries them direct to a vehicle running alongside.

Among other American machines, the International is equipped with a sorting belt and a trailer which takes the beet to the headlands, and there transfers them by means of a power-driven elevator to a truck. The John Deere has adjustable conveyors, which deliver four rows of tops and eight rows of beet into single windrows. A separate beet, loader is available for picking up the beet from the windrow and loading them into trailers.

Multi-purpose Root Harvesters.

The amount of capital invested in equipment steadily increases, and one of the chief problems confronting farmers is the extent to which further mechanization can be economic. In some instances, on account of the small acreages of particular crops grown, the saving in labour or increase in output can hardly justify the high cost of specialized equipment. This applies particularly to root harvesting machinery, where small acreages of several different crops may be grown. Many farmers are, therefore, interested in the development of "universal" root harvesting machines which, with minor adjustments, can be used for a range of crops such as sugar-beet, potatoes, swedes and turnips.

The development of satisfactory and cheap beet topping machines has made it. practicable to omit the requirement of accurate topping from the specification of a universal root harvester, and some promising machines are now available.

Some potato harvesters, for example, fitted with a beet-lifting share, will lift and clean previously topped beet satisfactorily and deliver it to a trailer alongside. A man can ride on the machine and deal with any incorrectly topped beet or other rubbish.

A cheaper adaptation of a potato lifting machine is the conversion of an elevator type potato digger. Equipping of such a machine with lifting shares and with a cross conveyor at the rear enables it to do a good job of lifting, cleaning and windrowing previously topped beet in reasonably light soils.

Beet lifting shares can be obtained for one mounted potato spinner (Blackstone), while one of the P.T.O.-driven spinner type root harvesters (Fordson) can also be adapted for lifting potatoes. There is, therefore, every possibility that a variety of satisfactory multi-purpose root harvesters will become available.

FIG. 214.—SUGAR-BEET ELEVATOR. (COOK.)

Beet Loaders.

When beet is carted from the fields and dumped in heaps by the roadside, loading into lorries for transport to the factory is facilitated by the use of an elevator, driven by a small engine, mounted on the frame. It often knocks off an appreciable quantity of soil, and its use thus results in better dirt tares.

A general-purpose loader equipped with a hopper consisting of open bars will get rid of a considerable amount of soil, and the effect is improved if the elevator bed is also of the open grill type (e.g. Fig. 287). One type of loader (the Robot) has a rotary cleaning drum which is particularly effective in cleaning the roots before loading them.

Early harvesting and carting to the roadside is common in the Fens. The practice has the advantage that the crop comes up cleaner and is more easily carted from the fields, while ploughing for subsequent crops may be done in better season.

CHAPTER FIFTEEN

MACHINERY FOR LAND DRAINAGE, RECLAMATION AND ESTATE MAINTENANCE

"The reader will imagine a deep knife fixed under the beam of a wheel-plough; 'at the bottom of this knife a pointed piece of iron fixed horizontally."

Rudimentary Treatise on Agricultural Engineering. G. H. Andrews, 1852.

Drainage work on rivers and brooks is of vital interest to many farmers, but its execution does not directly concern them, since Catchment Boards or County Councils are generally responsible. The work on held drains and farm ditches, however, is the farmer's responsibility; and it is important that he should be aware of the great advances made in the mechanization of this work during the last 20 years. Almost all work on rivers, water-courses, farm ditches and field drains can now be carried out by mechanical devices of one sort or another.

Excavators for Cleaning Ditches.

Development work on drag-line excavators started long ago; but it was not until the 1920's that efficient, compact, self-contained and self-propelled machines appeared. A range of machines of varying capacities is now available; and many British excavators have been exported to work in all parts of the world. A typical small drag-line excavator with a scoop of 5 to 8 cubic ft. capacity costs over £2,000. Such machines are generally operated by contractors.

Side-drag-line equipment enables a small machine to clean narrow ditches and watercourses, even where there is a hedge running alongside. The drag pull on these machines (Fig. 215) is displaced from the foot of the jib to an adjustable point several feet away, by means of an arm and a sheave block attached to the jib members. The path of the scoop becomes along instead of across the drainage channel, and a ditch with a cross section no greater than that of the scoop itself can. be cleared.

It should, however, be remembered that the side-drag-line outfit, efficient though it is, is essentially a "bottoming" tool; and hand labour is needed to finish the sides, which should be battened to an angle of about 60 degrees to the horizontal for most purposes. One or two men equipped with long-handled spades generally work with the machine, so that the trimmings which fall into the ditch are picked up by the machine as it works. Under favourable conditions, a good driver is able to excavate up to a chain of small ditch per hour, with a 20-inch bottom width and a depth of about 3 feet. One small side-drag-line outfit can use specially tapered buckets giving a bottom width of only 12 inches. Some drag-line excavators have been mounted on widely-spaced tracks so that they straddle the ditch being cleaned. This, however, requires very careful use, and there is difficulty where ditches intersect. Blade graders attached to the front of tracklaying tractors are widely used in the United States, but have not been found suitable for British conditions.

FIG. 215.—SIDE-DRAG-LINE EXCAVATOR CLEANING OUT A FEN "DYKE". (PRIESTMAN.)

Trench Cutting Machines.

Some small drag-line excavators can be fitted with a special trenching attachment for preparing trenches suitable for laying tile drains. An example of this is the "Teredo" attachment for the Priestman "Cub" excavator. A special jib and scoop are fitted for this work, the scoop being held rigidly on a pair of arms pivoted to the top of the jib. The machine stands centrally along the line of the proposed drain and draws the scoop directly towards it, retreating from the work as the trench is completed. The excavated material is deposited at the side. A tooth at the bottom of the scoop cuts a groove for the pipes, to suit the size of drain required. Two men generally work with the machine, one helping to side up the trench and the other giving the final touches with a "bottoming" scoop, and laying the pipes. It is advisable to mark out the drain by drawing a shallow plough furrow before starting with the machine. Outputs of up to 20 chains per 8-hour day, with a trench 1 ft. or 1 ft. 3 in. wide at the top and 2 ft. 6 in. deep, have been achieved in good conditions.

FIG. 216.—EXCAVATOR FITTED WITH TRENCHING
ATTACHMENT FOR TILE-DRAINING. (PRIESTMAN.)

Some of the most successful trench-cutting machines employ a system of endless buckets, with small cutters attached either to a wheel or to an endless conveyor. Examples of these include the Neal excavator, the American Buckeye excavator, and the Rotehoe trench digger. The outfit is self-propelled and has the cutter mechanism power driven. Such machines work well in soil where there are not many stones, and are particularly suitable for tile draining and for putting in tiled mains for mole drainage systems.

There are many trench-cutting machines which are pulled like a plough, being operated either by a powerful tracklayer or by a power-driven winch.

The trencher generally consists essentially of a pair of knives which cut the sides, and a share followed by an incline which cuts the bottom and raises the spoil.

The Guthbertson drainage machines consist essentially of large trench-cutting ploughs with arrangements for easy transport, and adjustments for depth control. The upper edges of the trench are cut by discs of 3 ft. diameter, and the lower cut is made by a share, behind which is an inclined mould-board. The depth of the drain is controlled by the angle of the main mouldboard. Pressure of the slice on the mouldboard exerts a thrust on the heel, which is shaped to give the desired profile to the trench bottom. The main mouldboard lifts and twists the spoil, and a trailing mouldboard pushes it to one side. A powerful tracklayer is needed to operate such trench cutters, and special machines with wide tracks, suitable for crossing bogs, have been developed to facilitate the drainage of hill land.

Narrow, deep drains suitable for tile-laying can be cut, and a tile-laying attachment can be fitted to some models. The tiles are fed into a chute by two men standing on a platform at the rear of the mouldboard. For continuous work it is necessary to have a trailer carrying tiles running alongside, with a man handing tiles continuously to the two feeding the machine.

The Barford agricultural drainage machine is a trench cutter suitable for the laying of drainage tiles, and designed for use in conjunction with the Fordson Major tractor. The machine cuts a trench) I in. wide at the top, either 5 or 6 in. wide at the bottom, and up to 27 or 30 in. deep. The tractor P.T.O. drives a winch through a reduction gearbox, and the winch pulls on a wire rope which is anchored up to 65 yds. in front of the tractor. The ordinary tractor gearbox is in neutral, and the machine is slowly moved forward by the pull on the cable. The trench is cut by a ratchet-driven rotor carrying 6 blades. Attachment of the machine to the tractor is a fairly lengthy operation, and this, like most other trench-cutting devices, tends to be a contractor's machine.

FIG. 217.—TRENCH EXCAVATOR. (R.H. NEAL & CO.)

FIG. 218.—HILL DRAINAGE PLOUGH PULLED BY "WATER BUFFALO". (CUTHBERTSON.)

Mole Drainage.

Mole drainage consists essentially of drawing a bullet-shaped steel mole or cartridge through the subsoil. It is practicable only on very heavy land in a moist condition. The process has been employed for 150 years. It was normally carried out by steam cable tackle before suitable tractor power became available.

Many theories of drainage and rules for drainage practice have been formulated from time to time, but most of these break down with mole draining. The important principles of drainage in heavy or clay soil have been shown to be broadly as follows:— (1) heavy soils on which, alone, mole draining can be successfully performed, are almost impermeable to water in the undisturbed lower layers during the drainage season. (2) Drainage from these soils is essentially surface drainage and takes place by the passage of water through major interspaces in the soil rather than by percolation through it. The permeability of such soils may be improved by suitable cropping and tillages; deep cultivations when the soil is dry assist materially in the drainage of such land.

It is possible to carry out mole draining at a comparatively low cost on heavy soils where there is a natural fall, and where the subsoil is uniform undisturbed clay to within about a foot from the surface. A fall of 1 in 500 may be sufficient for short lengths of mole; but it is desirable to have more where possible, and good results cannot be expected with falls of less than 1 in 200. The surface of the ground needs to be free from irregularities of contour, and best results can only be obtained if the subsoil is free from pockets of sand, gravel or pebbles. When a mole channel is drawn through clay in which sand or gravel occurs, the walls are left rough, and soon collapse when drainage begins.

Moles range in diameter from 2 to 4 in., and are generally pulled through the ground at a depth of from 12 to 24 in. The mole cartridge squeezes the plastic clay outwards in its passage through the soil and leaves a channel which forms the drain. Well-constructed moles have been said to function for fifty years, but the normal life is from 5 to 15 years. Factors making for success are careful planning of the system and the construction of suitable tiled and bushed or tiled and clinkered main collecting drains, with well-protected outfalls. If the mains are well constructed the life of a system may be indefinitely prolonged by re-drawing the mole channels at intervals of a few years. The moles are generally drawn from 2 to 6 yds. apart; but where the land is laid up in ridge and furrow they should run along the furrow bottoms. Their length should not exceed about 1 o chains, and there should be a gradual fall throughout the length.

Methods of Mole Draining.

Steam cable sets were the most popular source of power until recent times, but they have been almost entirely displaced by tractors working by direct haulage. Powerful tracklaying tractors can do good work by direct haulage of the mole plough, and if the conditions are such that the tractor can secure good adhesion, even a medium-powered

wheeled tractor can usually pull a $2\frac{1}{4}$-in. mole about 15 in. deep.

Pulls of 5,000-10,000 lb. are required for pulling 3-4 in. moles 20-24 in. deep, and for this work it may be necessary to equip the tractor with an anchoring device and winch which enable the plough to be pulled indirectly, through a wire cable. Such a tractor proceeds up the field along the proposed track of the drain, paying out cable as it goes, and leaving the plough at the starting-point. The winding drum is then put into gear and the pull causes the anchor to dig into the ground. The plough is pulled up to the tractor, the winding mechanism is disengaged, and the tractor then moves on to repeat the operation. The advantages of this method are that the tractor can work effectively in slippery conditions, and that high drawbar pulls (at low speeds) are attainable. The disadvantages are that the method is not as simple as direct haulage, and the equipment is rather expensive.

It has been shown that there are good reasons for the old-established practice of drawing all moles uphill and running the machine light when returning it to the lower part of the field. Moles drawn uphill have been found to run more quickly and to silt up less readily than those drawn downhill.

Drainage should be performed when the subsoil is moist, for the draught is much lower in wet soils than in dry. For example, a $2\frac{1}{2}$-in. diameter mole, which has a draught of 2,000 lb. when pulled at a depth of 15 in. at 2 m.p.h. through a wet soil, may require double this pull when the soil is drier. Moreover, the drains are more likely to last well if the soil is moist and plastic enough to leave the walls of the mole smooth. But in deciding on the best time to operate, other considerations are involved. The surface must be dry enough to carry the tractor, and this is especially important where direct haulage is employed; for the high drawbar pulls necessitate a good grip by the wheels or tracks. The right conditions of surface soil and subsoil most frequently coincide in the spring, when the surface has dried out and the subsoil is still wet. The early winter is also sometimes favourable

It should be understood that movement of water and the accumulation of excess water in heavy land take place in the surface layers, and that drainage depends on giving this water access to the drainage channels. In mole drainage, this is mainly via the slit and the disturbed ground in its neighbourhood, and it is doubtful, therefore, whether there is any advantage in drawing the drains very deep. It is, in fact, unnecessary to go deeper than the minimum required to prevent cultivations and weathering influences from destroying the channel. When the drains are drawn the ground is fractured and opened up to a distance of 1 to 2 ft. on each side of the mole slit. The slit and major fractures persist for a long time, as is very obvious in dry summers on some fields. Slits have been observed open in grassland during a drought sixteen years after the moles were drawn.

In drainage work the tendency is to expect the system to last for many years, and with tile drainage on light land, something approaching permanency can be achieved. But the heavier the land the less likely are either mole or tile drains to continue to function indefinitely. Generally speaking, mole drains work best in the first five years of their life, and it is gradually becoming recognized that drainage of heavy land should be regarded more as a cultivation than as a permanent improvement. Where a semi-permanent tiled main is installed, redrawing the moles costs little and may be repeated every five years or so.

Types of Mole Plough.

The mole ploughs used by steam cable tackle sets are very heavy 3-wheeled implements with a large mole, 3 to 4 in. in diameter, capable of working at depths of up to 30 in. They are fitted with a seat, a steerage mechanism and a device for regulating the depth of work.

Mole ploughs for use with tractors are of three main types. One of these, the beam mole plough, is a simple implement consisting of a steel beam, fitted with skids and carrying a disc coulter and the mole (Fig. 219.)

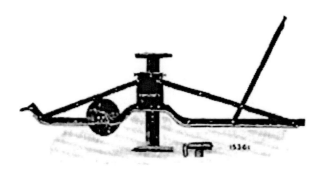

FIG. 219.—BEAM TYPE MOLE PLOUGH. (RAN-SOMES.)

The mole cartridge on this type of plough is generally $2\frac{1}{4}$ to $2\frac{1}{2}$ in. in diameter, and 12 to 24 in. long. It may be set parallel to the beam or, more usually, with a slight pitch downwards. An expander or bob, $\frac{1}{4}$ or $\frac{1}{2}$ in. greater in diameter, is often attached behind the cartridge to assist in smoothing the walls of the mole channel. The depth of the cartridge may be adjusted either by means of a series of holes in the knife or by a screw mechanism. The mole is raised to the surface at the ends by a sideways pull on a lever attached to the beam.

Another type is a wheeled implement with a powerful self-lift. Such an implement may generally be used as a subsoiler or heavy cultivator by the substitution of suitable attachments in place of the mole. A third type has a *floating* beam carried on a heavy 2-wheeled cultivator-type frame, fitted with a powerful self-lift device. Such a machine combines the advantages of a one-man self-lift outfit with a long, sliding beam. Fig. 220 shows one such machine of heavy construction, suitable for use with a very large track-layer. The lift on this machine is rather unusual. Star wheels attached to the axle engage a winding drum, and wind up a steel wire rope that is arranged to lift the end of the beam. A brake on the drum enables the beam to be lowered gently.

All types of plough should be constructed so that they enter and leave the soil without the necessity of digging holes. It is, of course, impossible for the cartridge to start work at the full depth of the mole channel without either digging eye-holes or starting in the ditch. The best practice is to start as near the ditch as possible with an implement which automatically draws itself into the soil and attains its full depth before passing through the porous layer above the tiled main drain.

Tractor-Mounted Mole Drainers. Mole drainers suitable, for mounting on the hydraulic 3-link system have been developed for most medium powered British tractors. The action of these simple mounted drainers is very satisfactory in most conditions, the 3-link hitch permitting the implement to float smoothly even if the tractor passes over rough ground.

*FIG. 220.—HEAVY FLOATING-BEAM TYPE MOLE
PLOUGH WITH POWER LIFT. (MCLAREN.)*

The power lift is a great advantage, and the equipment
is so cheap to operate that moles can be renewed frequently
at no great cost.

Mole-tile Drainage.

Research has been carried out in Germany and
Holland with the object of perfecting a cheap method of

draining soils that are unsuited to ordinary mole drainage. The methods tried include (1) laying tiles by means of a special type of mole plough; (2) laying a continuous length of porous concrete pipe by means of a machine which feeds the concrete coating in behind a special mole plough; and (3) placing inside the mole mild steel tubing, wooden drains, etc. The first partially successful experiments in these directions were carried out in England in the middle of the last century. No details of recent machines can be given here, since none can yet be considered to have passed beyond the experimental stage. Some of the methods give promise of ultimate success.

Machinery for Land Reclamation.

Since 1939 a great deal of land which was formerly derelict has been brought back into cultivation. There were several types of derelict land in 1939, the most common being clay land covered with scrub consisting of hawthorn, brambles and wild roses, with enormous ant-hills as well in some districts. Such land with good drainage, cultivations and phosphates, is generally capable of growing good crops of either wheat and beans or grass. Other types of land, potentially fertile but badly neglected, were certain parts of the Fens— where the difficulties consisted mainly of bad drainage and bog oaks—and derelict orchards. In addition there were large areas of upland covered with bracken, gorse and other shrubs. All these types of land have been successfully reclaimed, and machinery has played a vital part in the work. Considerable progress has been achieved in the technique of reclamation, and several new machines developed for the work.

The best method to employ for clearing scrub depends on the size and density of the bushes. For isolated bushes or trees there is probably nothing better than pulling them out with a wire rope, power being supplied by a hand-operated monkey winch, a tractor with or without a winch attached, or a steam engine, according to the equipment available and the number and size of the bushes to be dealt with. Most bush clearing must be done with the ordinary farm tractor, and devices such as bush pullers speed up the work. The bush puller consists essentially of a light, steel draught-pole attached to the tractor drawbar', and having at the free end a V-shaped guide and jaws designed to grip the trunks of bushes as the tractor advances, and to release them when it reverses. One man in addition to the driver is needed for rapid work. With such devices land having up to about 500 small bushes per acre can be cleared fairly economically; but where the bushes are more numerous the cost of pulling them individually is prohibitive.

FIG. 221.—PULLING OUT A TREE, WITH EASILY ATTACHED WINCH AND SPRAG FITTED TO LIGHT TRACTOR. (FERGUSON.)

"Bulldozers" have frequently been used with success for bush and hedge clearing; but they often take too much soil with the bushes and so make burning difficult.

Various kinds of tractor-mounted bush diggers, pushers and sweeps have also been built and used successfully by both farmers and contractors.

Bull-Dozers, Angle-Dozers, Earth Scoops and Graders.

The equipping of most medium-powered British tractors with hydraulic lift systems has facilitated the development of cheap earth-moving devices which are of great value for many purposes.

The term Bull-dozer is applied to a blade which is set square across the front of the tractor, while "angle-dozer" refers to a blade which is set at an angle. The same blade and equipment can often be used for both purposes, by using different methods of attachment between the blade and the framework which connects it to the tractor.

Both bull-dozers and angle-dozers are usually more effective on tracklayers than on wheeled tractors, but provided that a medium-powered wheeled tractor equipped with a dozing blade is intelligently used it can do a great deal of useful work on such jobs as removing unwanted banks and hedges, filling in drainage trenches, etc. A half-track conversion can do slightly heavier work than a wheeled tractor, on account of its extra push.

Earth scoops which are made for mounting at the rear of the tractor, on the hydraulic 3-link hitch, are useful tools for small earth-moving jobs such as making silage pits. Some are designed to be fitted with the tractor moving forwards, and others with the tractor in reverse gear. A narrow scoop makes it difficult to work up to a side wall, and necessitates hand-work, or the use of some other implement such as a plough, in making a silage pit with steep side-walls. The types with blades as wide as the tractor do not suffer from this disadvantage, though it may be necessary to use other implements for loosening the soil. Working depth is simply adjusted by altering the length of the top link. Small trailed types of earth scoops are available, but they are necessarily more expensive than those designed for mounting on the tractor's hydraulic lift system, and less easily manœuvred. The bucket is usually arranged to pivot about a horizontal axis, and is emptied by releasing a catch which allows it to roll over.

None of the earth-moving devices mentioned above is really suitable for accurate levelling, or for the final shaping of road surfaces. For this purpose a grader or leveller with a fairly wide and easily adjustable blade is required. The grader blade consists of a curved body with a hard steel cutting-edge. Graders suitable for use on the 3-link system of British tractors are available and can do good work. In one instance the grader is attached directly to the hitch, while in another it is designed for attachment to the earth scoop (Fig. 222). The scoop provides weight to ensure that die blade keeps down to its work, and additional ballast can be easily added when greater penetration is needed. When grading a road, the leading end of the blade is run at the road side, and this is set lower than the trailing end by means of the levelling screw on the hydraulic linkage. This causes earth to be excavated from the side and pushed up to the crown, the same setting being used on the return journey to deal with the other side of the road.

FIG. 222.—GRADER BLADE MAKING A CROWNED ROAD. (TASKERS.)

Tractor Winches.

For work where very heavy pulls are needed, such as trench cutting, pulling trees, pulling a threshing drum out of a boggy field, etc., it is often desirable to be able to increase considerably the maximum pull of a tractor by fitting a winch.

Winches vary considerably according to whether they are a subsidiary or a major feature of the tractor concerned. For example, the winch on a rubber-tyred tractor used for contract threshing work will probably consist of a small, compact winding rope, built neatly on to the rear of the tractor and driven by the power take-off. Some light winches can be used without interfering with the operation of the tractor's hydraulic lift system. On the other hand, some winches for tree-pulling, etc., equipped with anchors and other refinements, almost hide the tractors around which they are built.

FIG. 223.—TRACTOR WINCH. (COOKE.)

One other distinct type of winch is the trailer machine that is driven from the power take-off. The advantage of this is that it can easily be driven by any make of tractor equipped with a power take-off. In many respects, however, it is not as handy as the built-on type.

Whatever type of winch is used, it is essential, for satisfactory service, that the gears should be of high quality, and that the winding drum should be so designed and constructed that the cable winds on reasonably uniformly.

Bracken Cutting Machines.

Where land can be ploughed, bracken eradication is most easily and reliably accomplished by ploughing and thorough cultivations. This method is not, however, applicable to most of the hill land that is in need of improvement, and various types of bracken cutting and bruising machines have been developed for this purpose.

One common type of machine (e.g. the Bissett, Collins, etc.) has a horizontal rotating knife which is driven by gearing from the land wheels. Such machines are unsuitable for dealing with rough land, or with very dense, tough bracken.

FIG. 224.—MEDIUM POWERED TRACKLAYER (DA-
VID BROWN) PULLING LARGE BRACKEN CRUSHER.
(CUTHBERTSON.)

The Henderson-Collins P.T.O.-driven cutter is a modern machine designed for attachment to the 3-link system of a medium-powered tractor. The tractor drives through bevel gearing two swinging horizontal bars which are hinged both at the point of attachment and half-way along. These bars rotate at 260 r.p.m. and have their working height controlled by a dome-shaped stool, liven the toughest bracken is not only cut, but also split to the ground below the cut. The machine normally works a 6 ft. swath, and can operate safely on rough land, wherever the tractor can go. Smaller engine-driven 2-wheeled models operated by a man walking behind are also manufactured.

The Holt bracken bruiser is a cheap attachment designed for fitting to a rear-mounted tractor tool-bar. It consists essentially of three heavy rectangular rollers which have reinforced and hardened bars attached to the corners. The bars roll down and crush the bracken as they rotate, and cover a swath about 7 ft. wide.

A machine suitable for a wide range of rough conditions is the Cuthbertson, which has V-shaped cutters attached to the rims of eight large-diameter independently mounted wheels. The wheels roll down the bracken, and the V-shaped cutters chop it through in several places, including approximately at ground level. In this way the young shoots are cut as well as the old (Fig. 224.)

It is essential for success to deal with the bracken at the height of the season. The work must be done before the bracken is too old, and must be repeated when new growth appears, and also in subsequent years.

Hedge-cutting Machines.

Hedge cutters suitable for a wide range of farm and horticultural uses are now available, and they may be classified according to their capabilities as follows:

Tractor-mounted or trailed machines employing high-speed rotating cutters, capable of cutting wood up to 2-6 in.

Tractor-mounted, trailed and self-propelled machines employing reciprocating cutter-bars with fingers 3 in. apart, capable of cutting growth up to about 1 in.

Horticultural trimmers with short cutter-bars which are held in the hands of the operator, and are capable of dealing with shoots up to about $\frac{1}{4}$ or $\frac{1}{2}$ in. thick. These are dealt with in Chapter Twenty-three.

Rotary Cutters. Rotary cutters include tractor-mounted (e.g. Foster, McConnell, etc.) and P.T.O.-driven trailed machines, both types being capable of dealing with overgrown and neglected hedges, and reducing them to reasonable proportions. These are usually contractors' machines. The revolving blades can be adjusted to cut at various heights and angles, but these machines cannot be set to cut the far side of a hedge. Guards are provided to protect the operators from flying pieces of wood, and it is essential that these be used and that due regard is paid to the somewhat dangerous nature of the cutting mechanism.

On most machines the blade is like that of a circular saw, while that of the Fisher-Humphries has scimitar-shaped knives attached to the periphery of a thick metal plate (Fig. 225. The saw-type blade is needed for very thick work.

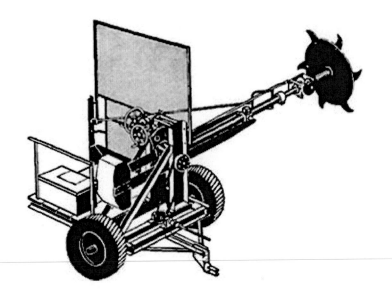

FIG. 225.—P.T.O.-DRIVEN TRAILED ROTARY HEDGE TRIMMER WITH SCIMITAR-SHAPED CUTTING BLADES. (FISHER-HUMPHRIES.)

Reciprocating Machines for Farm Work. Hedge cutting machines should generally be regarded as devices for carrying out an annual trimming, rather than as machines for dealing with the results of years of neglect. Hedges which are regularly trimmed by machine can be steadily improved, and one of the great advantages of a hedge cutting machine is the considerable saving of labour and effort involved in keeping the hedges in good order. The ideal is to have a hedge properly laid and then keep it in order by trimming. Where annual trimming is carried out, the reciprocating type of cutter-bar is most suitable. There are several distinct kinds of reciprocating attachments, viz.:

(a) Machines having a long beam which is mounted on a pivot on the tractor itself. The cutter-bar is at one end of the beam and the engine which provides the drive at the other (e.g. Bomford, McConnell-Gilmour, etc.).

(b) A somewhat similar attachment which is mounted in place of the usual bucket on a hydraulically-operated front loader (e.g. McConnell, Fleming, etc.).

(c) An attachment mounted on a trailer, or on its own two wheels behind the tractor (e.g. Fuller, Blanch, etc.),

(d) An attachment pushed in front of the tractor (e.g. Baker & Hunt).

(e) A machine independent of the tractor and self-propelled (e.g. Baker & Hunt).

FIG. 226.—ENGINE-DRIVEN TRACTOR-MOUNTED HEDGE TRIMMER CUTTING THE TOP. (BOMFORD.)

FIG. 227.—HEDGE TRIMMER (MCCONNELL) MOUNTED ON HYDRAULIC TRACTOR LOADER. (HORNDRAULIC.)

The cutting mechanism of all these machines is similar, and the main differences are in the time taken in getting the machine ready, manœuvrability, reach in various directions, and in details of construction.

An important requirement in this type of cutting mechanism is a "thin" cutter-bar, with only a small distance between the knife itself and the bottom of the knife back. The uncut "stubble" at the top of a hedge will not bend over like grass or corn stubble does after cutting, and the result of using a thick cutter-bar is that each successive cut is at a higher level, making it impossible to keep a level top without having the cutter-bar pitched downwards at a steep angle. Short, snub-nosed fingers are used, and it is an advantage if all nuts and bolts are countersunk.

Points to study in comparing the manœuvrability of different machines are the ease or difficulty of dealing with the corners of fields, and of getting past obstructions such as fencing posts and trees. Machines such as the Bomford and the Fuller, which need two men to operate them, are in many ways easier to operate than machines which have flexibility only in one plane. The ease of making adjustments for different angles and heights of cut should also be studied.

Independent engine drive is advantageous in providing, in conjunction with belt slip, an adequate safety factor. A 2 H.P. engine will stall when an attempt is made to cut growth that is too thick for the machine.

Some machines can be satisfactorily used for cutting the sides and bottoms of ditches. For example, the Bomford machine can be fitted with a short extension cutter-bar which can be set at an angle to the main bar and used for cutting reeds growing under water in ditch bottoms.

Machines which are mounted on hydraulic front loaders are very easily put on and removed, and the outfit is a one-man machine which is easily controlled on straightforward work. When cutting is finished, the manure bucket or buckrake on the front loader can be used for collecting up the trimmings.

Operation of Hedge Trimmers. It is usually advisable first to brush out the hedge bottom with a horizontal cut as near the ground as possible. In making a square hedge the next operation is to cut off the top at the correct level. The near side is next cut with the cutter-bar vertically downwards, and then the far side is trimmed, either by reaching over the hedge or by working from the other side.

FIG. 228.—P.T.O.-DRIVEN TRACTOR-MOUNTED POST-HOLE DIGGER. (ROBOT.)

Some farmers prefer to cut a tapered hedge, and in this case only two cuts are needed for hedges with sides no longer than the cutter-bar. If labour can be spared it often speeds up the work to have an extra man clearing up trimmings as the work proceeds. Alternatively, a buckrake can be used to collect most of the trimmings after cutting is completed. The overall rate of work with a tractor-mounted trimmer usually ranges from about 1 to 5 chains per hour, according to the number of cuts necessary and the nature of the work.

Post-Hole Diggers.

Power-driven post-hole diggers which are now available for attachment to the hydraulic power lift of most British tractors save a great deal of time where many holes are required. The digger consists of an auger which is driven through bevel gears by the tractor power-take-off. Auger points and the leading blades are made replaceable, and different shapes of point are available for dealing with different types of soil. When the borer is fitted it takes only about $\frac{1}{2}$ minute to bore a 6-inch hole 3-feet deep. Holes up to 12 in. diameter can be bored by fitting suitable augers.

REFERENCE

(25) Nicholson, H. H. *Field Drainage.* Cambridge University Press.

CHAPTER SIXTEEN

BARN MACHINERY

"Though the utility of having grain reduced by machinery, before it is employed in the feeding of horses or other animals, has been disputed by some, on the supposition of its not being so intimately combined with the saliva of the animals, on account of less chewing being required, there can hardly be any doubt but that, it will go considerably further when partially broken by such means. . . ."

The Farmer's Companion, vol. i, 2nd ed. R. W. Dickson, 1813.

This chapter contains a brief account of many machines used in or about the farm buildings. The variety and efficiency of such machines have been increased as oil engines, tractors and electric motors have become more commonly used on farms. For example, the grinding of cereals for stock feeding, an operation that used to be most commonly done by contractors using wind or steam power, is now generally done on the farm with the help of a stationary engine, tractor, or electric motor.

Grain Cleaning and Grading Machinery. Modern finishing threshers are capable in most circumstances of producing, with proper adjustment, samples of corn and seed that are sufficiently well cleaned and graded for immediate sale, or for use as stock feed; but in many instances the sample of grain delivered from the threshers can be much improved by a further dressing in special cleaning and grading machines. When corn is required for seed, the advantages of sowing a clean, uniform sample are obvious, and a further dressing after threshing is usually essential. The removal of weed seeds reduces cultivation costs and gives increased yields, while drills function much better when clean and properly graded seed is used.

The price received for grain or other seed that is for sale, particularly in the case of crops such as malting barley, may often be substantially increased by a relatively insignificant expenditure on cleaning and grading.

When combine harvesters are employed, samples of grain frequently contain green leaves and thistle-heads, as well as chaff, pieces of straw, a good deal of dust, and often many insects. In such cases it is usually essential to subject the grain to a cleaning process before it is stored. If the grain needs drying it is often advisable to pass it through a pre-cleaner before drying and through another cleaning and grading machine afterwards. The first operation removes those coarse impurities which are likely to interfere with the drying process. After drying, impurities are much more easily removed, and a final dressing can produce a clean and well-graded sample. The term "pre-cleaner" is correctly applied to a machine which carries out a preliminary operation, rather than to any particular type of cleaning machine.

The objects of using a cleaning machine are to separate from the grain all chaff, straw, weed seeds, broken and inferior seeds, dust and other rubbish; and many machines also grade the grain according to size into the various samples required.

Seed cleaners range from very simple riddling or winnowing machines to complex dressers which treat the grain in many ways. The processes commonly employed on farm machines are as follows:

(a) Screening over a sieve with large holes to remove stalks, stones and other large roughage.

(b) Separation by air blast, either with a simple winnowing device or by aspiration.

(c) Use of nests of sieves or rotary screens to separate grain according to its width.

(d) Use of pockets (e.g. indented cylinders) to separate grain according to length.

Seed merchants use special methods for grass and clover seed, and for the removal of particular types of impurities from all kinds of crops.

The simplest form of cleaner may carry out only one of the above processes. E.g. a simple pre-cleaner may consist of only a single reciprocating sieve, or may be a simple aspirating cleaner. The most complicated cleaners may employ all the processes detailed above.

Sieve Gleaners. Sieve cleaners which achieve separation without using an air blast are uncommon, but are sometimes installed as pre-cleaners in conjunction with tower-type grain driers. Their chief disadvantage is a tendency to blocking of the sieve or sieves—a fault which necessitates frequent inspection and attention when dirty samples of grain are being deal with. A steep sieve angle helps to prevent such troubles.

Winnowers. In the simple winnowing machine the blast of air from a fan which rotates rapidly inside a suitable casing is directed on to the grain as the latter falls in a thin stream from a hopper at the top of the machine, on to a reciprocating sieve. The blast blows chaff and other light material through the open end of the machine, while small corn, weed seeds and dirt pass through the lower sieve. The dressing may be controlled by adjustment of the sieves, the strength of the air blast, and the rate at which the grain passes through the machine.

Winnowers in which corn was thrown into the air past a blast produced by a crude fan were used from very early times. Machines similar to the modern simple winnower were first produced by James Meikle soon after 1710, as a result of a visit by him to Holland for the purpose of studying "the perfect art of sheeling barley". The design of the fan was crude in the early machines, but in the nineteenth century more efficient fans, with curved blades, and mounted in a suitable housing, began to be employed.

The chief disadvantage of the winnower, apart from its limitations, is the fact that unless special arrangements are made for dealing with it, the dust blown out may be an intolerable nuisance. This disadvantage applies not only to the simplest winnowers, but also to many more complex dressing machines which employ the winnowing process. The remedy, which must be borne in mind when planning any cleaning installation that employs such machines, is to arrange to blow the cleanings into a large settling box which has a wire-covered outlet of adequate size on the outside of the building, and is arranged so that it is easy to clean out. Unless this requirement is borne in mind at the outset, expensive modifications may become necessary.

Combined Winnowing And Grading Machines. Machines which employ the winnowing process often have a nest of grading sieves suitable for dealing with almost any kind of seed. Some deliver the cleaned grain to bagging-off spouts by means of a built-in bucket elevator.

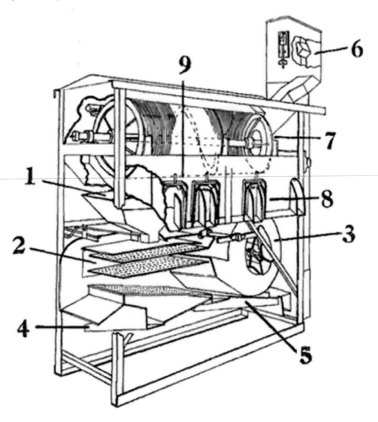

FIG. 229.—COMBINED WINNOWER AND GRADER WITH ROTARY SCREEN. (PENNEY AND PORTER.)

1, Feed hopper; 2, sieves; 3, fan; 4, chute for sand and small weed seeds; 5, grain delivery chute leading to bucket elevator 6; 7, rotary screen; 8, tail corn sacking spout; 9, head corn sacking spouts.

One type of machine that is in common use employs a winnowing blast and a pair of sieves for the primary cleaning, and then elevates the grain and passes it through a rotary wire screen for the final width grading. One of the advantages of this combination of components is that use of the built-in elevator provides for bagging off with the machine standing at ground level, whereas many other complex machines need to be raised about 6 ft. above the floor on which sacking-off is carried out. The rotary screen is similar to those employed in finishing threshers. Combined Aspirating And Sieving Machines. Many of the complex cleaning and grading machines now used on farms, and also by seedsmen, employ a combination of sieves and aspiration. The essential difference between aspiration and winnowing is that winnowing employs a more or less horizontal blast, whereas an aspirating machine employs a controlled vertical air blast which operates in a confined space termed an aspirating leg. This permits a finer "weighing" of the impurities in the air stream, and is more selective than winnowing. It also has the advantage that the escaping dust-laden air can be conveniently delivered by suitable trunking either directly out of the building, or to filter socks or a collecting cyclone. The sieves in such complex machines are automatically cleaned by brushes which move to and fro beneath them.

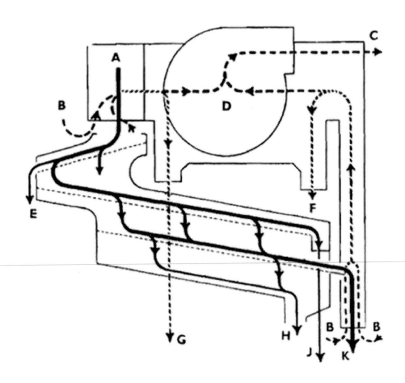

FIG. 230.—*FLOW DIAGRAM OF SIEVING GRAIN
CLEANER WITH DOUBLE ASPIRATION. (TURNER.)*

A, feed; **B**, air enters; **C**, dust and air exhausted to
atmosphere or collector; **D**, fan draws air through grain at
feed and delivery ends; **E**, rubble; **F**, liftings from delivery
end; **G**, liftings from feed end; **H**, dirt and small seeds; **J**,
large seed, straws, etc.; **K**, clean grain.

Indented Cylinders. For accurate grading, indented
cylinders are finding increasing use. These are long, slightly

inclined cylinders with indentations punched, cast or drilled on the inner surface. They are driven by gearing or belting, and as they revolve, grains which fit into the indentations are carried over and dropped into a chute inside the cylinder. Cylinders with various sizes of indentation will produce very accurately

FIG. 231.—COMBINED GRAIN CLEANING AND GRADING MACHINE WITH SIEVES, ASPIRATING LEGS AND INDENTED CYLINDER. (KAYBEE.)

graded samples. For example, a sample of wheat, after being well cleaned in a winnower or an aspirating and sieving machine, may be passed through a cylinder with small indentations to pick out the small and broken grains and small weed seeds (e.g. "goose-grass"). The grain may then be passed through a cylinder which picks out all the wheat except the very largest grains, and leaves the seed free from large weed seeds (e.g. the wild onion). This type of machine is extensively used in Germany and Sweden. Its main disadvantages are that the rate of grading grain is slow, and the machine is an expensive one for individual farm use.

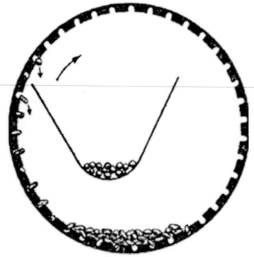

FIG. 232.—*DIAGRAM ILLUSTRATING PRINCIPLE OF INDENTED CYLINDER SEED CLEANER*

Pneumatic Separators. Some simple cleaning and grading machines achieve separation by aspiration alone (Fig. 233). The air blast in the aspirating leg is strong enough to "weigh" the grain in the air stream, the heavier fractions falling to the bottom, light fractions being taken off at an intermediate position, and the lightest materials being blown out with the air stream. The factors governing separation are a combination of weight and air resistance. Such machines do not block, so are suitable as pre-cleaners, but cannot be expected to achieve such fine cleaning and grading as the more complex and expensive machines which employ a number of different separation processes.

Operation of Seed Cleaners. The efficient operation of complicated cleaning and grading machines requires skill, for grain Samples are so variable that a setting of the machine that is suitable for a particular sample is unlikely to be correct for the next sample dealt with. If the principles of operation are understood, however, the machine can be correctly adjusted with practice.

The air blast can be controlled by adjustment of the fan blinds and sometimes by varying the speed of rotation. In many machines the slope, and the length and frequency of vibration of the sieves can be varied; and the sieves themselves are interchangeable, with a series of various sizes and shapes of mesh for different purposes.

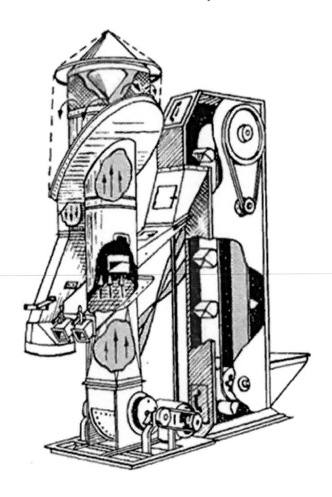

*FIG. 233.—DIAGRAM SHOWING MODE OF OPERA-
TION OF A SIMPLE ASPIRATING CLEANER, (SCIEN-
TAIRE.)*

Where it is necessary to clean and grade seed very
thoroughly, advantage must be taken of many kinds of
differences between the desired seed and the contaminant. In

addition to the devices already mentioned, seed merchants often employ disc separators, which perform a function similar to that achieved by the indented cylinders already mentioned, velvet band "dossors" for cleaning seeds such as clovers, and spiral separators for vetches. Flat seeds such as linseed may be separated by using special corrugated riddles with slots at the bottoms of the corrugations.

Seed Dusting Machines (Chemical Dressers).

The practice of dressing seed corn with an organic mercurial dust is a sound insurance against a risk of severe attacks of bunt of wheat, covered smut of barley and the leaf stripe diseases of barley and oats. The machine employed for this purpose usually consists essentially of a barrel in which the grain and the dusting powder are very intimately mixed. The aim is to cover each grain with a thin film of powder. For small quantities of corn an old disused butter churn can be converted into an effective dresser. It should be filled only two-thirds full of grain, and rotated with the right amount of powder for two or three minutes. It is not sufficient to turn the churn only about half a dozen times. In the best machines operation is continuous, the dusting powder being gradually measured out at a constant rate as the grain flows through the barrel. With large, power-driven machines, fifty to eighty bushels per hour can be put through.

FIG. 234.—CHEMICAL SEED DRESSER. (BARCLAY, ROSS AND HUTCHISON.)

Care needs to be exercised in carrying out this work. It is important that the machine be well designed, so that the dusting powder is efficiently used and so that the operation of the machine may not be dangerous to human life and health owing to the escape of poisonous powder from the machine.

FIG. 235.—LARGE CHAFF CUTTER WITH SIFTING AND BAGGING ATTACHMENT. (COVER REMOVED TO SHOW FLYWHEEL.)

Chaff Cutters.

The advantages and disadvantages of feeding whole or chopped hay and straw aroused much controversy in this country from about 1925-30. One expert on the feeding of livestock declared the folly of using the chaff cutter in the following terms: "The chaff cutter has sent more cows away in the knacker cart than any other invention of man." A result of the controversy was that the use of chaff cutters declined. Nobody who has studied the question can believe that the chaffing of bad hay can make it into a good foodstuff, and there is sound common sense in the argument that when hay is fed whole, animals are better able to select the good and reject the bad material. But this is no argument against the chaffing of good hay or oat straw, and there are circumstances where the use of the chaff cutter is desirable. When working horses live largely on good oat straw it is advantageous to chaff most of it.

Cato (234-149 B.C.), one of the earliest agricultural writers, mentions chaff as the food for oxen; but apart from a trough and a knife attached to the end of a lever, no machines for cutting it appeared until the end of the eighteenth century. The development of the chaff cutter was then fairly rapid, and by the middle of the nineteenth century cutters employing the essential principles of modern machines were in use.

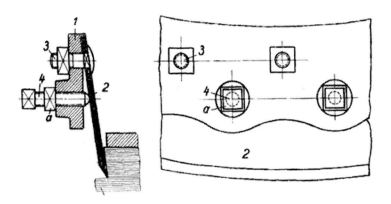

FIG. 236.—*DIAGRAM SHOWING METHOD OF AD-JUSTING KNIFE OF CHAFF CUTTER*

1, flywheel; 2, knife; 3, fixed screws; 4, set-screws.

On modern power-driven machines the hay or straw is fed on to a conveyor which runs along the bottom of a feed trough. At the throat of the conveyor the straw is gripped between two toothed rollers, and these pass it forward to a shear plate, where it is chopped into short lengths by rotating knives. The knives are mounted on a heavy flywheel, carried on the main shaft. It is important to "set" the knives carefully; the backs of the knives are bolted to the flywheel about $\frac{1}{2}$ in. from the shear plate, and the cutting edges should be given such a pitch (by the adjustment of set screws) that they nearly touch the edge of the shear plate along their whole lengths. The drive for the feed rollers and conveyor is taken through bevel gears, with two dog-clutches arranged to provide forward, neutral and reverse gears, to a countershaft running at right angles to the main shaft. The countershaft and the shaft which drives the conveyor have a set of change-speed pinions usually designed to provide three different conveyor speeds. Adjustment of the length of cut is made by varying the speed of the feed relatively to that of the knives, the high gear producing chaff about $\frac{3}{4}$ to $1\frac{1}{2}$ in. long, the medium $\frac{1}{2}$ to $\frac{3}{4}$ in., and the slow $\frac{1}{4}$ to $\frac{3}{8}$ in. The coarse cut is usually employed for bullocks, the medium for horses, and the short for sheep.

The drive to the upper feed roller is usually through a universal joint in order to enable this roller to "float" so as to accommodate varying quantities of fodder. A lever attached to the floating upper roller automatically disengages the dog-clutch or puts the machine into reverse gear if the throat becomes choked. Whether such a safety device is fitted or not, inexperienced men should be warned of the danger of putting their hands too near the throat, or of attaching the heavy flywheel to the mainshaft insecurely.

On the larger machines, self-feeding, sifting and bagging attachments make it possible to produce a clean chaff at fairly high speed. The approximate power required and the output of various sizes of machine are given below.

TABLE V.—APPROXIMATE POWER REQUIREMENTS AND OUTPUTS OF CHAFF CUTTERS.

Width of throat (in.)	Number of knives on flywheel	Approx. quantity of $\frac{1}{2}$-in. chaff cut (cwt. per hour)	Power required (b.h.p.)
9	2	4	$1\frac{1}{2}$
11	3	12	4
13	3	20	6
15	5	40	8

Root Cutting Machines.

There is little advantage in feeding chopped roots in preference to whole ones, except when it is desired to feed large quantities, or where the animals have bad teeth. This view has led to the abandoning of many root chopping machines. Slicing of the roots is sometimes desirable for horses or for old ewes with poor teeth, but for dairy cows, fattening bullocks and young sheep it is seldom really essential.

The root cutter usually has a V-shaped hopper, one side of which is formed by the cutting disc. The cutter is a steel disc in which are a number of openings, surrounded by sharp-edged lips that face in the direction of rotation.

Rotation of the disc forces the projecting lips against the roots, and slices are cut off and pass through the openings to the back of the disc. The size and shape of the slices depend on the shape of the projections on the disc, and are largely a matter of personal choice. Some machines are fitted with a cleaner consisting of a rotating wire cage with spiral cleaning bars. The whole roots are fed into the upper end of the cage, and the dirt is scraped off them as they pass slowly down to the cutter.

FIG. 237.—DISC-TYPE ROOT CUTTER AND CLEAN-ING CAGE, (BAMEORDS.)

Another type of root cutter has a cutting cylinder with knives mounted spirally on the periphery. The cylinder rotates at the bottom of the feed hopper, and cuts the roots against a stationary shear-plate (Fig. 238).

Grinding, Kibbling and Crushing Machines.

The practice of grinding food on the farm is becoming more general. This is due to a greater interest in the correct feeding of livestock, and to the greater availability of suitable mechanical power for the work. The grinding of grain has been practised since very early times, when a device resembling a pestle and mortar was employed in the production of meal for human consumption. The first mills were modifications of this device, in which grain was fed through an opening in a disc-shaped stone which was caused to rotate upon another. The gradual development of this type of mill during thousands of years has led to the evolution of the modern buhr-stone mill.

FIG. 238.—ELECTRIC MOTOR-DRIVEN CYLINDER-TYPE ROOT CUTTER WITH CLEANING CAGE. (BEN-TALL.)

The objects of grinding grain for stock are to increase the digestibility or palatability, and to facilitate mixing with other constituents of the ration. In this country little attention has yet been paid to the very debatable question of how fine various foodstuffs should be ground for the various classes of stock. In the United States, a general result of intensive research on this subject has been that for most classes of stock in that country grain is now being ground to a coarser grade than it was previously. For example, it has been shown that maize and wheat for laying hens should be kibbled as coarsely as possible, though for chicks a finer grinding of maize is recommended. Oats, owing to the fibre in the husk, should be ground fine for poultry. Coarse grinding is generally recommended for beef cattle, sheep and horses, while a rather finer grade is desirable for the high-producing dairy cow. In regard to pigs, the requirements are not yet clear. It seems likely that a fairly fine grinding is desirable for young and fattening pigs, but it should be remembered that the power requirement for very fine grinding is much greater than that for a medium grade.

Types of Mill.

Three main types of mill are now used in the preparation of food on farms, viz. buhr mills (including plate mills), hammer mills and roller-crushing mills.

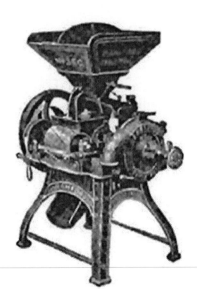

FIG. 239.—COMBINED CRUSHING AND GRINDING MILL. (BAMFORDS.)

The BUHR-STONE mill is so termed because of the grooves (buhrs) which are cut on the grinding faces of the two disc-shaped stones. As one stone revolves upon the other, grain fed in at the centre passes towards the periphery, being gradually ground in the process. These mills can do excellent work, producing a fine meal without much heating; but they are somewhat cumbrous, and require a powerful engine to drive them. They have been largely superseded on farms by CHILLED IRON PLATE mills. The plate mill is a modification of the buhr-stone type. The plates are cast with corrugations which force the material outwards as in the buhr-stone type. The plates are made of a suitable iron to produce, by chilling, the maximum hardness with adequate toughness. They are easily interchangeable, and are usually made with corrugations on both sides so as to be reversible.

The fineness of grinding is controlled by the nature of the corrugations and by the clearance between the two plates, which are mounted on a horizontal spindle. The corn is fed from the hopper over a small oscillating screen which removes any large foreign bodies, and thence, by way of an auger-like device, to the centre of the inner plate. The outer plate is usually feather-keyed to the rotating shaft, and may be set at a variable distance from the fixed plate by means of a hand wheel. A spiral spring between the plates prevents them from touching when no grain is passing through, provided they are properly adjusted. A wooden break-pin, a strong spring, or a quick-release device is provided as a safety measure in case hard objects are fed into the mill. This type of mill performs coarse grinding or kibbling of wheat, barley and maize very efficiently, and the addition of a pair of crushing rollers makes up a combined unit that is cheap, reliable, and capable of fulfilling most of the food-grinding requirements of many small farms.

FIG. 240.—ROLLER CRUSHING MILL WITH SMALL PLATE GRINDER. (BAMFORDS.)

The CRUSNRNGMILL consists essentially of two cylindrical rollers between which the grain is rolled. Grain is generally delivered to the crushing rollers by a small fluted feed roller, through a variable slide in the bottom of the hopper. The large roller is keyed to the main driving shaft (which also carries the grinding plates in a combination mill) and is carried in fixed anti-friction bearings. The smaller roller, which is usually ungeared, is carried in sliding bearings, and can be pressed against the driven roller, through the medium of strong spiral springs, by the operation of a hand wheel. The spring mounting provides a safety device in case hard foreign bodies are introduced between the rollers. Oats are commonly crushed for horses, in order that the digestive juices may not be prevented, by the unbroken husk, from acting upon the interior of the grain. With a combined crushing and grinding mill it is possible to perform both operations at once if sufficient power is available. The approximate power requirements and capacities of various sizes of plate and crushing mills are given below.

Very small electrically operated crushers designed for automatic operation are now available. The mill may be mounted on top of a bin into which the crushed corn falls (Fig. 241).

FIG. 241.—SMALL ELECTRTIC MOTOR-DRIVEN OAT CRUSHER MOUNTED ON STORAGE BIN. (BEN-TALL.)

Hammer Mills. The hammer mill, though a comparative newcomer to British farming, is now very widely used. The mill consists of a number of steel hammers which rotate at high speed on a shaft or rotor set in a strong housing. As the material is fed from the feed hopper into the mill, the hammers strike it with great force and rapidly pulverize it. At a point close to the periphery of the hammers is a screen, and as the material is reduced in size it passes through this screen. It is then usually elevated by means of a fan to a collecting cyclone, where it can be bagged off. The fineness of grinding is regulated by the use of screens of different mesh. Hammer mills are simple in construction and cost little for the replacement of parts; but the initial cost of some types is high. Large capacity mills need an engine of about 20 b.h.p. to drive them. They are designed for use with the farm tractor. Outputs of 10-20 cwt. per hour with medium grinding can be obtained.

There is, however, much to be said in favour of smaller hammer mills designed for electrical operation. One popular type is an 8-in. diameter mill that requires only 3 b.h.p. With a suitable layout of feed bins and meal bins it is possible with such a machine to secure fully automatic operation. A time switch may be set to start the mill at a predetermined time at night, when a cheap rate for power is available; and grinding will continue until the bin is empty, when an automatic cut-out switches off the motor. The corn to be ground falls by gravity from a bin above the mill, while the meal is blown into a storage bin. Such an outfit has been found capable of dealing with 3-4 tons of mixed grain per week when used in the way outlined above.

TABLE VI.—APPROXIMATE POWER REQUIREMENTS AND CAPACITIES OF PLATE MILLS.

Diameter of plates (in.)	Speed (r.p.m.)	Approximate power required (b.h.p.)	Approximate Capacity (bushels per hour)	
			Medium grinding	Kibbling
6	600	1- 2	3- 4	10- 15
9	550	3- 4	5- 8	20- 25
12	500	5- 6	12-16	35- 45
15	500	8-10	15-20	50- 70
18	500	10-12	20-30	75-100

TABLE VII.—APPROXIMATE POWER REQUIREMENTS AND CAPACITIES OF ROLLER CRUSHING MILLS.

Diameter of rollers (in.)	Width of rollers (in.)	Speed (r.p.m.)	Appropriate power required (b.h.p.)	Approximate Capacity (bushels of oats per hour)
16 and 10	3	200	$1\frac{1}{2}$- 2	8-15
21 and 12	4	200	2 - 4	15-30
25 and 12	5	250	4 - 6	25-40
25 and 12	8	300	6 -10	45-75

Fig. 243 illustrates the method of operation and automatic control employed on a typical small mill. So long as grain flows from the hopper above the mill, the flap of the automatic switch is depressed, and the mill continues to

599

run. As soon as the grain hopper is empty, a spring causes the flap to move upwards, and this switches off the mill. In its passage from the hopper to the grinding chamber, the grain is sucked over a weir, heavy impurities which might damage the mill being retained at the bottom of the hopper.

FIG. 242.—INTERIOR VIEW OF SWINGING-HAM-MER TYPE HAMMER MILL. (MASSEY-HARRIS.)

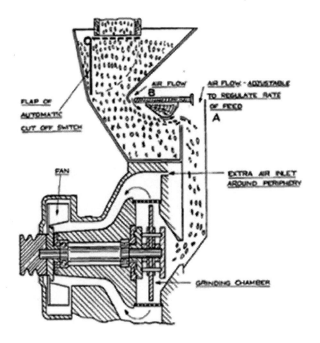

FIG. 243.—*DIAGRAM SHOWING FLOW OF GRAIN
AND MEAL THROUGH 3-H.P. AUTOMATIC HAMMER
MILL. (CHRISTY AND NORRIS)*

When used at full capacity on coarse grinding, such
mills can deal with 5-6 tons of feeding stuffs weekly.

Meal Bins. The 3-h.p. mill employs only a small fan,
and a satisfactory method of separating the meal from the
air is to blow it into bins fitted with filter-cloth tops. The
filter cloth is fitted loose, so that it balloons upwards when
the mill operates. When blowing stops, the filter cloth falls
back and jolts off much of the adhering meal. It is necessary

occasionally to beat the distended surface of the filter with a carpet beater or stick to assist the cleaning process.

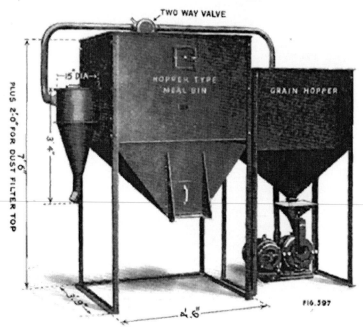

FIG. 244.—ARRANGEMENT OF 3-H.P. AUTOMATIC MILL WITH GRAIN HOPPER, MEAL BIN AND CYCLONE ON A SINGLE FLOOR. (CHRISTY AND NORRIS.)

When larger mills are employed, use of a cyclone is essential, but such mills can be made semi-automatic in operation by delivering the meal from the cyclone to storage bins. It is almost impossible to construct simple meal storage bins which do not give some trouble with sacking-off. Even if a very steep angle to the sacking-off spout is employed the meal will bridge when the bin is full. The most satisfactory form of construction is to have parallel sides and a fairly flat bottom leading to a large vertical bagging-off slide which permits use of a small rake to pull the meal forward when this is necessary. Bins should not be more than about 3 ft. wide and 4 ft. 6 in. from front to back. To prevent meal lodging hard against the slide it is advisable to construct a deflecting board above it. The top of the deflector should be about 6 in. above the top of the slide, and the board should slope inwards at an angle of 45°. It is advisable to provide a large inspection panel near the top of the bin.

FIG. 245.—TRACTOR-MOUNTED HAMMER MILL DRIVEN FROM BELT PULLEY. (FERGUSON.)

Choice of a Mill. For medium or coarse grinding or cracking of maize, wheat or barley, the plate mill is satisfactory; but it is less efficient for fine grinding. For the fine grinding of these grains, and especially of oats, the hammer mill is most efficient. On farms where a large head of stock is kept it may be economical to use a plate mill, a hammer mill and a crushing mill to meet the variable requirements of horses, sheep, beef and dairy cattle, poultry and pigs. On smaller farms the combined plate and crushing mill may be the most suitable equipment.

Much depends on the power unit to be employed. Where a tractor is used it is generally advisable to employ a mill of high capacity, which will get the work done quickly and provides the operator with a full-time job. There is a wide choice of hammer mills suitable for driving by tractors of 20-30 b.h.p. Where electricity is employed it is often essential to keep power requirements to a minimum, and a small hammer mill arranged for automatic operation may be best in every respect. A small electrically-operated crusher may also be needed where oat crushing is a regular need.

Cake Breakers.

Oilcake is usually delivered to the farm in the form of long flat cakes, which must be broken up before being fed to stock. Breaking the cake with a hammer is laborious and tedious, and a simple cake breaker soon pays for its cost in the saving of man-labour. The breaker consists of one or more pairs of spiked rollers mounted in a strong frame. The cake is placed in a hopper situated above the rollers, and as it passes down between them it is broken into pieces, the size of which may be varied to suit different animals by adjustment of the rollers. The broken cake is usually delivered over a screen which separates out the smallest pieces and the fine meal produced. The drive to the crushing rollers is geared down on both hand and power machines. With a suitable power-driven machine, a $1\frac{1}{2}$-to 2-h.p. engine will crush 30 to 40 cwt. in an hour.

Food-mixing Machines.

The mixing of foodstuffs to form a uniform ration is a serious problem on large stock farms; and where there is a sufficiently large head of stock the use of a mechanical mixer has much to commend it. The mixer usually consists of a steel cylinder tapering to a conical hopper at the bottom. The ingredients to be mixed are gradually fed in, and the mixing is performed by a large vertical worm which revolves continuously, carrying material from the bottom of the hopper and sprinkling it over the surface. The worm is usually driven at the top by means of bevel gearing and a short countershaft, fitted with fast-and-loose pulleys and a striking gear. An elevator that enables filling and bagging to be carried out side by side is generally desirable, but machines which automatically fill themselves from the bottom are now available, and these are very suitable where there is little head-room. Power required for a 10-cwt. mixer of this type is 3 to 5 h.p. The mixer must be run while it is being filled and for a few minutes after filling is complete. With all the materials handy one man can complete the mixing and bagging of $\frac{1}{2}$ ton of rations in 15 to 20 minutes with the help of a 10-cwt.-capacity machine.

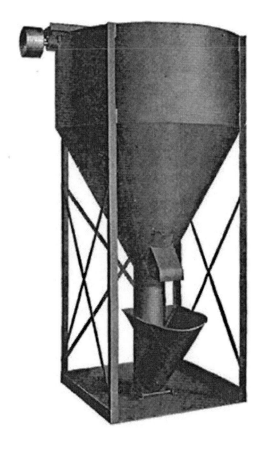

FIG. 246.—A FOOD-MIXING MACHINE WITH
WHICH MEAL IS BOTH FED IN AND BAGGED OFF
AT THE BOTTOM. (HOWES.)

Vertical mixers are generally used for dry materials, but are unsuitable for wet. Where wet mixtures have to be prepared a horizontal mixer may be used. These usually have a U-shaped trough with a mechanical agitator running along the bottom. Horizontal mixers are filled from above and discharged through a valve in the bottom of the trough. Capacities of wet-mash mixers range from 1 to 30 cwt. Power required is greater than that for dry mixers, being approximately 1 h.p. per cwt. capacity in the small sizes.

The installation of a mixer involves a rather heavy capital outlay, but this may well be justified by the increased efficiency of the mixing alone, while the saving of labour is considerable. The following table shows the approximate capacities and power requirements of various sizes of mixer of the usual conical shape, fed from above. Bottom-fed mixers operate at 800-900 r.p.m. and need more power for a given capacity.

TABLE VIII.—APPROXIMATE POWER REQUIRE-MENTS AND CAPACITIES OF LOW-SPEED TOP-FEED FOOD-MIXING MACHINES.

Overall height (ft.)	Overall diameter (ft.)	Capacity (cwt.)	Speed (r.p.m.)	Power required* (b.h.p.)
7	2·5	5	140 — 180	1 -2
9	3·0	10	140 — 180	1½-3
11	4·0	20	140 — 180	2 -4
13	4·5	30	140 — 180	3 -6
15	5·0	40	140 — 180	4 -7

* Machines which operate at higher speeds have higher power requirements.

The Lay-out of Barn Machinery.

The lay-out of barn machinery should be carefully planned with the object of avoiding all unnecessary labour and keeping operating costs at a minimum. An installation involving several machines requires planning, both in the selection of the machines and in their arrangement, if the money spent is to be a profitable investment.

The power for driving barn machinery may be supplied by fixed engines, electric motors or tractors. Where a fixed engine is used it should be accommodated in an engine-house shut off from the barn. If several machines are to be driven the engine must drive a countershaft from which drives are taken off, through fast and loose pulleys, to the various machines. The shafting should generally run at about 250 to 350 r.p.m., and should be in such a position that drives may be taken to machines situated on both the first and ground floors in a two-storey building.

609

Where electricity is available, individual electric motor drive to each machine is almost invariably preferable. This enables machines to be sited in the most convenient places, eliminates the maintenance of shafting, permits use of efficient V-belt drives, and is normally little more expensive than a countershaft system.

As a general rule, heavy mills should be mounted on the ground floor. They need a firm foundation, and may cause damage to the building if this is not provided.

Two-storied buildings are no longer a necessity, but where available they can be put to good use. The first floor can be conveniently used for storing corn in bins, whence it may be fed by gravity to the mill. It is often convenient to have several flat-bottomed holding bins and one or two hopper-bottomed bins to which the corn is transferred for grinding. It is usually desirable to provide some kind of elevator or power-driven hoist for getting grain up to the first floor. Such devices have a value in making work congenial as well as in saving labour.

When a milling and mixing plant is to be arranged in a building with only a single floor, it is necessary to construct an elevated grain hopper which can be filled by a suitable conveyor, and to employ a mixer of the bottom-feed type. A hammer mill can often conveniently be set in a shallow pit in order to reduce the height of the grain hopper.

Sheep-shearing Machines.

Sheep-shearing machines have now displaced hand shears on a great number of sheep farms all over the world. Their advantages over hand shears lie in the saving of labour and in the fact that they enable comparatively unskilled men to do work equal to or better than that of the best hand labour.

The operating principle of the shearer is the same as that of the cutter-bar of a grass mower. The cutter, which corresponds to the mower knife, reciprocates between the comb and the forks of the shear head. It is driven at high speed (up to 3,000 strokes per minute) by a flexible shaft which may either be driven by hand through suitable gearing, or better, by a $\frac{1}{2}$-h.p. electric motor or small engine. The knives move over several closely-set teeth of the comb in order to produce short and even cutting. For good cutting the combs and cutters must both be kept sharp; grinding may be effected by means of a revolving disc covered with emery cloth.

REFERENCE

(26) *The Small Automatic Hammer Mill.* British Electrical Development Association. London.

CHAPTER SEVENTEEN

PUMPS AND SPRAYING MACHINERY

The raising of water is an essential part of the routine on most British farms, and agriculture itself in some districts, such as the Fens, is wholly dependent upon pumping. An ample supply of pure water, properly distributed, is necessary for technical efficiency on stock farms, permitting maximum productivity from controlled intensive grazing of leys, and saving labour about the buildings. A modern dairy farm must have a plentiful water supply. Milk yields are greatly influenced by free access to drinking water, while shortage of a good water supply is one of the chief single causes of unsatisfactory milk quality. A piped water supply also greatly facilitates all other kinds of livestock enterprises. Few farms are able to draw upon a public water supply, and even where such a supply is available it may be economic to instal a pumping plant if a suitable source of water is available.

During recent years an extensive range of small and effective pumping sets at reasonable prices has been developed, and these help to make a supply of power-pumped water available to small farms as well as large. The electric motor is an ideal power unit for driving small modern pumping sets, especially as it is easily fitted with an automatic control. Where supply conditions are favourable, a small electric set may be installed cheaply, and the running costs of such sets are low compared with the cost of a public supply. For example, one investigation showed that the power used by five typical modern sets with total suction and delivery lifts ranging from 28 to 84 ft. varied from 0.66 to 1.89 electrical units per 1,000 gallons of water delivered. It is rarely that a public supply outside towns costs less than is. 6d. per 1,000 gallons, and the actual cost of pumping with the small sets mentioned above is generally much less.

The pumping outfit installed must be suited to the form of supply. The ideal supply is from a true artesian well or from a natural reservoir at a sufficient height to provide a service without pumping. Such supplies are rarely available, but open springs, lakes, streams and wells of various kinds from which water may be pumped are more common. There is, of course, a danger of pollution in many such sources, but the water may be suitable for stock if not for human consumption. Wells and springs should always be thoroughly tested before embarking on any extensive water supply scheme.

The water requirements of farms vary greatly. In arriving at an estimate of the need, the following consumption figures may be used as a rough guide: dairy cows in milk, 15 gallons per day; bullocks, 10; horses at work, 10; pigs, 3; sheep, 1; poultry on range, $\frac{1}{20}$; poultry indoors, $\frac{1}{8}$ Water used for milk cooling may generally be re-used for washing-down, or for watering stock.

Domestic requirements average 12-15 gallons per head per day in unsewered houses, and 20-25 gallons in fully sewered.

Glasshouses need up to 6,000 gallons per acre per day, and cold frames about half this quantity.

It is usually necessary to provide a storage reservoir near to where the water is chiefly required. If the topography is suitable, an elevated tank may be used. This should be at least 20-30 ft. above the taps, milk coolers, etc., at the farm.

Where high pressures are needed (e.g. 40 lb. per sq. in. for a power jet or for irrigation), use of a gravity tank may be impracticable, since it would need to be go ft. or more above the point of use. (A pressure of 1 lb. per sq. in. is equivalent to a static head of 2.3 ft.) The advantage of employing a large-capacity gravity tank is that it provides an adequate reservoir against peak demands, which may exceed the capacity of the pump. Thus, a cheap pump with a capacity of only 200 gallons per hour may be adequate for dealing with a peak demand of 400 gallons per hour if it delivers into a 1,000-gallon reservoir.

There are, however, many situations where use of a gravity tank is impracticable. Where electricity is employed for pumping, a good alternative is a pressure tank with an automatic control switch. This has the great advantage that the pressure tank can be situated anywhere convenient near the pump. The pump delivers the water through a non-return valve into the closed pressure tank, which is automatically maintained about three-quarters full of air. When a tap is turned on anywhere in the delivery system, water is drawn from the tank, and after the pressure has fallen to a certain level, the pump is automatically switched on. In this case it is uneconomic to have a very large tank, and the output of the pump must be equal to the peak demand. The whole outfit can work very satisfactorily, with a minimum of maintenance.

The type of pump required depends mainly upon the "suction lift" and the "delivery head". Almost any kind of pump may be used for raising water short distances (up to about 20 ft.); in such cases the simplest and cheapest types of pump may be chosen. But where water has to be raised from a deep well, one of the many types of pump specially designed for this service must be chosen. Deep wells are those in which either the pump must be placed below ground-level in order to be within suction distance of the water, or a special type of deep-well pump must be employed.

The Lift pump is one of the simplest and commonest types, and is widely used as a hand machine for raising water from shallow wells. It should be observed that the supply of water to the suction side of this and of all other pumps situated above the water supply depends upon the fact that the pressure exerted by the atmosphere (i.e. the barometric pressure) is capable of supporting a vertical column of water about 30 to 34 ft. high in a pipe exhausted of air. In the lift pump a suction pipe leads to the source of supply, which should not, in practice, be more than 20 to 25 ft. below the delivery spout. The pump cylinder has a one-way "foot valve" situated at its junction with the suction pipe. Operation of the pump causes a plunger to move up and down in the cylinder, a water-tight fit on the up-stroke being secured by "cup leathers" which press against the smooth cylinder wall. The plunger has, at its centre, a transfer port and valve through which the water can pass. When the plunger moves downwards the plunger valve opens, the foot valve closes, and water passes through the plunger to the upper side. As the plunger rises, this water is pushed out through the spout, and the partial vacuum created in the barrel below the plunger causes the foot valve to open and water to rise into the barrel from the supply, owing to the excess of the pressure of the atmosphere over that in the barrel. If the suction lift is greater than 20 to 25 ft., the action of the pump is unsatisfactory because the water, owing to its inertia, does not follow the plunger quickly on the up-stroke.

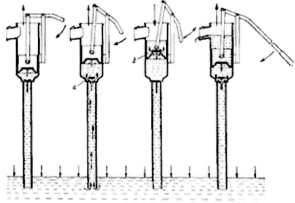

FIG. 247.—*DIAGRAM ILLUSTRATING ACTION OF LIFT PUMP A, FOOT VALVE; B, DELIVERY VALVE.*

Modern pumps for operation by internal-combustion engines or electric motors include the following three main types which are in general use on farms:

(*a*) Plunger pumps, suitable for deep wells.

(*b*) Piston pumps, suitable for shallow wells,

(*c*) Centrifugal pumps.

Other types include rotary, semi-rotary, diaphragm and chain pumps.

Plunger Pumps. The conventional type for deep wells is a "plunger" pump, which is placed in the well and is operated through a system of rods by a crank or lever above the ground. Operation of the crank or lever may be by a windmill, an engine or an electric motor. The pump chamber, which may either be immersed in the water or placed on a platform just above, consists of a vertical brass cylinder. The plunger is also usually of brass and is fitted with cup leathers, transfer ports and a delivery valve as in the lift pump. This type operates exactly like the lift pump, except that the chamber is connected to the rising main and a water-tight gland surrounds the pump rod. The rising main is usually screwed to the top of the pump chamber, extending it to the top of the well; the pump rod then works inside this pipe, and as the pipe is of slightly greater diameter than the plunger the latter may be easily withdrawn for occasional inspection or repair without the necessity of going down the well.

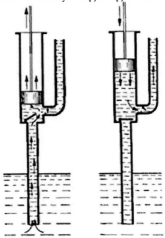

*FIG. 248.—DIAGRAM ILLUSTRATING ACTION OF A
PISTON PUMP*

A modern electrically operated plunger pump has the wellhead gear immediately above the rising main. The motor drives a reduction gear by V-belt, and the plunger usually operates at 25 to 90 strokes per minute. The water can be raised 400 to 500 ft. above the plunger if necessary. A typical 1 h.p. pump operating with a $2\frac{1}{2}$ in. cylinder at 45 strokes per minute in a well 135 ft. deep against a total head of 225 ft. may be expected to deliver about 270 gallons per hour. There is a range of plunger pumps suitable for almost all common farm requirements.

Piston Pumps. High-speed electrically operated piston pumps are suitable for use in conjunction with shallow wells. They consist of a close-fitting piston reciprocating in a small cylinder which contains both outlet and inlet valves. The volume of the cylinder is small, displacement per stroke usually being only about $\frac{1}{16}$ pint, and the pump is almost invariably horizontal and double-acting. A typical installation with a $\frac{1}{2}$ h.p. motor will deliver 300 gallons per hour against a total head of 120 ft.

Low-speed high-duty piston pumps will deliver against very high heads. These need a reduction gear and are therefore more expensive than the high-speed types.

Centrifugal Pumps. Centrifugal pumps operate on the same principle as a propeller or fan. Simple types are not as "efficient" for small duties as piston or plunger pumps, but have the advantage of less complication and maintenance. In the simplest type, an impeller, consisting of curved vanes, is driven at high speed inside a disc-shaped casing. Water enters at the centre of the casing, and the motion of the impeller pushes it up the delivery pipe, which is attached tangentially to the outer rim of the casing. Such simple pumps are not self-priming.

Modern centrifugal pumps are usually of a modified design, in which there are inclined vanes on the inside of the pump chamber, and the rotor consists of straight radial vanes. With this type, inlet is at the top of the pump beside the outlet, so the pump is never empty and is self-priming.

Multi-stage pumps have two or more rotors mounted on the same shaft in separate casings, water from the first in the series being delivered to the inlet of the next. By this means the working pressure is increased at each stage, and centrifugal pumps which will work against heads of up to 1,000 ft. can be constructed.

Submersible Electric Pumps. A very neat and efficient type of pump that may be employed for deep wells where electricity is available is the submersible multi-stage centrifugal type. The unit consists of a motor and pump which are lowered into the water, attached to the rising main. The pump is mounted immediately above the motor. The motor itself is of squirrel cage induction type, with a thin alloy cylinder to protect the stator from the water. The rotor is of non-corrodible metal, and is immersed in the water.

Current is conveyed to the motor by rubber-sheathed cable. Control may be by a push-button starter, a float switch working on an open reservoir, or by an automatic pressure system.

A great advantage of the submersible type of pump is the ease of installation. The outside diameter of the unit is generally less than 6 inches, and it is easily lowered into the well at the end of the rising main, and withdrawn when necessary for servicing.

Portable Centrifugal Pumps. A handy type of pump for general farm use is the portable self-priming centrifugal type. The best examples of this type are fully automatic in their self-priming action; and with a small internal-combustion engine attached they are easily portable by one man.

The capabilities of such pumps are indicated by the example of a tiny engine-driven outfit, weighing less than $\frac{1}{2}$ cwt., when built of aluminium, which is capable of pumping 3,000 gallons per hour with a suction of 15 ft. and against a total head of 30 ft. Such pumps are able to deal satisfactorily with water containing up to 25 per cent, of solid matter, and are suitable for handling liquid manure.

Rotary Pumps, in their simplest form, consist of a pair of spur gear wheels, which mesh together in a suitable housing and carry the liquid round in the spaces between the gear teeth and the outer casing. Such pumps are not generally used for water supply or for pumping gritty fluids, but are commonly used on internal-combustion engines for pumping oil from the sump to the principal bearings, and on spraying machines. Other rotary pumps work on a similar principle, but have the rotating parts of various shapes.

One type of rotary pump that is used for many purposes, including water pumping and spraying, has a single rotor which revolves inside a fixed stator made of rubber. The stator has the form of a double, internal helix. The rotor is also helical, but of half the pitch. The rotor maintains a seal with the stator, and this seal travels continuously along the stator as the pump operates. Rotary pumps give a positive displacement.

The Semi-Rotary Pump is a compact, efficient, double-acting type that is popular for the hand pumping of water, oil, etc. Its lightness and portability make it useful for such work as the filling of field spraying machines.

The Diaphragm Pump is a type suitable for short suction lifts. It has a rubber or chrome leather diaphragm in place of the piston of the lift pump. The diaphragm has a valve in its centre, and as it is moved up and down by operation of the handle, the delivery valve opens and closes alternately with a foot valve, as in the common lift pump.

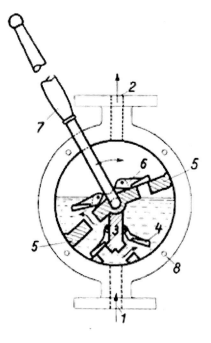

FIG. 249.—DIAGRAM OF SEMI-ROTARY PUMP

1, inlet; **2**, outlet; **3**, stationary partition with **4**, valves; **5**, semi-rotating partition with **6**, valves; **7**, handle; **8**, casing.

The diaphragms are easily and cheaply renewed, and the complete absence of sliding friction permits a fairly high efficiency and a capacity for dealing satisfactorily with muddy water.

Chain Pumps consist of a set of chilled cast-iron cups or discs connected by an endless chain which runs over grooved wheels. The chain is driven at high speed through a pipe leading from the well or cistern to the surface. These are sometimes used for pumping liquid manure from a storage cistern to the distributing cart, since they will deal with thick fluids or with liquids containing solid material. Wearing of the discs causes a loss in the efficiency of chain pumps.

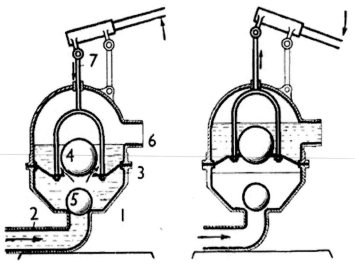

FIG. 250.—DIAPHRAGM PUMP

1, casing; 2, inlet pipe; 3, diaphragm; 4 and 5, ball valves; 6, delivery pipe; 7, pump rod.

Capacities and Efficiencies of Pumps.

The theoretical capacities of reciprocating pumps may be calculated in gallons by multiplying the product of the length of stroke (ft.), the area of the cylinder (sq. ft.) and the number of strokes per minute by (gallons per cubic ft.). The work done per minute may be calculated if the total head and the amount of water lifted per minute are known. Then but losses of energy due to friction in the working parts and pipes, leakage of the valves, etc., may cause very low efficiencies. Thus, whereas carefully designed and constructed pumps may have efficiencies up to 90 per cent., the efficiencies of small farm models, owing to their imperfect construction, generally range from 20 to 75 per cent.

$$\text{theoretical h.p.} = \frac{\text{foot-lb. of work per minute}}{33,000},$$

The limiting height or pressure at which water can be delivered by pumps is determined by (1) the power available; (2) the absence of leaks in the valves; (3) such details as the fit of the piston in the cylinder; and (4) the strength of the materials. In theory the quantity of water delivered is independent of the head; but in practice the "volumetric efficiency" (i.e. the ratio of the actual volume discharged to the plunger displacement) falls off at high heads, and it is advisable when selecting a pump to know its capacity at various heads.

The manometric head is the sum of the static head (the vertical distance in feet from the lowest pumping level to the highest point of delivery) and the friction head, representing all friction losses. The friction losses per 100 ft. of various sizes of piping are given in Table IX. Where there is an appreciable bend in the pipe, each bend produces a friction loss equal to 25 ft. of straight pipe of the same diameter. Where a pressure tank is used, about 20 ft. must be added to the head.

TABLE IX.—FRICTION LOSSES IN PIPES.

FEET LOSS PER 100 FEET OF PIPING.

Inside dia. of Piping	Volume in Imp. Gallons Per Hour.						
	200	300	400	500	600	800	1,000
1 in.	2	4	7	11	17	26	37
1¼ in.	0.6	1.9	2.2	4	6	8	13
1½ in.	—	—	—	1.6	1.8	3.2	5

The Hydraulic Ram.

The hydraulic ram is a device by which the energy of a quantity of water with a small head may be used to elevate a *proportionate* quantity to a higher level. The working of the ram is illustrated by Fig. 251. Water flows from the source, which may be a flowing stream, along the drive pipe and through the waste valve. The rapid flow of water through the open waste valve then causes the latter to close. When this occurs, the momentum of water in the drive pipe forces a proportion of it through the delivery valve, thus compressing the air cushion in the air vessel and causing a continuous delivery to the storage reservoir. When the pressure in the drive pipe and the air chamber become equal the water in the drive pipe recedes, causing the waste valve to open, when the cycle is repeated. For successful operation, a continuous flow of water must be available. The driving head is the fall from the surface of the supply to the waste valve of the ram. Where the delivery tank is not more than forty times as high as the driving head, efficiencies up to 67 per cent, may be secured. A rule for computing the capacity is as follows: "to estimate the quantity in gallons per minute delivered to the reservoir, multiply the fall (in feet) by the gallons of supply per minute and divide by twice the height in feet to which the water is forced". If plenty of fall is available rams can be worked by springs yielding as little as one gallon of water per minute. If, on the other hand, an abundance of power water is available, falls as low as 18 in. can be used. They can be made to force any distance if provided with suitable delivery main. Where there is a small supply of pure water and plenty of impure, the power of the impure supply may be harnessed to pump the pure by using a compound ram. Thus, water from a spring may be led by a small pipe down to a brook, and the brook water used as a driving force to deliver the spring water to a farm on the high ground. Maintenance expenses are low, and in suitable conditions the ram is an ideal mechanism for providing a water supply.

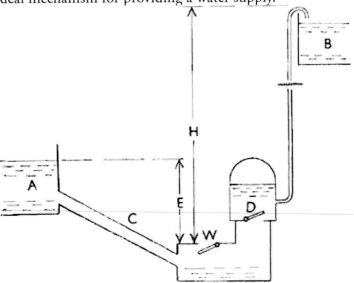

FIG. 251.—*DIAGRAM ILLUSTRATING ACTION OF HYDRAULIC RAM*

A, supply; B, delivery tank; C, drive pipe; D, delivery valve; W, waste valve; E, driving head; H, delivery head.

Choice of Pump.

For farm pumps the absence of operating troubles and cheapness in first cost are usually considered to be of more consequence than the highest mechanical efficiency.

For shallow wells the choice generally lies between the horizontal piston pump and the simple centrifugal type, and for deeper wells, either the vertical plunger type or a submersible centrifugal one may be used. Piston pumps usually cost slightly less to run than centrifugal ones, but the latter are less subject to mechanical troubles and are generally to be preferred where there is sand or grit in the water.

When a new pump is installed on a medium-sized dairy farm it ought, if possible, to be capable of delivering up to 400 gallons per hour at a pressure of about 50 lb. per sq. in. It is often expensive to secure a good pressure by use of a high storage tank, and a pressure tank with automatic control is sometimes preferable where an electric drive is employed. Electric motors are ideal power units because of the ease of arranging an automatic control, but where electricity is not available a Diesel engine or a windmill may be used. The windmill has strong claims for consideration in situations where it is convenient to draw an alternative supply from mains during calm periods.

Methods of Installing Pipes, etc.

Modern methods of installation enable farm water supply schemes to be completed more quickly and easily than hitherto. With iron water pipes one modern method is to join them together and draw them into the ground by using a mole drainer, fitted with a special device to connect one end of the pipe to the rear end of the mole cartridge. The mole plough is pulled by a tracklayer, and on suitable land up to 500 yds. can be pulled in to a depth of 2 ft. in one length. By this method as much as 2,000 yds. of I-in. pipe has been laid by one outfit in a day. Where there are rocks or tree roots this method is, naturally, not as satisfactory.

In recent years the use of asbestos cement for public water supply pipes has developed considerably, and there are some circumstances where they may be used with advantage in farm schemes. They must, however, be laid on a reasonably level surface, and they cannot be drawn into the ground by using a mole drainer in the way that iron pipes can.

Other materials, e.g. plastics, may also be used. Pipe lines should run as straight as possible to the main points of use, but should avoid rock outcrops, marshes and wooded areas, in order to keep down the cost of excavation. They should be laid with a continuous fall in order to avoid air locks at high points or the collection of sediment in depressions. Wherever possible they should be buried 2 ft. deep in order to avoid trouble from frost, or mechanical damage.

The design of drinking troughs has been improved in recent years. Concrete troughs with no corners to hold dirt, and a good concrete base for the animals to stand on, are probably best for use out in the fields in present circumstances.

The supply pipe should be brought underground into the bottom of the trough, and each trough should be fitted with a separate emptying tap and stopcock, protected with a concrete box and cover. Where galvanized iron troughs are used they should be supported on wrought-iron legs built into concrete, the trough being kept clear of the ground to prevent the bottom from rusting. The service pipe, fitted to the ball tap in the bottom of the trough, may be protected from frost by covering it with a 4-in. asbestos pipe.

Where stand-pipes are fitted outdoors they are best protected by a galvanized iron waterproof frost box which covers the pipe and all but the high-pressure type nose of the tap.

Spraying Machinery.

As the application of science to farming progresses, spraying is employed for an ever-widening variety of purposes. The chief of these are (1) the application of herbicides in order to reduce the competition from weeds; (2) the application of protective fungicides to minimize the effects of fungus diseases; and (3) the use of insecticides to control various kinds of insect pests. In addition, spraying is employed for the application of micro-nutrients such as manganese or boron; for burning off excessive "flag" on cereal crops in order to reduce the danger of lodging; and for destroying potato haulm before mechanical harvesting equipment is used. The economic advantages of such operations as spraying for weed control are experienced in the form of increased yields and in future benefits due to prevention of seeding. In this, as in most other spraying operations such as spraying with fungicide against potato blight, benefits are proportional to the degree of infestation that is present or threatened.

FIG. 252.—HILLER HELICOPTER SPRAYING CORN FOR CONTROL OF CHARLOCK. (PEST CONTROL LTD.)

Many kinds of modern spraying machines are suitable, with appropriate accessories, for spraying both field crops and orchards, while other machines are designed for only a limited range of uses.

SPRAY PUMPS.

Many kinds of pumps are employed in spraying machines, but the two kinds most commonly used are high-pressure piston pumps, and rotary pumps, mostly of the gear type.

The piston pumps used may have from one to four cylinders. On some machines the plungers are of corrosion-resistant metal and reciprocate in suitable pump packings. More usually, there are plunger cups made of rubber, leather or fabric composition, that reciprocate in cylinders which are porcelain-lined, or have stainless steel or monel metal liners. Valves are usually of the stainless steel ball type and may fit on renewable monel metal seatings. This type of pump is used on most orchard sprayers and modern "high-volume" field sprayers. In modern machines such pumps are capable of working at pressures of up to 600 lb. per sq. in. Any excess liquid pumped is returned through a relief valve which can be set to operate at any desired pressure. An air vessel is fitted close to the delivery manifold in order to produce a more even discharge.

Many low-volume and universal sprayers employ gear pumps, made of gun-metal or phosphor bronze, and mounted on stainless steel shafting. The gear pumps used on some machines operate at pressures of only 30-40 lb. per sq. in., but well-made large-capacity gear pumps can operate at higher pressures and are suitable for use on "universal" machines.

Spray Regulators. The regulation of pressure is an important requirement of a spraying pump. The original method was to use a relief valve, which prevented the pressure exceeding a set amount by passing the excess liquid back from the delivery pipe into the supply tank. This method tends to keep the pump working at full capacity, regardless of the amount of liquid used, with a consequent waste of power and unnecessary wear and tear. On modern pumps, the discharge pressure acts against a spring-loaded diaphragm, and when the pressure exceeds a set (adjustable) amount, this opens a ball valve from the pump discharge pipe, allowing the spray to pass directly into the supply tank. The pressure in the delivery pipes and air chamber is held by a separate check valve, and when the pressure in the air chamber falls below the set amount the relief valve closes, and the liquid is forced out through the regular discharge line (Fig. 253). With a fully automatic regulating device of this type, shutting off all the nozzles merely causes the engine to idle. For good operating conditions, it is not wise to load the pump to the limit of its capacity, but rather to allow some liquid to by-pass the regulator while spraying is in progress.

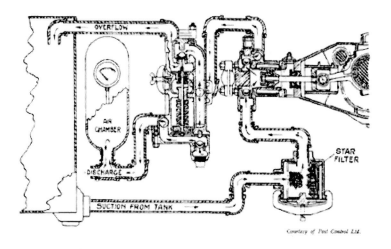

FIG. 253.—DIAGRAM OF MODERN HYDRAULIC
SPRAYING MACHINE.

The liquid is sucked from the tank through the filter to the
inlet valves of the pump, and then forced by the compression
stroke of the piston to the pressure regulator. From there it
passes to the discharge nozzles; or, if the discharge is blocked
and the overflow valve at the top of the regulator is open,
back to the tank.

Low or moderate pressures (30 to 150 lb. per sq. in.)
are suitable for spraying ground crops, but high pressures are
needed for certain types of orchard work. If a single machine
is required to do both types of work, therefore, it must be
capable of operating at high pressure.

Spray Nozzles. Atomization of spray fluids is usually achieved by discharging the liquid through an orifice under pressure, but a second main method of commercial importance achieves atomization by breaking up the jet of liquid with a blast of air. Nozzles which do not employ air blast may be broadly divided into two main types, viz. (1) swirl nozzles and (2) fan nozzles.

Swirl Nozzles, which are usually made of brass or stainless steel, consist of a swirl plate, a swirl chamber, a nozzle disc and a cap (Fig. 254a). A filter is also often incorporated in the nozzles. The spray fluid is given a whirling motion as it approaches the orifice, and the result is a cone-shaped spray pattern. The nature of the spray may be controlled by varying the number and size of holes in the swirl plate, the depth of the swirl chamber (thickness of washer), and by fitting discs with holes of various sizes.

Pressure also has a considerable effect on nozzle performance. For a given size and design of nozzle and depth of swirl chamber, the higher the pressure, the higher the throughput, the wider the spray angle and the finer the spray droplets.

Fan Nozzles may be made of metal (Fig. 254b) or of ceramics (Fig. 255). These are usually designed for "low volume" work, but by selecting nozzles with apertures of suitable size, quite high rates of application can be obtained. Individual filters are normally fitted to each nozzle.

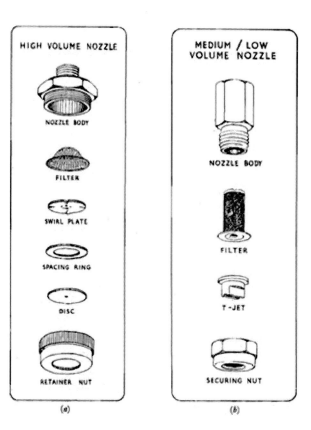

FIG. 254.—DIAGRAM SHOWING (A) TYPICAL SWIRL NOZZLE, APPROXIMATELY HALF ACTUAL SIZE; (B) METAL FAN JET, SLIGHTLY LESS THAN ACTUAL SIZE.

SPRAY MATERIALS AND APPLICATION RATES.

Many spray materials can be easily dissolved in water to give solutions, or may be mixed with water to form stable emulsions. Such materials, once mixed, are easy to apply, and can be effectively sprayed through fine nozzles by "low volume" machines at rates as low as 5 gallons per acre. Other materials, such as some of the copper fungicides, are suspensions. These may easily be mixed with water but need continuous agitation during work if a spray of constant strength is to be applied. Such materials must usually be applied by machines employing fairly large nozzles, i.e. either "high volume" sprayers which apply 60-100 gallons per acre, or by "atomizer" (air-flow) type sprayers.

The type of "cover" required from a spraying machine depends on the purpose for which the crop is being sprayed and the type of spray material. For example, sulphuric acid, which destroys the leaf tissue on which it alights, needs to cover as completely as possible the leaf surfaces of weeds that are being sprayed. With "growth regulating" weed killers, on the other hand, the toxic spray material is absorbed and translocated if only a part of the leaf surface is wetted. A thorough cover is needed for insecticides which act through the stomach or by contact with the insect's cuticle, and also for fungicides of all kinds.

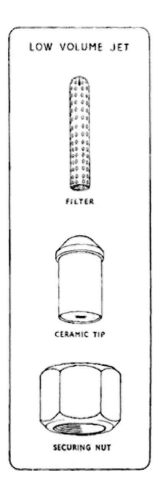

FIG. 255.—FAN JET WITH CERAMIC TIP. (SLIGHTLY ABOVE ACTUAL SIZE.)

Some herbicides (e.g. D.N.B.P., which is employed in weed-killers used for spraying peas) are more selective in action if applied with a large volume of water.

A partial wetting may be adequate for systemic insecticides.

With "low-volume" spraying, the small quantity of spray applied per acre is split up into very fine droplets, and the spray which strikes the plants adheres to the surface in individual specks. With "high-volume" spraying the drops are larger, and tend to coalesce and cover the whole surface, any excess running off. High volume spraying is, therefore, more suitable where a complete cover is required. A further advantage of applying high volumes is that the large droplets, delivered at high pressure, have more penetrating power than the small droplets delivered by low-volume sprayers, and are therefore more effective in achieving a complete cover in dense crops. Moreover, there is less danger of damage from drifting of the spray on to neighbouring crops.

The advantages of spraying at low volume when this is practicable are fairly obvious. Machines designed only for low-volume work can be relatively light and cheap, and owing to the small amount of water needed, rates of work are high, and labour costs low.

Table X gives typical figures for the volumes of spray fluids commonly applied.

Types of Field Crop Sprayers,

HIGH-VOLUME sprayers are usually trailed machines, driven either by the tractor P.T.O. or occasionally by an independent engine, and capable of delivering a coarse, drenching spray at fairly high pressures at a rate of up to 100 gallons per acre. A large spray tank is necessary, and an injector system or other power-driven pump is needed for quick filling. Such machines are larger, heavier and more expensive than low-volume sprayers. Most modern high-volume machines may be employed for both orchard and field crops. The spray tank is usually fitted with a mechanical agitator, and any kind of spray material except perhaps sulphuric acid may be used.

TABLE X.—COMMON SPRAY MATERIALS AND AP-PLICATION RATES (GROUND CROPS).

Spraying Compound	Purpose	Approximate quantities per acre (gallons)	Notes
M.C.P.A	Weed killer	5–100	No agitation necessary after mixing.

D.C.P.A (2-4D) Amine salt	Weed killer	5–100	No agitation necessary after mixing.
D.C.P.A (2-4D) Ethyl ester	Weed killer	10–100	No agitation necessary after mixing.
D.C.P.A (2-4D) Sodium salt	Weed killer	25–100	Difficult to dissolve at lower volumes.
Sulphuric acid	Weed killer	10–100	Corrosive. Few machines suitable. Care necessary in mixing.
Vaporizing oil or White Spirit	Weed killer	50	For spraying carrots, etc.
D.N.O.C Amine salt	Weed killer	25–100	Poison. Difficult to dissolve at lower volumes.
D.N.O.C Suspensions	Weed killer	100	Poison. Continuous agitation essential.
D.N.B.P Ammonium salt	Weed killer	100	Poison. More selective at high volume.
Bordeaux mixture	Fungicide	100	Continuous agitation essential.

| Proprietary Fungicides for potato spraying | Fungicide | 10–100 | Agitation usually necessary at low volume. |
| D.D.T Emulsions | Insecticide | 5–100 | No agitation necessary after mixing. |

Low-Volume sprayers are light, simple, inexpensive machines designed to apply small quantities of highly-concentrated material in a fine, misty spray at low pressure. Owing to the fine nozzles employed, only solutions or emulsions or very fine suspensions may be used. A small tank of about 50 gallons capacity is adequate, and this is often tractor-mounted. Fan-type nozzles are usually fitted. Low-volume sprayers can only deal with a limited range of spray materials. Agitation of the liquid in the spray tank is usually achieved by employing a pump of higher capacity than is needed for supplying the nozzles, and directing the surplus back into the bottom of the tank.

SPRAYING MACHINE CIRCUIT.

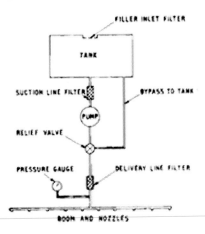

FIG. 256.—DIAGRAM SHOWING LAYOUT OF SIM-PLE FIELD SPRAYER.

"Universal" Sprayers. Many spraying machines are capable of spraying either at high volume (up to 100 gallons per acre) or low volume (5-10 gallons per acre) or at intermediate rates. This is achieved by having pump control mechanisms which are capable of a wide range of pressure adjustment, and by having a range of nozzle sizes. The tank needs to be of reasonable size for the high-volume work (100 to 200 gallons), and the machine is usually trailed on account of this. In order to provide for the low or moderate pressures (30 to 100 lb. per sq. in.) needed for field crop spraying and also for the high pressures (300-500 lb. per sq. in.) needed for orchard work, some universal machines have two pumps—a low-pressure one for ground crops and a high-pressure one for orchard spraying. Such machines can undertake a great variety of spraying jobs.

Pneumatic Sprayers. Pneumatic sprayers employ an airtight spray tank and a compressor which pumps air into the top of the tank. When the required pressure has been reached, a control valve is opened, and the spray is forced out of the nozzles by the air pressure. A safety relief valve in the delivery line prevents excessive pressures. This type of sprayer is suitable for handling sulphuric acid, since the acid only comes in contact with the barrel and the delivery line, all of which can easily be made of acid-resistant materials.

Air-Flow Sprayers. In air-flow sprayers, the spray liquid is delivered to the nozzle at low pressure, and there it meets a powerful high-speed air blast which breaks it up into droplets and conveys it to the target. This type of sprayer normally applies 10 to 25 gallons per acre, and because the nozzle has no fine orifice it is capable of dealing with many types of suspensions. The foliage of the crop is blown about by the air stream, and this helps to secure an effective cover.

FIG. 257.—TRAILED P.T.O.-DRIVEN AIR-FLOW ("AT-OMIZER") FIELD SPRAYER (RANSOMES.)

Spray Booms for Ground Crop Sprayers. The machine is usually provided with a spray boom 15 to 35 ft. or more in length, nozzles being fitted, at intervals depending on their characteristics. The boom is often mounted in three sections, the central part being fixed and the outer sections being hinged and supported both vertically and laterally. The outer sections will fold upwards or forwards for transport. Some short booms have a single section, and some have two sections hinged at the centre. Some long booms have castor wheels at the ends to assist in maintaining the correct height on undulating land. Many modern machines incorporate a suck-back device to prevent drip from the nozzles when the control lever is operated.

Fig. 258 shows a boom with hoods designed to prevent drift of the spray. For spraying crops such as potatoes, nozzles may be attached to the bottoms of more or less vertical drop-legs in such a way that the lower surfaces of the leaves are sprayed by jets directed upwards (Fig. 259).

FIG. 258.—TRACTOR FITTED WITH GAS-PROOF CAB, HAULING 500-GALLON CONTRACTOR'S HIGH-VOLUME SPRAYER EQUIPPED WITH ANTI-DRIFT SPRAY BOOM. (PEST CONTROL LTD.)

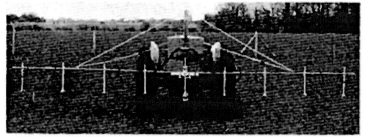

FIG. 259.—TRACTOR-MOUNTED SPRAYER FITTED WITH UNDERLEAF SPRAYING EQUIPMENT. (ALL-MAN.)

Adjustment of Application Rates on Ground Crop Sprayers. Most modern field sprayers are either engine driven or P.T.O. driven, and with a given set of nozzles and adjustment of the pressure, the output of spray per minute remains constant, and the quantity of spray applied per acre is inversely proportional to the forward speed. Most manufacturers supply a chart showing the application rates with various nozzles at different pressures at a standard speed of 4 m.p.h., and advise operators to vary the forward speed slightly in order to make minor adjustments of application rate.

If it is essential to apply an exact quantity per acre and there is doubt about the setting, the sprayer may be calibrated as follows: Calculate the distance to be travelled in spraying an acre.

$$\text{Distance (ft.)} = \frac{4,840 \times 9}{\text{Effective spraying width (ft.)}}$$

Fill up the tank with water to a known mark, spray an acre at the recommended speed, and then refill the tank to the original mark, measuring the quantity of water added. Usually, a slight variation in speed will achieve any necessary adjustment, but if a major alteration is necessary, nozzles must be changed and the calibration repeated.

Operating speeds should not normally exceed about 5 m.p.h.

Operation of Ground Crop Sprayers. All spraying machines are fitted with a number of filters, and it is essential before starting work to see that all are fitted and that all are clean. The tank should be flushed out with the drain plugs open, and the action of the nozzles should be tested and any defects remedied. The boom must be adjusted for height so that the spray patterns of adjacent jets just overlap where the spray meets the target, and care must be taken in driving to ensure a slight overlap between adjacent bouts. Care must be taken in turning near trees and fences, to avoid damaging the boom. The headland should be sprayed after the body of the field, and when weed-killers are being used it may be advisable to reduce the dosage or concentration on the headland, because of the damage that occurs where the crop has been badly bruised.

It is essential to wait for suitable weather for spraying. Windy weather makes good work impossible, and may result in harmful drifting of the spray, especially when spraying with fine droplets at low volume. The toxicity of some herbicides to both weeds and crops varies with temperature, and no attempt should be made to use dinitro compounds in very hot weather.

When spraying is finished, spraying machines should be thoroughly washed down both inside and out, and particular care must be taken to remove all traces of growth regulating sprays before the machine is used for spraying susceptible crops. After using dinitro compounds, growth regulating substances, etc., it is advisable to wash out with a suitable detergent. It is a good plan to leave the nozzles of low-volume sprayers to soak in a detergent overnight.

MAINTENANCE OF SPRAYING MACHINES.

Many sprays are corrosive, and after spraying with materials such as sulphuric acid, at least 25 gallons of clean water should first be put through the machine. All metal parts should then be washed with a solution of washing soda (1 lb. to 10 gallons of water), and some of this solution left in the machine for 24 hours. After this all detachable metal parts, including filters, should be removed and oiled or treated with an anti-rust compound before storage. All rubber parts should be well washed and stored in a cool place.

SAFETY PRECAUTIONS.

Some of the spray materials now employed are poisonous and highly dangerous to the operators unless very great care is taken in handling them. Particular precautions must be taken in handling the dinitro weed-killers (D.N.O.G. and D.N.B.P.) and the insecticides T.E.P.P., H.E.T.P., Parathion and Schradan. Sulphuric acid is dangerous on account of its corrosive action, and the best method of handling it is usually to pump the concentrated acid direct from the carboys into the sprayer barrel. Whenever the above materials are handled the advice given in the Ministry of Agriculture's Advisory Leaflet No. 374 and in the leaflet "Take Care When You Spray" should be strictly followed. Small amounts of poison taken into the body may produce no ill effects at first, but being cumulative, may suddenly give rise to serious poisoning. Protective clothing should be worn, and operators should avoid absorbing poison through the mouth, nose or skin.

The sprayer should always be stopped when nozzles are to be cleaned or changed, and spraying in windy weather should be avoided. The dinitro compounds are particularly dangerous in hot weather. When mixing poisonous sprays, care should be taken to avoid splashing them on to the skin.

When using dilute sulphuric acid, the concentrated acid must always be added to water, never water to acid, and stirring should proceed as mixing occurs. The strength of the mixture may be tested by using a hydrometer. Finely dispersed tractor vaporizing oil is inflammable, and precautions against fire should be taken when this is used. Some dry D.N.O.G. compounds will ignite spontaneously in hot, dry weather, and this is a further reason for washing down the outside of the sprayer, as well as the inside.

Orchard Sprayers.

The most popular type of orchard sprayer is a 2-wheeled, tractor-trailed, P.T.O.-driven machine having a tank of 200-300 gallons capacity and a multi-cylinder pump with an output of about 20 gallons per minute. In addition, there are larger and smaller machines of the same general type, with tanks ranging from 100 to 500 gallons, and outputs of 10 to 50 gallons per minute. All these machines employ reciprocating pumps working at pressures of up to 550-600 lb. per sq. in., and are, in fact, often the same multipurpose machines as have already been mentioned in dealing with high-volume and universal ground-crop sprayers. In addition, there are many kinds of small "portable" sprayers, with engines of about 3 h.p., and also hand-powered machines suitable for small growers and for countries where mechanical power is not yet available.

With stationary or portable plants, steel mains and flexible connections may be used to transmit the spray to hand lances or guns, each lance being attached to about 60 to 120 ft. of rubber hose. With mobile machines, the rubber hose can be short and is attached directly to the sprayer. The tractor pulls the hose along when necessary, and so saves the operator a great deal of work. On some mobile machines the operators work from platforms on the machine itself (Fig. 260), and in such conditions can easily control a multi-nozzle lance. The advantage of a mains system is that work can be done on hillsides and on soft ground where the use of mobile equipment is difficult; but the high cost of maintenance of pipe-lines, and the loss of time in moving them, combine to make mobile machines more attractive to most growers.

A typical mobile sprayer that will deliver 20 gallons per minute would, in the past, have kept 6-10 men busy with single-nozzle lances. To-day, such a machine normally supplies two triple-nozzle lances, each of which delivers 7-8 gallons per minute. The output of two good men with such equipment can be equal to that of many more men with low-capacity single nozzles, and such an outfit is capable of spraying up to 5 acres a day at 400 gallons per acre.

The type of spray pattern produced by modern nozzles is easily varied by changing swirl plates and discs, and by altering the thickness of the rubber washer which alters the depth of the swirl chamber. A shallow swirl chamber gives a wide angle, while a deeper one gives a narrow cone and a greater effective spraying distance.

For a very long throw, needed to reach the tops of tall trees, swirl plates having a fairly large hole in the centre must be fitted, and large-size discs which will give a throughput of 8 gallons per minute with a 3-nozzle lance are needed.

*FIG. 260.—MOBILE TYPE OF ORCHARD SPRAYER
WITH POWER DRIVE AND MULTIPLE-NOZZLE
LANCES*

With uniform orchards which are suitably laid out, the operators of mobile machines may be replaced by a range of fixed or oscillating nozzles. With a 20 g.p.m. machine of the type described, the nozzles are usually arranged on one side of the machine only, being set on an incline from a point high above the tank to one just above the top of the wheels. Six nozzles set 9 in. apart, and properly adjusted can do satisfactory work on low trees. In general, the uppermost nozzles should be as high as the tree tops. Larger machines, which are capable of an output of 40 gallons per minute, may have the nozzles arranged in the form of an inverted V and can spray both sides at once. One firm employs a series of oscillating nozzles (Fig. 261) which are kept in motion up and down by means of eccentrics.

FIG. 261.—AUTOMATIC HIGH-PRESSURE ORCHARD SPRAYER, WITH BANKS OF OSCILLATING SPRAY GUNS. (PEST CONTROL LTD.)

Some mobile sprayers employ a high-speed air blast for providing atomization and delivering the spray to the target. The nozzles may be arranged in a semi-circle, or in a complete ring across the air stream, or in a line along a kind of fish-tail. Some recently developed machines have no nozzles as such, but deliver the liquid into the air stream from between large-diameter revolving metal plates which are connected to the axis of the aerofoil-type fan. It seems likely that sprayers of this type will be increasingly used in the form of automatic or semi-automatic machines, as orchards which are really suitable for such equipment are planted.

FIG. 262.—AIR-FLOW TYPE AUTOMATIC ORCHARD
SPRAYER. (K.E.F.)

Dusting Machines.

The dusting machine consists of a hopper with feed rate control and an agitator, a blower, distributing pipes and one or several nozzles. The equipment is usually either mounted on a trailer or tractor or on a frame which can be placed on a lorry. The blower is operated either by power take-off, by a small petrol engine, or by a sprocket chain drive and suitable gearing from the wheels.

There are small hand-operated dusting machines on the market, which, in use, are worn across the chest. They also consist of a small fan, a hopper and dosage mechanism.

For dusting potatoes, eleven- and thirteen-row power machines are available, and these are capable of working at a rate of 6 to 7 acres per hour (Fig. 263). The position of the pipes may be adjusted by operating a handle on the machine, and they may be raised into an almost vertical position for transport. If the spraying is carried out very early in the morning or late at night, when there is a good dew on the leaves, control of blight by this method may be fairly effective. The ease with which the operation is carried out makes it possible to dust several times each season, if necessary.

*FIG. 263.—ENGINE-DRIVEN TRACTOR-MOUNT-
ED DUSTING MACHINE TREATING POTATOES
AGAINST COLORADO BEETLE. (ALLMAN.)*

Dusting is often preferred to liquid spraying because it dispenses with water transport; but for the application of fungicides and insect stomach poisons a dust deposit on the plant is less rain resistant than a liquid spray deposit. For contact insecticides, which produce an instantaneous kill, dusting is as effective as spraying. It is widely used against Colorado beetle.

Dusting should not be carried out if the wind velocity exceeds 6 m.p.h. The use of a dragsheet which is trailed behind the machine to prevent wind drift and to confine the dust over the plant, is often advisable. Dusting with fungicides and stomach poisons gives best results when carried out in the early morning and late evening. When contact insecticides are used, the time of application should be chosen in accordance with the activities of the insect, as it is often necessary to dust when the temperature is high and the insects active. Nicotine dusting has to be carried out when the temperature is above 60° F. Dusts are often carried away by wind, and unless sufficient care is taken, this may cause damage to neighbouring crops.

FIG. 264.—ELEVEN-ROW ENGINE-DRIVEN HORSE-DRAWN POTATO DUSTER. (GRATTON.)

Irrigation.

Overhead irrigation is now successfully employed by many vegetable growers, to assist in the production of high-value crops such as cauliflower, lettuce, onions, tomatoes, etc. It is used for watering-in the plants, and for maintaining a luxurious growth in times of drought.

Experiments have shown that where water supplies are readily available, the irrigation of farm crops such as sugar beet may be economic. It seems likely that if portable equipment were really cheap, several farmers in the drier parts of Britain could profitably employ irrigation. Experience of irrigating riverside meadows shows that substantially increased yields are obtainable, and there must be many thousands of acres of grassland which would show a worthwhile response. The main factor which often restricts the development of irrigation is, of course, lack of a suitable water supply. There are many situations where making water available is either physically almost impossible, or likely to be impossibly expensive.

Where a good water supply is available the equipment used generally consists of a fairly powerful engine (a tractor is often used, but an electric motor in conjunction with a pressure tank and automatic switch gear is ideal), a good pump (often of the multi-stage centrifugal type), delivery pipes, and a spray line. The object is to apply the water evenly and not too fast, giving the equivalent of about one inch of rain in 3-4 hours. It is always advisable to give a good soaking—not less than half-an-inch.

The delivery pipes may be portable, of light metal, with quick-coupling connections. Some farmers prefer to lay permanent underground mains to every suitable field, and to instal stand-pipes at. intervals along the headland, so that only the spray lines need to be moved. The spray lines normally consist of $1\frac{1}{2}$ to 2 in. diameter pipes, carried on light steel chairs, and having jets every 2 ft. along the line. The spray line may be automatically oscillated through an angle of about 90 degrees by a device fitted at one end, and operated by the water flow. Some users, however, prefer to throw the water all in one direction and to operate the line by hand. A pressure of 30-50 lb. per sq. in. must be maintained at the jets, and with the oscillating line a lateral throw of about 14 yds. is generally obtained. The length of spray line that can be operated naturally depends on the capacity of the pump, size of the delivery pipe, distance from pump to spray line, etc. A length of 200-300 yds. is fairly common.

In the United States, where overhead irrigation has lately spread rapidly throughout, humid areas as well as arid ones, a common method of application is by means of rotary sprinklers spaced at 30-40 ft. intervals along a portable pipeline. Each sprinkler covers a circle about 90 ft. in diameter, and the line is moved 60-80 ft. at a time.

The success of such a system depends on having very light aluminium pipes which are connected up or disconnected very rapidly by means of lever-operated fasteners. Rubber seals are provided in the pipe ends, and these allow flexibility which permits the pipe-lines to run easily over undulating land. The sprinklers need to be simple, light and cheap.

The rotary sprinklers which have hitherto been most popular for commercial use in Britain are usually of very large size, a single "rainer" taking in many cases the whole output of a large pump. Smaller rotary sprinklers are, however, now available, and it is possible that portable systems similar to those common in North America may become more widely used in Britain.

REFERENCES

(27) "Potatoes." Bulletin No. 94 of the Ministry of Agriculture and Fisheries.

(28) "Commercial Fruit Tree Spraying." Bulletin No. 5 of the Ministry of Agriculture and Fisheries.

(29) "Irrigation." Bulletin No. 138 of the Ministry of Agriculture and Fisheries.

CHAPTER EIGHTEEN

MACHINERY FOR MILK PRODUCTION

Of about 157,000 farmers in England and Wales who sell milk to the Milk Marketing Board, about half use milking machines. Many of the remaining herds are very small, and there are now few large herds which do not employ milking machines and milk cooling devices. Most of the larger herds also employ utensil sterilizing equipment. In recent years British manufacturers have developed a vast range of equipment for all these purposes, and only a brief outline can be given in this chapter of the main features of the most common kinds of equipment.

Milking Machines.

During the 19th century, many attempts were made to milk cows by inserting small metal tubes into the teats. These were a complete failure, partly owing to inflammation of the udder caused by inefficient sterilization. There were also many attempts to imitate hand milking, but such machines failed to produce satisfactory results. A patent for employing a vacuum to draw out the milk was taken out in 1851, and by 1889 a machine operating on the continuous vacuum principle was marketed. These were found unsuitable, and the principle of intermittent suction and release, now universally employed, was first tried about 1895. The other important principle which is employed in all modern milking machines—use of a double-walled teat-cup which provides for both suction and squeezing by a rubber liner—was introduced in Australia in 1903. Later developments of importance included the application of releaser-type machines to portable milking bails by A. J. Hosier during the nineteen-twenties, and the incorporation of refrigerated cooling in the milking bails in 1928. The most important recent developments are concerned mainly with the development of various types of milking parlours, and of portable machines suitable for milking small herds.

The increasing difficulty of getting hand milkers, together with cost studies which show that at present wage rates even herds of under 10 cows can be more economically milked by machine than by hand methods, are combining to accelerate the adoption of milking machines even on very small farms.

The Essentials of a Milking Machine.

An electric motor or a small engine drives a vacuum pump, producing a suction which is transmitted by a pipeline to the milking units. This suction is continuous, but the teat cups are double-walled, with metal bodies and rubber linings; and a device called the pulsator alternately connects the space between the metal walls and the rubber linings, first with the atmosphere and then with the suction. When the space between the "liners" and the metal walls is at atmospheric pressure, the liner is compressed and the suction is for a short time cut off, thus producing the intermittent suction and release that is essential.

FIG. 265.—TYPICAL PAIL MILKING UNIG (GASCOI-GNE)

The intermittent release from suction is necessary owing to the manner in which the milk is delivered to the teats. The milk is secreted in the alveoli in the udder, and passes down ducts to the teat cisterns, which are situated in the lower part of the udder, just above the teats. Suction draws off this milk but causes constriction of the cisterns and ducts, thus interfering with the blood circulation and preventing milk from passing rapidly from the alveoli into the cisterns. The release period allows the cisterns to become filled, and assists in this by a light massaging action.

The vacuum pump may be of rotary or reciprocating type. It sucks air from the vacuum reservoir and pushes it out to atmosphere. In general, rotary pumps work more smoothly and have fewer moving parts. It is necessary to have a vacuum tank in order to provide a reserve, which helps to keep the pressure constant. The pump usually maintains a vacuum of 15 in. of mercury, or about 7lb. per sq. in. (i.e. about one-half of the normal atmospheric pressure). A vacuum regulator or relief valve is placed in the pipe-line, in order to ensure against an excessive vacuum, which might be harmful, to the cows. In addition, a vacuum gauge and moisture traps should always be incorporated in the pipe-line.

The pulsator is an air valve so constructed that suction and release impulses alternate in the air line leading to the space surrounding the rubber teat cup liners. Pulsators may be driven mechanically or electrically, but are generally driven by the vacuum itself operating alternately on two diaphragms or on the two ends of a piston. It is, in effect, a two-way tap. In most machines it is set to give from 45 to 65 pulsations per minute. There is much variety in the construction of different makes of pulsator, and the buyer should make sure that this vital part is reasonably simple and free from trouble.

There is much difference of opinion on the requirements of pulsators. Most manufacturers aim to have suction and release periods of the same length, but some consider that a long period to extract the milk and a short release period to enable the teat cistern to refill is desirable, and design their machines accordingly.

The teat cups are an important feature of a milking machine. The type now generally accepted has cylindrical metal cups with rubber liners, but there are wide variations in details of design. The two main types of liner employed are the stretched liner and the moulded type. The relative merits of these two main types are hotly debated, each having some advantages. Some manufacturers attempt to combine the advantages of both. As in the case of the differences of opinion on requirements of pulsators, there is little scientific evidence to uphold one view or the other. One essential is that the cups should be easy to clean and assemble.

Types of Milking Plant.

BUCKET PLANTS. In bucket plants, the bucket is connected to the vacuum pipe-line by a rubber hose, and forms a sealed container from which the air is extracted by the vacuum pump. The pulsator is usually attached to the lid, and operates exactly like a vacuum-operated windscreen wiper. It alternately connects the space between the metal teat-cup wall and the rubber liner to the vacuum and to atmospheric pressure. On most machines, one pair of teat-cups alternates with the other; but some machines have all four teat-cups working together. The bucket usually stands on the floor, and the milk passes into it from the teat-cups by way of a metal claw and a long rubber milk hose. Several manufacturers produce suspended-pail units which are supported partly by a surcingle over the cow's back and partly by the pull on the udder. An advantage claimed for this type is the easy variation of the pull on the teat-cups. The teat-cups lead directly into the pail, and this eliminates the claw and the milk hose.

Portable Bucket Plants. The bucket type plant is easily made portable by mounting the power unit, vacuum pump and regulator on a trolley which can be pushed along behind the cows. If an electric motor is used, a trailing cable and one or two conveniently situated sockets are required. A pair of jib arms are usually provided to carry the vacuum line to the buckets, and these make it possible to milk several cows without moving the trolley.

Releaser and Recorder Plants. In releaser milking plants, pails are not used, and the milk is drawn into a pipe-line and conveyed directly to the cooler and churns. The pulsator is carried on the standing. An early disadvantage of this type of machine was the difficulty of cleaning long milk pipe-lines, but the modern tendency is to employ a small milking-shed, and keep pipe-lines short. The cows are milked in rotation.

Where the milk has to be weighed it is passed first into a glass jar suspended on a spring balance. After the weight of milk is recorded the turning of a master tap lets air into the jar, and this forces the milk along the milk pipe-line to the cooler. This type of plant is called a "recorder". Both releaser and recorder plants employ a device termed a "releaser" which delivers the milk from the pipe-line to the cooler without upsetting the vacuum.

Portable Recorder Plants have the power unit, vacuum pump, churn and weighing apparatus all carried on a trolley or on an overhead gantry rail which runs the length of the cowshed. As with the stationary releaser type plants, one of the advantages is elimination of the task of carrying milk in pails. With this type of plant, some form of in-churn cooling is generally employed.

The Milking Bail.

The application of the principle of "continuous" milking to British farming owes much to the pioneer work of Mr. A. J. Hosier, who, about 1925, commenced to manage his cows on what has come to be known as the "Bail" or "Open Air" system. It soon became apparent that some of the advantages of the system practised by Mr. Hosier were of wide application.

The modern bail consists of a movable shed, with an engine, vacuum pump, water pump, dynamo and boiler at one end; a cooler and milk store at the other end; and half a dozen cow stalls in between. The cows are fed as they are milked, the mangers for cake being built into the doors in front of the cows. When milking is finished the doors are opened by means of a lever, and the cows walk out to make room for the next batch.

The bail normally has a dynamo and a battery lighting set, with lights in the stalls and also in the compartments at the end of the bail. A water pump and cooling tank in the roof may be used to provide water for the cooler; but a refrigerator is much more satisfactory.

The advantages claimed in favour of the bail system are healthy cows, no manure spreading, no long walks for cows with full udders, no buildings to erect or repair, saving of labour and improvement of the land. It has certainly been proved to be a satisfactory method of running a dairy herd on poor downland.

FIG. 266.—A MODERN MILKING SHED EQUIPPED WITH AUTOMATICALLY RECORDING MILKING MACHINES

FIG. 267.—MOBILE RELEASER-RECORDER TYPE MILKING MACHINE WITH TWO MILKER UNITS. (GASCOIGNE.)

The chief disadvantages of the bail system are the difficulty of controlling dates of calving; the fact that the method is suitable only for herds of moderate yields; and the bad conditions under which the men have to work during the winter months.

The Operation of Milking Machines.

In the operation of milking machines it is essential to be systematic in everything that is done, if efficient working is to be secured. Such operations as oiling the pump and motor, keeping the pulsator at the correct speed and keeping the vacuum correct, should be attended to from time to time. The cows should be treated gently, and washing the udder, putting on and removing the teat cups, stripping, etc., should all be carried out in a thoroughly systematic way. Perhaps the most important requirement of all is keeping the machine clean.

The maintenance of high quality in milk production where machines are employed depends partly on cooling, but primarily on having all the milking machine parts and utensils properly cleaned and sterilized. The procedure to be adopted varies in detail according to the type of machine used, but the following general rules apply to all. First, thoroughly rinse in cold water before the milk film has a chance to dry on. Then wash well in hot water (120° F.) to which a detergent has been added. Next rinse in clean hot water, and finally sterilize metal parts at 210° F. for 10 minutes. Machines should be completely dismantled weekly, and thoroughly cleaned. Where chemical sterilization methods are employed, teat-cup clusters may be sterilized in $\frac{1}{2}$ per cent, caustic soda or a proprietary solution containing chlorine, by using a rack which keeps the cups filled to their brims.

The Economics of Machine Milking.

There is little to be gained to-day by arguing whether hand milking or machine milking is cheaper. Most farmers with more than 20 cows in milk simply cannot do without a machine, owing to the impossibility of obtaining skilled hand milkers. The time may be approaching when farm workers willing to do hand milking will be almost unobtainable.

In the past, the use of machines has certainly not been invariably successful; indeed, there were so many disappointments and failures among earlier models that the unfavourable opinions created still often militate against the use of machines; but it should be understood that the modern equipment is very different from that of twenty years ago, and there is to-day every prospect of satisfactory results, if it is installed in suitable conditions and is carefully operated. Machines are sometimes blamed for transmitting udder troubles throughout the herd, but a study of most cases where this happens reveals that the fault lies in the operation, rather than in the machine itself.

A Cambridge study has shown substantial savings of labour when machine milked herds are compared with hand milked; and releaser type plants were found, as expected, to require less labour per cow than bucket plants. Farms with herds of only 8 to 10 cows were found to be effecting economy by the use of a milking machine.

The effect of a machine on yield and depreciation of the cows must, of course, be considered. Many farmers consider efficiency of milking more important than small variations in the cost. Milking should be performed "quickly, quietly, gently and thoroughly". While a machine, with adequate attention, may milk a herd quickly, it does not generally milk individual cows more quickly than does a good hand milker. As regards quietness and gentleness, the modern machine has a distinct advantage, especially in hot weather when flies are troublesome. So far as thoroughness is concerned, the machine is at a disadvantage. Stripping is generally necessary after the machine if maximum yields are to be obtained.

Decreased yields may result from keeping the machine on too long and drawing the last pint in driblets. A valuable feature of modern outfits is a glass tube or vessel which reveals when milking is complete. Unfortunately, the amount of milk available in the different quarters, and the rate at which it can be removed, vary. Usually, the hind quarters give more milk, but the fore quarters have bigger teats and can be milked more quickly, so that the fore teats are often finished an appreciable time before the hind ones. One of the requirements of a perfect machine would seem to be that the machine should stop drawing at each teat automatically as the milk ceases to flow from that quarter.

In conclusion, it may be stated that although milking machines are not perfect, and will never be foolproof, they have been improved so much that any farmer who regularly milks a herd of more than fifteen cows would do well to consider whether the use of a machine would improve the efficiency of his farming. If management of the machine is good, production costs may be reduced and no reduction of yield or extra depreciation need occur.

Milk Cooling.

When milk is to be marketed it is generally necessary to cool it as rapidly as possible to a temperature of about 45° F. Cooling may be effected by causing the milk to flow over a special cooling surface, or by treating it in the churn. The medium used for cooling has in the past usually been water just as it comes from the farm supply; but in recent years there has been a rapid development of the use of refrigerated coolers.

*FIG. 268.—COMBINE RELEASER AND MILK COOL-
ER. (ALFA-LAVAL.)*

Surface coolers are used by the vast majority of British producers. They are normally made of heavily-tinned copper, and the milk flows down by gravity over the outside of the cooler, while the coolant or refrigerant passes through the inside from the bottom upwards. The tubular type of cooler consists of a series of parallel tubes sweated together and connected at the ends in such a way that the coolant must pass to and fro across the inside of the cooler. The corrugated type is a cheaper form of construction, two spaced corrugated sheets with sealed ends being substituted for the tubes. The corrugated type is slightly less thermally efficient, but has the advantage of simplicity and of providing a surface which is very easy to clean.

A perforated trough at the top distributes the milk over the cooling surface, and it is important that this should be level, so that the whole cooling surface is efficiently utilized. After long use, when the tin is worn off the copper on the cooling surface, the cooler must be re-tinned, since milk will attack and dissolve the copper. With an adequate flow of water through the cooler (3 gallons water to 1 gallon of milk), the milk can be cooled to within about 5° F. of the water temperature.

With churn-immersion cooling, the churns are immersed up to their necks in a tank through which water is flowing. Bacteria tend to collect near the surface with the cream, and it is important that these upper layers should not be cooled too slowly. Results should therefore be judged by the time taken to cool the milk in the tops of the churns. With a well-distributed water flow of 30 gallons per hour per churn the temperature of the milk should be within 10° F. of the water temperature in 2 hours. Advantages of the simple churn immersion method are that there is a minimum of equipment that comes in contact with the milk, and there is no cooling equipment that must be cleaned and sterilized regularly. Disadvantages are the difficulty of handling full churns into and out of the tank unless lifting equipment is provided. The work is facilitated if the tank is placed partially below a loading dock, and the churns may then stand in the tank until they are collected.

Coil immersion coolers employ a coiled tube made of stainless steel, which is connected to the water supply by rubber hose, and immersed in the churn.

Refrigerated Cooling.

In hot weather, cooling by water alone may be unsatisfactory, especially if the supply of water is limited. In this case it may be necessary to employ a refrigerator. The principle on which refrigerators work is, briefly, as follows:— An engine or motor drives a compressor which compresses a refrigerant, such as methyl chloride, ammonia or freon. Heat is developed in the compression of the refrigerant, and this is dissipated by cooling with water which may have a temperature of up to 70° F., or by air. In another part of the system the compressed refrigerant is allowed to expand, and in doing so it absorbs heat.

FIG. 269.—AIR-COOLED ELECTRIC CONDENSING UNIT FOR SMALL REFRIGERATOR. (PRESTCOLD.)

With DIRECT-EXPANSION coolers, the refrigerating liquid passes through an expansion valve and then boils at a temperature of about 30° F. as it passes through the coils of the cooler. The advantage of this method is that there is only one heat-exchange process, the refrigerant acting directly on the milk. The disadvantage is that the cooler cannot be dismantled for cleaning. It must be designed for easy washing, and must be sterilized *in situ* by providing covers which form a steam jacket.

The direct-expansion process is thermally efficient, but the refrigerating unit must be of higher capacity than storage-type coolers which use chilled water or brine as an intermediate cooling agent, and can build up a store of coolant in the intervals between milkings.

With CHILLED WATER surface coolers, the expansion coils of the refrigerating unit are immersed in water in an insulated tank; and in the period between milkings, a fairly large volume of water below 40° F. is built up. Frequently, the aim is to build up a block of ice round the coils. During milk cooling, the chilled water is pumped through a surface cooler and usually returned to the tank.

The advantage of building up an ice block is that the melting of a small volume of ice will absorb a large amount of heat. A small disadvantage is that care must be exercised to ensure that the size of the ice block is kept within the correct limits.

The chilled water is frequently employed in a 2-stage surface cooler—mains water being used as the coolant in the upper part, and the chilled water in the lower part only.

The great advantage of chilled-water cooling compared with other methods is that there is no difficulty about dismantling, cleaning and sterilizing the cooler. Brine may be used instead of chilled water, the advantage being that a concentrated solution can be cooled to very low temperatures without freezing. The use of very cold liquid in a cooler is not, however, without its disadvantages; for unless flow rates are very carefully controlled the milk will freeze on parts of the cooler, and operation becomes unsatisfactory. It seems likely that the chilled water method will become increasingly popular. Chilled water may be satisfactorily employed in insulated churn-immersion coolers, and also in coil immersion systems and other methods of in-churn cooling. With the churn-immersion method, the expansion coils may be arranged either round the walls of the tank or at one end. Agitation of the water is essential, and care must be taken to see that the ice block is kept to a reasonable size, so that there is no difficulty in getting the churns into the tank.

A method which has many advantages is that illustrated in Fig. 270 whereby the chilled water flows first through a cylindrical metal cooling head which is immersed in the neck of the churn, and then out and over the outside of the churn in a substantially unbroken sheath. The churns stand on a rack over the tank which contains the ice-bank, and a high-capacity pump has the double function of delivering chilled water over the churns and of circulating the water in the tank to ensure sufficiently rapid melting of the ice. Introduction of the coldest water into the milk at the top of the churn produces strong convection currents in the milk. As a result of this, circulation of 100 gallons per hour of chilled water at 40° F. effects cooling of the complete churn to 50° F. in as little as 30 minutes.

Rotating cooling heads which stir the milk are available, but the static type of head which relies on convection currents has the advantages of simplicity and cheapness.

Sterilizing Equipment.

If milk of good keeping quality is to be produced, all utensils for handling it must be kept clean and sterile. This means that after thorough washing, equipment such as milking machines, coolers, churns, and, in fact, everything that comes into contact with the milk, must be regularly sterilized. It has been found by practical tests that the best method of sterilizing utensils on farms is to place them in a box in which a temperature of 210° F. is maintained for not less than ten minutes.

Milk producers normally need some kind of steam generator, a sterilizing chest, nozzles or jets to which teat-cup clusters or milk pipe lines can be connected, and a churn stool on which churns or buckets can be steamed. In the past, the usual arrangement has been to have a solid-fuel boiler and separate sterilizing chest, but where electricity is employed for providing heat there are advantages in having combined units.

FIG. 270.—CHILIED-WATER IN-CHURN COOLER EMPLOYING "CONVECTION" TYPE COOLING HEADS. (CASCOIGNE.)

Solid-fuel boilers which burn coal or coke are available in a range of sizes designed to evaporate upwards of 30 lb. of water per hour. Working pressures need not exceed 30 lb. per sq. in., since a pressure of only 1 lb. per sq. in. is adequate to drive steam into sterilizing chests, pipe-lines, etc., where there is a free exit. What matters is the ability of the boiler to evaporate the water at the desired rate, so that a temperature of 210° F. can be maintained. The disadvantage of using solid-fuel boilers is the work involved in laying the fire and lighting up, and the need to stoke carefully to get an even fire when the steam is required.

Several manufacturers now produce oil-fired equipment for their dairy boilers, and this has many attractions. Fuel cost is usually little higher than with solid fuel, because the burner can be switched off as soon as steam is no longer needed. The labour of stoking is saved, and automatic or semi-automatic control can be arranged where necessary.

Electric Sterilizing Equipment.

Though heating by electricity looks expensive if it is compared with solid fuels and oil in terms of cost per B. Th. U., the fact that the heat can be much more efficiently applied to the job, and the great advantages of cleanliness and ease of control, combine to make it the most attractive method to a steadily increasing number of farmers. Three main types of electrical sterilizing equipment are employed, viz.:

Boilers with immersion elements.

Electrode boilers in which the water itself is the conductor.

Heat storage units where the heat is stored in a cast iron block or in water that is kept at a high pressure.

The first two methods involve fairly high electrical loads, whereas equipment of the third type operates at low loadings.

Boilers which employ immersion elements can be obtained with loadings ranging from tinder 10 to over 30 kW. Working pressures are normally up to 30 lb. per sq. in., and control is either manual or by thermostat. Safety devices are provided against the risk of letting the water get too low.

Electrode boilers usually have iron electrodes mounted on a float, and a typical 20 kW. unit will deliver 60 lb. of steam per hour at 15 lb. per sq. in. Heat storage boilers are somewhat more expensive, but are becoming increasingly popular. The most common method is to employ a loading of 1 to 2 kW. to heat up a heavy cast iron block to a temperature of 600 to 1,000° F. The block is well insulated, and is enclosed in a steel casing. When steam is required, water can be introduced into the casing, where it immediately forms steam, the rate of steam production being regulated by the flow of water. Where water-heating is also provided for in a hot-block storage steam raiser, it is usually provided independently by means of a small immersion heater. A second type of storage unit employs a large volume of hot water under a pressure of 90 lb. per sq. in. One model which employs a 3-kW. heater raises $67\frac{1}{2}$ gallons of water to a temperature of 331°F., and this is sufficient to provide 80 lb. of dry steam at atmospheric pressure.

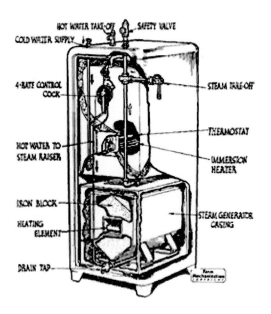

FIG. 271.—*ELECTRICALLY-OPERATED STORAGE TYPE STEAM RAISER WITH AUXILIARY TANK FOR HOT WATER SUPPLY. (WOOLEY.)*

Sterilizing Chests.

Sterilizing chests are usually square boxes made of steel, galvanized after manufacture. A large door is provided, together with various racks on which the equipment can be stacked. Free drainage outlets are provided for the water that condenses, and these ensure that no pressure is built up inside the chest. After the door is closed and steaming commences, it usually takes 20-30 minutes before the temperature of the whole chest reaches 210° F. Steaming should continue for a further 10 minutes, and after this the door is opened slightly to allow the steam to escape and the utensils to dry. Sizes of chests commonly range from 16 to over 40 cubic ft., the 27 cubic ft. size being widely used. Teat-cup clusters and churns may be sterilized on special fittings outside the chest.

Self-contained Electric Sterilizing Chests.

Self-contained electric sterilizing chests may be either of the immersion heater or hot-block heat storage type. The chest itself, unlike most of those which are independent, is thermally insulated, having a double-walled construction. The steam passes directly into the chest without any bottleneck, and the unit is thermally efficient. It occupies only a small floor space, and can provide a hot water supply for washing. Such units are not suitable for recorder plants.

Milk-bottling Machines, etc.

On many farms the milk is filled into bottles ready for retailing, and there exist many useful devices for filling, sealing and washing the bottles. Bottle washers usually consist of a galvanized steel trough, fitted with electrically driven rotating brushes and with rinsing jets. Such outfits are almost essential where there is a large retail round, for they not only save labour but also do the work more efficiently than it can be done by hand methods.

The Cream Separator.

On farms where a trade for cream exists or where butter is made, the cream is usually separated by mechanical means. Efficient separation of the butter fat from milk is an important matter, for the total fat is normally less than 4 per cent., and a small loss of cream is serious. Separation is effected by taking advantage of the fact that the specific gravity of the fat (0·930) is lower than that of the rest of the milk (average 1·036). The old methods of separation were by gravity alone, the chief being the shallow pan method, in which the cream rose to the surface and was skimmed off by hand after about twenty-four hours. This method is slow, and results in skim milk a day old. On average, about 0·45 per cent, fat is lost, and the method is now obsolete.

The centrifugal separator accomplishes in a few minutes a separation which is much more efficient than can possibly be achieved by any method relying on the force of gravity alone. The whole milk is rotated at high speed in a steel bowl, and the heavier elements of the milk pass to the periphery, leaving the lighter cream near the centre. The process is continuous, and with a separator properly adjusted it is possible to reduce the loss of fat to 0·02 per cent.

Construction of the Cream Separator.

The whole milk is placed in a large receiving tin, and from this it passes through a float-controlled valve into the centre of a bowl which is rotated at 6,000-9,000 r.p.m. The bowl is generally driven via several pinion gears and a final worm gearing, in which a large wheel drives a worm connected to the bowl spindle. On modern machines the gears are enclosed in an oil bath, and ball bearings are provided throughout. The bearings and gearing must be carefully constructed to ensure smooth running and a low power requirement. Where a power drive is employed, provision must be made for attaining full speed slowly, for the heavy bowl has considerable inertia.

The bowl is normally supported on a vertical spindle. It must be perfectly balanced, otherwise a dangerous vibration may occur at high speeds. The whole milk passes down through the hollow centre of the bowl, and through the holes in the base of the "distributor" to the spaces between a series of conical plates. The conical plates separate the liquid into thin layers with the object of ensuring that the speed of the bowl is rapidly transmitted to its contents. During separation the cream passes to the inner edges of the plates, and is forced upwards by the pressure of the incoming milk, along grooves in the outside of the distributor, to the top of the bowl. There it passes through the cream screw, a hollow regulating screw which projects into the central space. The purest cream is nearest the centre of the bowl, and the farther the screw is moved inwards, the richer the cream skimmed. In practice, a cream containing about 35 per cent, butter fat is usually skimmed. The separated milk passes to the outer edges of the conical plates, and moves upwards round the outside of the false top to the skim-milk outlet. The skim-milk and cream are thrown into separate tinned covers, which convey the two constituents to separate receptacles. The bowl should be constructed in such a manner that it is easily taken apart, cleansed and reassembled. All parts which come into contact with the milk should either be well tinned or made of stainless steel.

Care and Operation. Separation should commence while the milk is still warm, as it is then easier and more efficient than at temperatures below 85° F. If the milk has cooled below this temperature, it should be heated to 90° F.

The flow of milk through the separator should not commence until the bowl has attained the requisite speed, when the tap should be turned on and separation performed at uniform speed and with no break in the process.

Special attention should be paid to cleaning the bowl. When separation is completed, a gallon of hot water (150° F.) may be passed through before the bowl is taken apart, washed clean in warm water and rinsed in boiling water. All tinned parts should be left absolutely dry. The bowl parts must be handled with care, for if they become bent the balance may be upset.

REFERENCES

(30) Dixey, R. W. *Open Air Dairy Farming.* Oxford Univ. Agric. Econ. Res. Inst., 1942, 2s. 6d.

(30a) "Modern Milk Production." Bulletin No. 52 of the Ministry of Agriculture.

(30b) A. J. and F. H. Hosier. *Hosier's Farming System.* Crosby Lockwood.

CHAPTER NINETEEN

CROP DRYING EQUIPMENT

"A future generation may witness the utilization of large areas of grassland for the sole purpose of production of protein concentrate."

H. E. Woodman *et al. Jour. Agric. Sci.,* 1927, xvii, 242.

The artificial drying of herbage and of grain on farms has been one of the most noteworthy developments in the technique of farming during recent years. Grass, lucerne and grain are now regularly preserved by drying.

The Principles of Crop Drying.

Although moisture may be extracted from various substances in a variety of ways, the method universally adopted in farm crop driers is "vaporization", with removal of the moisture in a stream of drying air. Vaporization does not necessarily involve boiling the water and converting it to steam; in fact, many farm driers operate at temperatures well below boiling point, and some at atmospheric temperature. The picking up of moisture by a stream of drying air depends on the fact that the moisture in the crop tends to come into a state of equilibrium with the drying air. The power of the air to dry the crop depends on the difference between the amount of water vapour it already contains and the amount that it can hold if saturated. The amount of water that a given volume of air can carry increases rapidly with temperature. For example, 1 cubic foot of saturated air holds 2 grains of water at 32° F., 20 grains at 100° F., and roughly 200 grains at 200° F. It may be seen, therefore, that if there is a given amount of water to be removed from a material by an air stream, this may be achieved either by a small volume of air leaving the drier at high temperature, or by a larger volume exhausted at a lower temperature. It would, of course, be useless merely to increase the temperature of air used for drying if this air remained saturated with water vapour. Heating a given volume of atmospheric air which contains a definite amount of water vapour in fact reduces its relative humidity—i.e. the ratio of the actual amount of water vapour that it contains to the amount that it could hold at the given temperature if the air were saturated. Heating the drying air is, therefore, a common method of speeding up crop drying processes, and incidentally of increasing thermal efficiency. Drying temperatures employed are usually as high as they can safely be, without causing damage to the material being dried, or leading to inefficient operation in other ways.

The germination of grain is easily ruined by use of too high drying temperatures, while burning of the most valuable parts of the crop may occur if too high temperatures are employed in grass drying. Methods of reducing the relative humidity of air without increasing its temperature include drying it by chemical means—a method mentioned later in connection with the drying of grain in ventilated silos.

Crop drying processes cannot normally be carried out at unlimited speed. For example, in grass drying, the moisture in the leaf readily passes out through the innumerable stomata in the leaf surfaces, but moisture in the stemmy parts of the crop takes a long time to move to the surface, where extraction can take place. Similarly, surface moisture on grain may be removed in seconds, whereas moisture from the middle of the berry may take minutes or even hours to diffuse to the surface. It is for reasons of this kind, as well as on account of damage to the crop, that drying must proceed at a moderate speed in many types of driers, and that grass bruising and similar processes which assist rapid drying are of commercial interest.

Most grass and grain driers employ some kind of furnace for heating air, and a fan for blowing it through the drier. Both green crops and grain are usually dried by "direct" heating methods, in which hot gases from a furnace, and a controlled volume of cold air, are mixed together and passed through the crop. In grain driers, temperature of the mixture may be controlled at from below 100° F. up to 180° F.; in low-temperature grass driers, temperatures of 220° to 300° F. are common; while in high-temperature grass driers initial temperatures may be well over 1,000° F. In the case of a continuous grain drier, drying may take an hour or more, whereas with high-temperature grass driers the light fractions may be completely dried in a few seconds. At the other extreme, grain at the top of a ventilated silo may take 10-14 days to dry.

When the capacities of various types and makes of driers are compared it is important to specify working conditions. With grass driers, the "standard" operating conditions are drying from an initial moisture content of 80 per cent, to a final moisture content of 10 per cent. With grain drying the standard conditions at which capacity should be stated are for drying grain from a moisture content of 21 per cent, to 15 per cent., at a hot air temperature of 150° F. Statements of output which do not define the initial and final moisture contents are usually either meaningless or misleading.

Grass Drying.

Moisture Content. The moisture content of grass that is growing in the field depends on the type of herbage and its maturity, but in general it is around 70 to 85 per cent, (wet-base). In favourable drying weather, drying of the cut grass for a short time in the field can reduce moisture content to about 65 per cent, or even to below 60 per cent, moisture. The effect of this field drying or "wilting" is important. If grass having an initial moisture content of 80 per cent, is dried down to 10 per cent., $3\frac{1}{2}$ cwt. of water have to be evaporated to produce 1 cwt. of dried grass. At 70 per cent. initial moisture content, only 2 cwt. of water has to be removed to produce 1 cwt. of dried grass, while at 60 per cent. the amount of water to be removed is less than $1\frac{1}{2}$ cwt. Thus, a small difference in moisture percentage at high levels has a great effect on the amount of water to be evaporated. The figures serve to emphasize the value of "wilting" where this is practicable.

It should be noted that grass is not normally completely dried. If it is to be baled there is no point in drying it to below 10 per cent. moisture content. In fact, if really uniform drying were possible, 15 per cent, moisture content would be low enough for baling, since the grass will arrive at some such figure by re-absorbing moisture from the air after storage. The reason for not aiming at an average of above about 10 per cent. is the danger of having damper patches which would go mouldy. When grass is to be milled it pays to dry it down to about 6 per cent., otherwise the mill requires an excessive amount of power, and blockages may result.

Drier Capacities and Efficiencies.

The output of a drier in terms of dried grass produced per hour fluctuates widely according to initial moisture contents. The best measure of a grass drier's capacity is, therefore, its water evaporative capacity under working conditions; and the best measure of its efficiency is its consumption of fuel in B.Th.U. per lb. of water evaporated. These figures are available for driers which have been officially tested by the N.I.A.E. A typical good conveyor-type low-temperature grass drier with arrangements for recirculating the air which finishes the drying process should have a specific consumption within the range 1,500 to 2,000 B.Th.U. net per lb. of water evaporated. In terms of more common usage, a good efficiency for a grass drier is 90 to 100 lb. of water evaporated per gallon of oil fuel. Very small cheap plants with no recirculation may have much lower efficiencies than this, and all driers represent some sort of a compromise between the requirements of high efficiency and low capital cost.

Fuels.

The fuel used in commercial grass driers is now usually either oil or coke, but coal is also used on some large machines which are equipped with automatic stokers. Compared on calorific value, gas oil is considerably more expensive than coke, and heavy fuel oil a little more expensive; but oil has advantages over coke in providing quick starting and allowing easy thermostatic control. It also has the advantage of needing little labour. A popular method of estimating the efficiency of drying is in terms of gallons of oil burned per ton of dried grass produced; but this, of course, varies considerably with wet grass moisture contents. With grass of 75 per cent, moisture dried to 10 per cent., an average figure of 90 gallons per ton is not unreasonable. In good weather, when full advantage can be obtained from field drying, figures of under 60 gallons per ton are possible. On the other hand, in a wet season, when the moisture content of the incoming grass often exceeds 85 per cent., well over 120 gallons of oil per ton of dried product may be required.

Types of Grass Driers.

Tray Driers. The simplest type of grass drier is a single-tray machine in which the hot air is blown upwards through a layer of grass placed on a perforated metal floor. In this simple form, a fairly thick bed of grass is employed in order to minimize the effect of the period towards the end of drying, when the drying air is exhausted in an unsaturated condition, with its drying potential not fully utilized. In operating a single-tray drier such as the Slade-Curran, the 2 ft. 6 in.-deep tray is first filled with grass, and this filling is partially dried. The next step is to top up the drier with a further supply of wet grass, and continue drying until this second layer is nearly dry. A layer of wire netting may then be placed over the grass and a third lot of wet grass added above it. The latter is removed before it is completely dried, the fully dried grass from the two bottom layers is unloaded, and the partly dried top layer is added to the next filling. Working in this way, two or three complete batches totalling about 10 cwt. of dried product may be dried in a day from 75 per cent, moisture content, at a fuel consumption of about 5-7 cwt. of coke. This type of tray drier, which in the small sizes has a very moderate capital cost, can produce its 10 cwt. of dried grass daily with only part-time attention by one man, and is suitable for small dairy farms. It can also be used for grain drying if necessary.

Most modern tray and conveyor driers achieve improved efficiency by having the air flow designed so that the relatively dry air leaving the grass in the final stages of drying is either returned to the furnace, or passed through further layers of grass which are not so dry.

Air is thus exhausted only from wet grass, which gives up moisture at a high rate, and so causes the air to leave the drier in a saturated condition.

For example, in the I.C.I, drier, 4 trays in a line are employed, and drying is carried out in two distinct stages. Wet grass is first loaded on to an outer tray to a depth of about 18 inches, and air at 300° F. is passed up through the mat. When this tray of grass is partially dried, the grass is tedded and turned into the inner tray, where drying is completed. The drying air which passes through the partly-dried grass in the inner tray is returned to the furnace by means of a steel hood which covers the tray during the drying period. The same 2-stage process is subsequently carried out using the other two trays of the drier, the method of operation being to make one pair of trays ready while the other pair is being dried. The steel hood slides sideways to cover either of the two middle trays. In this way, drying is almost continuous, and tests show that with careful operation a quite good thermal efficiency is obtainable (e.g. 90 lb. of water evaporated per gallon of oil burned in the furnace).

FIG. 272.—A BATCH TYPE GRASS DRIER. (I.C.I.)

Several makes of tray driers are available, with outputs ranging from 1 to 4 cwt. of dried grass per hour. Some employ recirculation, some use the technique of shaking out the grass half-way through the drying process, and some use both methods. Some makes of tray driers are portable (e.g. Opperman).

One type of batch drier, the Fewster, has six trays disposed in the form of a hexagon, and the whole assembly of trays can rotate in a brickwork pit. The two trays adjacent to the fan are covered by a sheet steel hood, which can be raised clear of the trays by means of hydraulic jacks. Four trays contain grass being dried, while one is emptied and one recharged with wet grass. Hot air passes up through the final drying tray, down *via* the hood through the grass in the next tray, and then up to atmosphere through the wet grass in the other two drying trays.

The labour requirement of batch driers is usually greater than that for other types, owing to the handwork involved in moving the grass to and from the trays. Efficiency depends to a great extent on the care that is taken to secure a uniform mat of grass. Advantages of batch driers are low initial cost and depreciation, and a low power requirement, compared with most other types. Most batch driers are multi-purpose, machines which can also be employed for grain drying, provided that temperature can be effectively controlled and distribution of air-flow through the trays is reasonably uniform.

Endless Belt Conveyor driers (e.g. Ransomes, Templewood, Petrie and McNaught, Wrekin, etc.) comprise a drying chamber or chambers through which run one or more endless conveyors. The most common type of endless conveyor consists of a series of well-fitting perforated plates carried on chains and driven by a sprocket at the end of the drier. Means are provided for easily altering the speed of the conveyor, frequently by means of a ratchet drive. Other types of conveyor are made of woven wire. While many driers employ only one conveyor belt, one widely used grass drier (Templewood) has two, one above the other, while some other machines have several, arranged so that the material from the upper conveyor falls on to the second, and so on.

FIG. 273.—ENDLESS-CONVEYOR TYPE DUAL-PUR-POSE GRASS AND GRAIN DRIER SHOWING (LEFT) GRASS SELF-FEED EQUIPMENT AND (RIGHT) OIL-FIRED FURNACE. (RANSOMES.)

The herbage must be fed on to the conveyor in a uniform layer, and practically all modern machines can be fitted with a self-feed device which will secure this with all but very long material, and which also saves labour.

A self-feed is almost essential to the efficient operation of conveyor type machines.

Practically all modern conveyor driers employ recirculation, the air which passes through the bed at the dry end being collected and led back to the furnace or to a mixing chamber which supplies part of the air to a second fan. Some driers employ two or three separate hot-air fans in addition to a cooling fan, with recirculation of air from one to the next until finally the air passes out to atmosphere through the wet grass near the inlet end of the machine. Drying temperature may be 300° to 350° F. at the inlet end, about 250° at the second stage and about 220° F. at the third stage.

The in-going material should be kept as uniform in moisture content as possible, and the bed should usually be maintained at the same thickness, control being achieved mainly by varying the speed of the conveyor. With full recirculation such driers can evaporate up to 100 lb. of water per gallon of fuel oil used.

Some types of conveyor driers can be efficiently used as both grain and grass driers. Securing a satisfactory performance from such machines demands not only a properly designed drier, but also a well-planned installation which takes account of the needs in handling both grass and grain crops.

High Temperature Driers are usually of the rotary drum type or may employ a pneumatic tower and a rotary drum. All forms of high-temperature driers require the material to be chopped into short lengths. Advantages claimed for high-temperature driers are *(a)* that their thermal efficiency is higher than that of low-temperature machines, and *(b)* that they require less labour to operate. It is true that high-temperature driers can have high efficiencies in terms of water evaporative capacity; and they are seen at their best in drying very wet materials such as sugar beet tops. Unfortunately, it is not always possible to get the best out of such machines when they are required to handle such materials as wilted grass. The labour requirement of various types of drier is determined by the feed mechanism and not by the temperature at which the drier operates.

In a typical rotary drum drier the chopped grass is fed into the slowly rotating drum directly in the path of the hot gases, which are sucked through the drum, by a fan at the discharge end. The temperature of the in-going gases is usually of the order of 1,000° F., and light, leafy material is dried in a few seconds, and is then carried along in the air current and delivered to a large cyclone. The heavier, stemmy parts of the crop take longer to dry and to become airborne. During its passage through the drum the grass is constantly picked up by vanes on the inner circumference of the drum, taken by them to the top, and dropped into the air stream. Staggered vertical baffles impede progress through the drum and increase selectivity. The whole drum is maintained at a slight vacuum, and a high exhaust temperature of about 225° to 240° F. is usually maintained.

Single-stage high-temperature driers tend to be inefficient due to their allowing insufficient time for moisture to diffuse out of the stemmy parts. They often present the equally unattractive alternatives of over-drying and scorching the leaf, or under-drying the stem. In order to remedy this, two-stage and three-stage driers are employed. In the first stage, very high temperatures are used and the exhaust gases go to atmosphere. In the second and third stages, lower temperatures are employed, and the exhaust gases are re-circulated.

An example of this type is the pneumatic tower and drum type (e.g. Aldersley-Pehrson). In the first stage, chopped grass is fed into an adjustable mechanical selector, which removes stones, etc., and also discards some of the grass if too much is accidentally fed in. The selector delivers the chopped grass into the main air stream, which passes up a tall tower and thence to a large cyclone, the exhaust passing directly to atmosphere. From the cyclone the grass passes on to two rotary drums, the first at about 500° F. and the second at a still lower temperature. Selection of dried material from wet continues at all stages of the process, the dry material passing rapidly to a position where it will not be burned.

It should be understood that in all types of drier in which really high temperatures are used, the gases and forage travel in the same direction (i.e. they are "co-current"); and by the time the material is dry enough to be inflammable, it should have travelled to a position where the temperature is low enough to preclude scorching or fire. So long as the herbage is moist, its temperature remains low, owing to evaporation; but it is rapidly spoiled if it remains at temperatures of 300° F. or over after it is dry. One of the chief disadvantages of high-temperature driers is that control is somewhat difficult, and needs the more or less constant attention of an experienced man. High-temperature driers are particularly difficult to manage if the material is wilted, so that parts of the leaf are already fairly dry before they meet the hot gases. Scorching and fires may then occur. For farm-scale drying in Britain there is much to be said in favour of batch and conveyor machines which employ initial temperatures not much above 300° F., and which can in many instances be used also for grain drying.

The Operation of Grass Driers.

The successful employment of grass driers on farms involves much more than the economical removal of the moisture from wet herbage. In the first place, suitable crops for drying must be grown, and the produce must be dried at the right stage of maturity. The productivity of permanent grassland varies widely, but good fields in the recognized grassland districts may yield 4-5 tons of wet grass per acre in May and June, 2-3 tons per acre in July and August and 3-4 tons per acre in the autumn, giving a total yield of *dried* grass of 2-3 tons per acre. It should, however, be observed that permanent grassland in the dry east of England cannot be expected to produce high yields during the summer months. In some districts, temporary leys or arable crops are more reliable sources of good quality herbage for drying. Where pastures are reserved entirely for drying, 3-4 cuts per year may be obtained. The question of pasture management for grass drying is a very important one; but it is essentially a husbandry problem and cannot be discussed here.

Collecting Grass for Drying.

The machinery used for collecting green crops for drying has been described in Chapter Ten, and all that need be added here is an indication of the size of the grass collection problem. A drier of moderate capacity, e.g. one which produces 4 cwt. per hour of dried grass from material of 80 per cent, moisture content, will need about 18 cwt. of wet grass per hour. This is not in itself a difficult task. The difficulty is to cut and collect the grass with a minimum of labour, and to keep the drier going for as long a day as is practicable. This usually means that some grass must be collected in the evening ready for an early start with the drier the next morning. As a general rule, cutting is not by any means a full-time job, and since the output of the drier may be greatly increased and the amount of fuel needed for removal of moisture reduced if the grass can be partially dried in the field, one of the chief problems is securing the right amount of field drying whenever weather conditions allow. In fine weather, 24 hours of field drying may be advantageous, and some kind of green crop loader is used for grass collection. In wet weather there is every advantage in using a machine such as a "Cutlift". Chopping long crops into 3-4 inch lengths appears advantageous, even for endless-conveyor low-temperature driers. As mentioned in Chapter Ten, grass bruising machines may also possibly be useful for speeding up the drying of stemmy crops, though the possibility that undesirable fermentation may develop if the crop is put into big loads after this treatment will need study.

Saving labour in unloading, and in feeding the drier is a matter which must be kept in mind when planning drier installations. One of the best methods for farm-scale work is to throw the grass from the trailers directly into the drier self-feed, or into the trays, and to have available enough trailers to collect the material for a day's drying in a fairly short period. In hot weather, care must be taken to avoid putting the grass into big loads or heaps if it has to stand over-night, otherwise serious heating may occur.

Dried Grass Storage.

Grass that is dried for home consumption may be conveniently stored in bales. Almost any type of baler may be used, provided that the baling is not done immediately after drying. It is a good plan to build the day's output into a large heap and allow it to stand overnight, so that moisture is re-distributed and the leaves become less brittle. The dried grass is easily handled, stacked and fed in baled form. The outsides of bales become bleached and there is some loss of carotene, but the insides retain their colour and feeding value fairly well over long periods.

Milling the dried grass requires fairly expensive equipment and more power than baling. If the moisture content were uniform at about 5 per cent., a motor of 10-15 h.p. would be adequate to deal with the output of a 4-cwt. per hour plant. In practice, however, damper patches are inevitable, and a mill of at least double this power is needed. A $\frac{1}{16}$-in. screen is usually quite fine enough. The meal may be stored in thick paper bags, and keeps its quality for long periods.

Many drying plants now employ cubing machines which add water or steam to the grass meal and compress it in a die which forms cylindrical cubes or pellets. Cubers are employed at most factory-scale plants where the dried grass is for sale, but it is doubtful if milling and cubing is better than baling for grass that is to be home-fed to cattle. Cubing machines are somewhat expensive to buy and to operate. Milling usually costs 30 to 40 shillings a ton and cubing 40 to 60 shillings a ton, even where the plant is fully employed.

Grass bales, meal or cubes should always be cool before being stored in bulk or sealed in bags.

Barn Hay Drying.

In the barn hay drying process, haymaking implements are employed to get rid of moisture as rapidly as possible in the field, and when the crop is half dry, having a moisture content of about 40-50 per cent., it is brought in and placed in a barn which has been laid out with suitable ducting and side-walls to ensure that an air velocity of some 20 ft. per min. can be maintained through the crop. The grass is carefully loaded into the drying chamber without trampling, and a depth of 6-8 ft. is put in at a filling, a further batch being added when this is partly dried. Axial-flow fans are usually employed, giving static pressures just over 1 in. w.g. Successful results have been obtained with unheated air, but there is more certainty of success where heat can be applied when atmospheric conditions make this desirable.

A few farmers have installed dual-purpose plants which are capable of drying hay and also grain in bags; and provided that fire risks are avoided, the limitations of the plants realized, and design requirements understood, it seems likely that such equipment may become widely used.

Grain Drying.

When corn is cut and threshed with the combine harvester in Britain, it is frequently necessary to dry the grain before it can be safely stored. Small quantities of grain containing only a slight excess of moisture may be dried slowly by repeated turning, but such methods are not applicable to large quantities, or to corn that is very wet. Grain, as harvested with the combine, often contains 20 per cent, or more of moisture, and it is necessary to reduce this to about 14 per cent, before safe storage is possible. Grain containing 16 per cent, moisture may be stored if it is turned occasionally, or if it is in sacks. Grain with 18 per cent, moisture will generally keep for a short period if stored in sacks which are well spread out and have their mouths open. The farmer who risks using combines without the help of a drier is at a disadvantage, for he must always wait for the grain to dry before he can start work. The use of a drier makes it possible to work the combine more hours per day, and this is important in a wet harvest.

Driers designed for handling grain include (1) continuous-flow driers, (2) batch driers including platform driers for bagged grain and (3) ventilated silo driers. These are briefly dealt with in the following pages.

Continuous Driers.

Continuous driers extract the moisture fairly rapidly, the usual time taken for individual grains to pass right through the machine being about an hour. In order to economize in size of plant but to maintain a good drying speed, fairly high temperatures are usually employed, and the grain is treated in thin layers. Heat is provided by an oil burner, a coke furnace, or sometimes by electricity, or steam heat exchanger. In tower driers, the grain is fed into a hopper at the top of the drier and moves continuously downwards through the machine, usually between perforated metal plates so arranged as to form vertical walls of grain, about 3-6 in. thick. The hot gases are blown or sucked through the grain, which gradually becomes drier as it falls to the bottom of the machine. The rate of flow of grain through the drier is variable, and is controlled by the speed of a conveyor, which removes the dried grain from the bottom. Samples of grain that are very wet must, of course, be passed slowly through the machine.

The grain walls in a tower drier may be arranged in the form of 2 walls along the long sides of a rectangle (Turner-Oxford, Mather and Platt), as 4 walls in a rectangular chamber (Kennedy and Kempe), in the form of a square (Porteus) or in a cylinder (Penney and Porter). Some driers do not employ perforated metal walls, but arrange for the air to flow through the grain from inverted V-shaped metal trunkings, into alternate rows of trunking that are connected to the exhaust system (e.g. Aldersley).

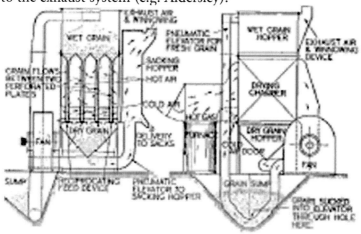

FIG. 274.—SECTION OF CONTINUOUS TOWER DRI-

ER WITH BUCKET ELEVATORS. (MATHER & PLATT.)

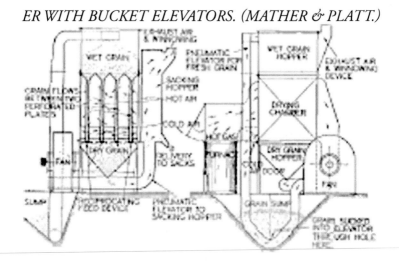

FIG. 275.—SECTION OF CONTINUOUS TOWER DRIER WITH PNEUMATIC ELEVATORS. (KENNEDY & KEMPE.)

The Jones-Mitchell is a thermally-insulated all-electric tower drier which operates on a contra-flow principle. As the grain flows downwards over a system of louvres, the air flows upwards and leaves the drier at the top. The electric heating is provided by means of thermostatically controlled elements arranged in two stages, part-way up the drier column. Cold air entering at the bottom first cools the grain which is about to be discharged. It is thereby heated to 100° F. At this point, its temperature is raised to 120° F. by the first heater elements. In its course up the drier the air temperature then falls to 80° F., by which time it reaches the second heater battery. This again heats the air to 120° F., and thereafter temperature falls until the cool moist air is discharged. This system results in a high thermal efficiency at the expense of a fairly high capital cost compared with many farm driers. As with many other driers in which the primary aim is high efficiency and not minimum cost with high output, indirect heating of the air by steam pipes can be employed if desired.

The Templewood-International is a continuous vertical-flow thermally insulated portable drier which employs an oil-fired burner with heat exchanger, and also makes full use of the waste heat from the power unit which drives the whole machine. The grain is elevated by auger and scraper-conveyor from an intake hopper to the top of the drier. It is distributed horizontally by an auger, flows downwards past longitudinal plenum chambers, and is drawn off by auger from the V-shaped bottom. The drying air is first preheated by being passed in part through the engine radiator and over the exhaust pipe. It is then drawn through the heat exchanger, the plenum chambers, and up through the grain by the main exhaust fan.

The engine drives, in addition to the fan, a $\frac{1}{4}$ kw. alternator to provide for local lighting and meter control circuits. An unusual feature is a moisture meter, in the grain conveying system, which effects recirculation of insufficiently dried grain.

Other types of continuous driers include those dual-purpose horizontal conveyor driers which may be used for both grass and grain, and also a number of other machines of miscellaneous type. Among the latter is the Butterley-Goodall, which is particularly suitable for drying grass seeds and other small seeds. The grain is moved over a series of open inclined trays by means of an adjustable knocking mechanism. The drying air is blown up through slots in the trays by a number of independent fans. Three, four or five trays may be employed in series, the last being used for cooling.

In the Wilmot-Blanch, the corn is elevated by wide buckets to the top of a fairly steep incline, down which it cascades. The drying air is blown through louvres in the floor of the incline. Three or more independent units may be employed in succession, the last being used for cooling.

Operation of Farm Type Continuous Driers.

For wheat or barley that is to be used for milling the temperature should not exceed 150° F. 180°F. is the maximum temperature that should be used for drying feeding grain. Great care must be exercised in drying corn that is to be used for seed or for malting. The critical temperature above which damage may result varies with the moisture content of the corn.

For barley and seed corn of up to 24 per cent, moisture content the upper limit should be 120° F., while for corn of over 24 per cent, moisture it is wise not to exceed 110° F.—a temperature which is safe for almost any cereal crop. With linseed and mustard, temperatures should not exceed 115°F. A temperature of 90° F. is the maximum recommended for some types of vegetable seeds.

Grain must be cooled immediately after drying, otherwise it will sweat. A separate cooling fan is usually employed to blow cold air through the grain columns during the final stages in the drier.

A continuous drier needs adequate storage capacity for damp grain, so that it can accommodate the production from the combines during the day, and continue with drying during the night if necessary. It is usually advisable to provide either a pit of about 10 tons capacity, or a smaller receiving pit and a separate pre-drying storage bin of adequate size. It is sometimes convenient to fill the pre-drying bin from the overflow of the elevator that feeds the drier. It is a considerable advantage to have a divided wet grain pit, as this facilitates changing over from one crop to another.

Where tipping trailers are used it is usually convenient to build a parapet about 12 in. high to increase the capacity and keep out water and rubbish; but with non-tipping vehicles a run-over pit covered by a grille is preferable. The sides of the pit should be made to slope at an angle of about 45 degrees.

When the grain is very wet, the discharge mechanism is adjusted to give a long drying period, and the moisture content of the dry grain must be checked to see that drying is adequate. If it is not, it may be necessary to dry the grain partially in the first operation and to re-dry it as soon as possible. In good harvesting conditions, on the other hand, the grain may be adequately dried if passed through the machine as fast as possible, or may not need drying at all. Temperature of the drying air with a coke furnace is controlled by a damper between the furnace and the hot-air fan, adjustment of this admitting varying proportions of hot and cold air to the fan. With oil-fired driers the intensity of the fire can be controlled by adjusting the amount of fuel burned. Where a coke furnace is used, this should be lighted while the drier is being filled, and all the gases must be diverted into the air until the fire burns clear. With all types of continuous driers, the grain that is in the drier at the beginning of work will not be properly dried. It should be returned to the receiving pit and mixed with incoming grain.

Batch Type Grain Driers.

When tray driers are used for grain harvested in sacks, the usual method of operation is to facilitate loading by laying a plank run-way for a sack barrow over the trays, and to level the grain at a depth of about 12 in. If much grain needs drying it will be worth while to fix up a grain hopper in front of the trays so that the dried grain can be easily raked off into it, and then transferred by mechanical or pneumatic conveyor to cleaner, and sacking-off spouts or storage as required. One small batch drier (Bentall) employs a rotary drum.

Platform Driers for Bagged Grain.

The platform drier consists essentially of a heater unit and a fan to provide warm air, and a platform with air ducts beneath and openings in the top, over which the bags of grain are laid. Warm air is blown into the air ducts and up through the grain to atmosphere. The drier normally operates at a temperature rise of about 25°—30° F., so drying temperatures can never be high enough to damage the grain. The usual size of platform has 40 to 50 apertures, and accommodates about 50 cwt. of grain at a loading. The most usual type of heater unit is an oil burner which has a consumption of about 1 gallon of Diesel oil per hour. In normal conditions, the rate of moisture removal is approximately 1 per cent, per hour.

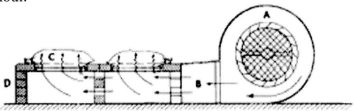

FIG. 276.—DIAGRAM SHOWING MODE OF ACTION OF PLATFORM DRIER FOR BAGGED GRAIN.

A, fan and heater unit; **B**, air duct; **C**, grain in bags; **D**, platform.

Electrical heating may be employed, and in this case it may be convenient to aim at a lower capacity. Thus, some small farmers find it satisfactory to have a 15 kw. unit operating in conjunction with a platform having 30 apertures, and removing only $\frac{1}{2}$ per cent, moisture per hour.

Coke-fired furnaces are also available for use with platform driers, and may be found advantageous in districts which have low night temperatures and high humidity.

The drier platform usually consists essentially of brick walls forming the air ducts, and pre-cast concrete slabs to provide the openings over which the bags of corn are laid. Where it is possible to excavate for the main duct it is usually convenient to lay the concrete slabs of the platform top so that they are flush with the surrounding floor. One of the chief requirements for easy operation is that there should be a good area of concrete at the same level as the platform, so that bags of corn can easily be moved to and from the drier on a sack barrow. Such an arrangement also permits easy storage of implements, etc., on the platform top during most of the year. Metal grids are placed over the apertures to help in keeping the grain layer fairly uniform in thickness, and to prevent accidents when walking on the platform. These should be removable to facilitate cleaning out any spilled grain or rubbish.

There must be easy access to the drier platform for unloading damp grain and taking off dry grain. Sometimes it is necessary to transfer the grain to a bulk store after drying, and account should be taken of all such handling problems in planning an installation.

The heater unit is preferably housed in a separate small shed adjacent to the drier building, in order to reduce fire risks and keep it clean. It is important to use suitable bags made of light hessian or similar material, and they should not be filled to capacity, otherwise it is impossible to get a good air seal when the bags are laid on the platform. The grain layer should be only 7-8 in. thick when the bags are on the drier, and the thickness should be as uniform as possible.

Uniformity of drying is improved if grain is collected from the field in a systematic manner, or if string of different colours is used to distinguish bags containing corn of different initial moisture contents. Provided that reasonable care is taken, uniformity of drying is quite satisfactory. It may be convenient to load the drier late at night with a batch of very damp grain, which need not be removed until the morning. One of the important advantages of the platform drier is the fact that a properly designed installation can be left unattended while the grain dries.

Other advantages of the platform drier include its ability to dry efficiently any kind of grass, root or vegetable seeds. Part loads can be dried when necessary by blanking off some of the sack apertures, but in this case the air flow will be cut down and care must be taken to ensure that the drying temperature does not become excessive.

Platform driers have potentialities as multi-purpose crop driers which are still not fully explored, and in view of their low capital cost it may be expected that their use on farms will steadily increase. Where grain storage in bulk is required, the grain which has been dried can be cleaned and conveyed to adjacent storage silos in the usual ways.

Ventilated Silo Grain Driers.

Where long-term grain storage is needed, and the grain can normally be harvested in a fairly dry condition, drying and storage in ventilated bins or silos may be the most economic system. A ventilated bulk storage system consists essentially of a fan and heater unit to provide a supply of slightly warmed air, ducting to convey the air to the grain, and some method of distributing the air so that it passes through the grain in the store. These essentials can often be quite cheaply provided, but where large quantities of grain are to be dealt with it becomes desirable to add such auxiliaries as a damp grain receiving hopper, grain clearing equipment, and conveying equipment to enable easy movement of the grain into and out of the store, or from one part of the system to another, The chief fundamental principle of operation of ventilated silo driers may be readily understood by reference to Fig. 277, which shows the effect of continuously blowing through corn air of varying degrees of relative humidity. It may be seen that for the production of grain of a moisture content of 14 per cent.—the figure necessary for prolonged storage in bulk—the relative humidity of the drying air needs to be about 60-65 per cent. The mean relative humidity of the atmosphere of the drier parts of Britain over 24 hours during harvest time varies from 75 to 85 per cent., and continuous ventilation with such atmospheric air will in practice produce grain having a moisture content of around 17-18 per cent. It is necessary to warm the air a little in order to produce grain of 14 per cent, moisture content, and the amount of warming needed varies with weather conditions. In cold, wet weather, the average temperature increase needed over the 24 hours is about 10° F., whereas 6° F. temperature rise is adequate in hot, dry weather. Heating equipment adequate to give 10°F. temperature rise must, therefore, be allowed for if the full drying capacity provided by continuous operation is to be obtained. If very little drying is needed and the fan only has to be operated in the daytime in good weather, less heating may be satisfactory; but farmers who plan to instal ventilated silos would be well advised to provide for full utilization where this is practicable. Heat is usually provided by electric heater elements, and the approximate requirement for 10° F. temperature rise is 3 kW. per 1,000 cubic ft. of air. Thus, if a fan working at maximum capacity delivers 7,000 cubic ft. per minute, the installed heater capacity should be 21 kW., and it should be possible to switch this on in stages, so that adjustment may be made for the amount of air being blown and for long-term weather changes.

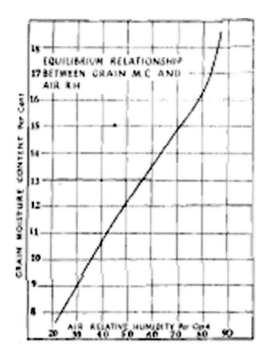

*FIG. 277.—GRAPH SHOWING EQUILIBRIUM RELA-
TIONSHIP BETWEEN GRAIN MOISTURE CONTENT
AND AIR RELATIVE HUMIDITY. (N.I.A.E. DRAWING.)*

Fig. 278 shows a typical layout of silos, ventilating ducting and conveying for a 150-ton farm plant consisting of six cylindrical silos. The conveying system allows for easy transfer of grain from one silo to another, and valves are provided to permit the ventilation of only one or several silos. As a general rule, a ventilating fan which is capable of effectively ventilating three or four silos simultaneously should be chosen. The rate of air flow through the silos needed for effective drying is a minimum of 10 ft. per minute, and preferably not less than 15 ft. per minute. The actual air speed will depend on the number of silos being blown and the depth of grain in them.

The silo floor may be constructed of porous foamed slag blocks laid over a honeycomb arrangement of bricks on edge, without any cement filling except where there are large crevices which would allow grain to run through. More expensive but more efficient alternatives include various kinds of wedge wire and perforated metal, and malt kiln tiles.

Drying proceeds from the bottom of the silo upwards, and after it has been in progress for a day or two there are three distinct layers in the silo, viz. dry grain at the bottom, a drying layer in the middle, and a layer where no drying has taken place at the top. The rate at which the dry zone moves upwards depends not so much on the temperature of the ventilating air as on the original moisture content of the grain. Thus, if only 1-2 per cent, moisture content has to be removed, drying should be complete in a 10-ft.-deep silo in 3 or 4 days; but if the grain originally contains over 20 per cent, moisture content it may be 10 days or more before the grain at the top is reasonably dry. Further heating of the air in such cases merely over-dries the bottom layers, and has little effect on the time taken to dry the top.

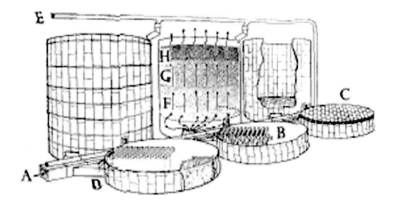

FIG. 278.—DIAGRAM SHOWING PRINCIPLE OF ACTION OF VENTILATED SILO GRAIN STORE.

A, ventilating air blown into main duct; **B**, beams over trench and bricks on edge to support floor; **C**, floor of porous blocks; **D**, valve in branch duct; **E**, grain conveyor; **F**, dry grain, zone; **G**, drying zone; **H**, damp grain.

The remedy for dealing with very damp grain is, therefore, to dry it in shallow layers. Grain of over 20 per cent, moisture ought to be first dealt with in layers not exceeding 5 ft. deep. It can be moved and bulked to full depth after a few days' drying. It should be emphasized that ventilated silo plants are at their best in areas where most of the corn can be harvested at moisture contents not exceeding 20 per cent.

Other methods of drying grain in bulk include distributing the ventilating air through a system of metal ducts which are built into silos or laid on barn floors. In the first case, the advantage is that the air has to flow only a short distance through the grain before it passes out through an exhaust duct, and so the drying of a full silo can be speeded up. In one type of plant, the air that is distributed through such ducts is dried by being passed over silica gel. The process is efficient, but rather high in capital cost, in relation to drying capacity. The advantage of the floor ventilating system is that little in the way of specialized equipment is needed, apart from the fan, heater unit and metal ducting.

Drying and storage bins may be circular or rectangular, and made of brick, concrete or other materials. Care must be taken to ensure that the walls are strong enough to withstand the pressures that will be developed when the bins are full. Where pre-cast concrete blocks are used, joints must be made air-tight. Successful installations range in size from single silos to plants having 8 or more silos arranged in two rows. Where multi-silo plants are installed, it is essential to ensure that the valves controlling the ventilating air flow are reasonably efficient. Heavy hinged doors, with oven-door type fasteners operated through a man-hole, or arranged to close on an inclined face and chain-operated from outside, can give a positive seal if a rubber strip is fixed round the edges.

The choice of a suitable ventilating fan needs careful study. The main types used are *(a)* centrifugal, and *(b)* dual-stage axial-flow. Both types have advantages for particular kinds of installation, and sound technical advice should be obtained on this and on many other aspects of planning comprehensive installations.

Cleaning and Turning. Dirty samples of grain should always be cleaned before being put into storage silos. Green material in the grain often collects in dense pockets in part of the silo, and these pockets may fail to dry out and may set up heating after ventilation ceases. For this reason it is always advisable to turn grain that has been dried from a high moisture content. One turning soon after drying is complete should be adequate to re-distribute any damp patches.

Choice of a Grain Drier.

Choice of a drier involves consideration of the climate of the area, the acreage to be harvested, and combine harvester capacity, since these all influence the moisture content of the grain. It also depends on the amount of capital available, and on whether long-term grain storage is required, as well as on such associated problems as grass drying and food preparing.

Continuous grain driers are usually needed where the drying problem is substantial, and where a fairly large amount of moisture normally has to be removed. Most farm driers of this type are rated at throughputs of from 1 to 4 tons per hour, for a reduction of moisture content from 21 to 15 per cent, at a hot air temperature of 150° F. Throughputs are lower at the lower drying temperatures needed for dealing with corn tor seed or malting. Nevertheless, a drier with a rating of 2 tons per hour should be easily capable of handling 100 tons a week, and is adequate for fairly large farms, except in the wettest areas, where much of the grain may need double drying. It should usually keep pace with two combines. The capacity of a continuous drier need not be equal to the maximum threshing capacity of the combine or combines. When the grain is wet the combine will always work short hours; and if necessary the drier can work extra time in order to keep pace. A small tray drier with an output of 2 cwt. per hour of dried grass can usually be expected to deal with about 10 cwt. of grain per hour. A typical platform drier will remove 6 per cent, from a 50 cwt. batch of grain in about 6 hours, and as it is easy to arrange for night operation it can deal with some 50 tons a week or more if necessary. The water evaporative capacity of a typical ventilated silo plant is small, but this can be partly made up for by continuous operation. In general, however, ventilated silo plants are unsuitable for farms where a large amount of grain is regularly harvested at moisture contents of over 20 per cent.

The cheapest insurance against damp grain being spoiled is a simple batch drier. Continuous drying machines are more expensive, but can be run with a minimum of labour. Where storage is needed and incoming grain moisture contents are not unduly high, ventilated silos may be most attractive. The cost of providing storage and drying need not be high, but a complete installation with elaborate conveying and cleaning machinery is usually fairly expensive.

Grain Drier Layouts.

With many plants incorporating continuous driers or ventilated silo driers it is necessary to arrange for complete flexibility so as to permit the grain to be passed through one or more of the following operations: (1) drying, (2) cleaning, (3) storing in silo, (4) removing from silo, (5) passing to food preparing plant, (6) sacking off for sale. In addition, it may be necessary to arrange for receiving grain in bulk and feedingstuffs in sacks, and also to plan easy loading out of corn on to transport vehicles. All this involves careful planning to ensure that the best advantage is taken of the abilities of the machines and of any differences in levels, so that the corn can be efficiently handled with the minimum use of mechanical equipment.

It must be borne in mind that the conveying equipment at typical farm plants is little used compared with that in provender mills, so it is necessary to keep expenditure on it within reasonable limits. Pneumatic conveyors have advantages in being completely self-cleaning and in being adaptable to irregular layouts. Disadvantages are a rather high power requirement, and sometimes excessive dust and noise. Bucket elevators are generally most satisfactory for straightforward vertical lifts, but portable scraper-conveyors which work at an angle may be preferable in some layouts. Where layouts are well-planned and regular, a combination of bucket elevators and horizontal mechanical conveyors may be most satisfactory. One of the simplest and cheapest types of horizontal conveyor is the shaker type, consisting of a wooden trough which is driven by an eccentric. Other types, which are less restricted in the length of run and can also easily be made reversible, include endless belts and chain-and-flight mechanisms.

REFERENCES

(31) Roberts, E.J. *Grass Drying.* H.M. Stationery Office, London, 1937.

(32) Hutt, A.C. *Combine Harvesting and Grain Drying.* London & Counties Coke Association, London, 1942.

CHAPTER TWENTY

ELECTRICITY ON THE FARM

The supply of electricity to country districts is important not only on account of its direct bearing on improved farming efficiency, but also because of its effects in eliminating drudgery from farm work and from the homes of rural workers. While this chapter deals mainly with the farming uses of electricity, it should be emphasized that in many instances the domestic applications such as lighting, heating, water-pumping, washing and ironing are of equal importance.

Many farms are, unfortunately, situated a few hundred yards up to half a mile or more from the supply line, and the cost of cable, whether it be placed overhead or underground, may be prohibitive. It has been estimated that 75 per cent, of the total cost of distribution of the supply to rural areas is accounted for by expenditure on the cables, etc. This raises the cost of the supply much above that in towns, where every house in a closely built-up street may be connected.

As is more fully explained later, electricity for most purposes must be available at a low voltage, e.g. 230 volts; but it is impossible to transmit power economically for long distances at low voltages, and the high-voltage lines now a common sight in rural areas convey it at 66,000, 33,000 or, for shorter distances, 11,000 volts. The transforming of power from high voltages to low requires expensive apparatus; and it is uneconomical to "tap" lines of 66,000 or 33,000 volts for the supply of isolated farms. It is, however, quite common to tap 11,000-volt lines, the cost depending to some extent on the amount of power required.

Electrical Terms and Principles.

It is quite impossible in a book such as this to explain, either in scientific terms or in simple ones, exactly what is meant by electricity. A fairly detailed study of physics is necessary for a proper understanding of the subject, and there are a number of books dealing with electricity and magnetism, and also with applied electricity. In this section, therefore, an attempt is made to describe, without detailed scientific explanations, those phenomena and terms which a farmer employing electricity is most likely to encounter.

Conductors and Insulators. All materials may be broadly divided into two classes, according to whether they contain electrically sensitive matter or not. Materials of the electro-sensitive class, including all metals, carbon, impure water, etc., are called CONDUCTORS, because an electric current flows easily through them. When an electrical pressure is applied across the two ends of a conductor the electrical particles therein contained are set in motion, thereby producing an electric current. Materials in the other class, such as wood, rubber, plastics, ceramics, air, oil and *pure* water, contain no electrically sensitive particles, and electric current cannot flow through them. Such materials are called non-conductors or INSULATORS.

Voltage and Current. The power station maintains an electrical pressure difference (VOLTAGE) between two or more conductors, which are normally separated by some form of insulator. (Wires which are laid underground or are fastened to the walls of buildings are covered with layers of insulating materials such as rubber, paper, cotton, etc., while overhead wires carried on wooden poles or metal towers are insulated by porcelain or glass bobbins and by the surrounding air.) Where the pressure available is to be put to useful purpose, e.g. a fire, or a lamp, means are provided to connect the two wires together by means of a path which includes the fire or lamp. The decision to use the fire or to light the lamp or start the motor is applied through a plug or a switch which is simply a convenient way of completing or breaking the conducting path and so allowing the electrical pressure to set the electrical current in motion or to stop it.

Any movement of electricity, or flow of CURRENT, through any conductor causes two things to happen:

(1) Heat is produced in the conductor.

(2) A magnetic effect is created.

These are the two main phenomena resulting from the movement of an electric current, and the whole end and aim of electrical engineering is to control the place, time and degree in which these two phenomena are expressed. For instance, the amount of heat developed in a cable carrying current *to* an electric fire must be kept to a minimum, whereas that developed in the wires forming the heating element of the fire must be sufficient to provide the heat required. Similarly, while the magnetic effect produced from a current in a single conductor may be little, this can be multiplied and concentrated by winding the conductor round and round to form a coil or bobbin; this produces the magnetic effect required for operating an electric bell, an electro-magnet or an electric motor.

It must be appreciated that the factor producing these phenomena and deciding their strength and effect is the electric current; the effect of electric pressure is only secondary in so far as it causes the current to flow or move and is therefore commonly called the ELECTRO-MOTIVE FORCE (or E.M.F.). The higher the E.M.F. operating on a given conducting path, the stronger will be the flow of current.

The E.M.F. is commonly measured in volts and referred to as VOLTAGE.

The current is measured in AMPERES or AMPS.

Electrical Power and Energy. The power available to do useful work is proportional to the product of the current (amps.) and the difference in electrical pressure between the ends of the wires (volts). It is measured in WATTS.

Power (watts) = pressure difference (volts) x rate of flow (amps.).

Since the watt is very small, so that many electrical machines require thousands of watts, the measure of power generally employed in practice is the KILOWATT (or kW.), equal to 1,000 watts.

The amount of energy conveyed by an electric current depends on the power and the time the current flows.

Energy=watts x seconds=volts x amps. x seconds.

The commercial unit of electricity by which the energy is sold is the kilowatt-hour or UNIT. One Unit is used if equipment taking 1,000 watts is on for one hour; if 100-watt equipment is on for 10 hours; or if motors, heaters, etc., taking io kilowatts are on for 6 minutes. Thus, a 40-watt lamp will run for 25 hours on a unit, while a 5-kilowatt motor will run for 12 minutes.

746 watts are theoretically equivalent to 1 h.p., but because of losses in the conversion of the electrical energy into useful work about 880 watts are normally requited for 1 h.p. with electric motors; and the use of o-88 Units provides for I h.p. hour of useful work. Thus a 5 h.p. motor takes 880 x 5 = 4,400 watts =4.4 kilowatts; and if it is run for 8 hours it may be expected to use 4.4 x 8 = 35.2 Units.

An electric fire dissipating 1,000 watts (1 kW.) in the form of heat generates heat at the rate of 3,400 British Thermal Units every hour. Thus if the heat requirements for a given job are known, the appropriate size of fire can be installed.

Resistance. All conductors offer some resistance to the passage of electricity (cf. pipe resistance in a water supply), and work must be done against this resistance in order that the current may flow. The work done is liberated in the form of heat, and the domestic electric fire is an example of a commercial application of this effect. In the electric lamp, the filament becomes heated to such a degree that it becomes incandescent, and some of the energy is liberated as light. The resistance of a wire is measured in OHMS and may be defined by the equation:

$$\text{Resistance (ohms)} = \frac{\text{voltage across the wire (volts).}}{\text{current flowing (amps.)}}$$

For any given material employed as a conductor, the resistance varies with the dimensions of the wire, being directly proportional to the length and inversely proportional to the cross-sectional area. The resistance depends on the metal of which the wire is made, a common conductor being copper, which has a very low resistance.

It has been pointed out that the power developed in an electrical machine is given by the product of the electrical pressure (volts) and the current (amps.). The power required can be produced either by a high voltage and low current or a low voltage and high current. In practice, however, there are two big objections to low voltages and high currents. Firstly, the value of the current determines the size (cross-section) of the conductor required to carry it; and heavy currents would require very thick cables. Secondly, the loss of energy in mains is proportional to the square of the current. High voltages are, therefore, employed for transmitting power long distances; but transformers are used to reduce the voltage to the consumer, in order to lessen danger to life and to simplify the design of equipment.

Alternating Current. We have so far considered mainly the case of direct or continuous current, where a pressure difference created by a dynamo or battery is transmitted to the equipment by two wires, and the current may be considered to flow from the high-pressure side of the dynamo, through the machine, and back to the dynamo. With the simplest type of alternating current there are also only two wires; but instead of the generator creating a pressure or voltage acting in one direction all the time, the direction of this pressure is changing rapidly from one conductor to the other. Thus if the two main conductors are joined by a third conductor or machine, the resulting current will flow backwards and forwards. In fact it will, in this country, flow 50 times in *each direction* in every second. Such an A.G. system, standard in this country, is said to have a frequency of 50 cycles per second.

It is quite easily demonstrated with a pocket torch that it is immaterial in which direction the battery is connected to the lamp—in either direction of current flow the lamp is lit. It is possible, too, but only to the craftsman, to rig up a device which will enable this direction of connection to be reversed rapidly, say 10 times a second, and it will then be seen that the lamp may flicker but it never really goes out. It is but a short step to realizing that with a reversal rate of 50 times a second the inertia effect will be sufficient to give the effect of continuous lighting. In a similar way, alternating current gives a continuous heating effect with fires and, indirectly, a continuous magnetic effect with such devices as electric motors.

Alternating current may have certain minor disadvantages as compared with direct current, but they are insignificant when compared with the advantages of the former. Without A.C., rural electrification would be impossible. This arises from the fact, stated earlier, that it is necessary to have high voltages so as to cut down the size of cables required to bring the energy from the power station, but that considerations of safety to life, etc., make it necessary to reduce the voltage to the user. A.C. solves this problem, since by means of a machine known as a transformer it is very simple to raise or lower the voltage in any part of the system. Thus it is possible to transmit large amounts of energy on comparatively light conductors at high voltages, using transformers to reduce the voltage to working value when required. The latest practice is to take high voltage lines right up to the farm and to provide a transformer in the farmyard.

Applications of Electricity.

Except for certain specialized applications such as electro-culture, electric fences, etc., electricity is used on farms for carrying out quite orthodox processes where heating, lighting and power devices are concerned. The general conditions applying in each case are dealt with in the appropriate chapters, and only those particular principles peculiar to electrical operation are dealt with here.

Lighting. The advantages of electric lighting in farm buildings, especially where dairying is carried on, are so obvious as to render detailed description unnecessary.

Special uses on farms include artificial lighting for laying hens during winter. With well-managed flocks it is found economic to extend the day to 12-14 hours, by providing light both in the morning and evening.

Heating. Practicable heating is produced by so designing apparatus as to concentrate in a comparatively limited space the dissipation of the appropriate amount of electrical energy. This heat may appear as "radiant" and visible heat as in an electric fire or radiator, or as "dark" heat as in the enclosed tubular heater. Apart from the psychological effect of the luminous bars and certain minor physical differences, the heating effect is similar from whatever source. The main criterion is the number of kWh. used per hour, remembering that every kWh. produces 3,400 B.Th.Units.

The drawback to electrical heating is, of course, that, whereas there are probably no rural areas in which the cost of a kWh. is less than $\frac{1}{2}$d. and several in which it is id. or more, I lb. of coal costing about $\frac{1}{2}$. is capable of producing some 12,000 B.Th.U.s or more. Why, therefore, is electricity used at all for heating? First of all, the straight "B.Th.U." comparison is not the whole story, since, while the conversion of the electrical energy into heat is possible at something approaching 100 per cent, efficiency, conversion of the coal's energy into heat when it is required is, in practice, probably rarely done at an efficiency higher than 33 3 per cent. This still leaves a gap which is met by the operational advantages of electricity—advantages on which it is difficult to place any money assessment, but which are of some considerable value.

These operational advantages are roughly as follows:

Cleanliness—an electrical sterilizer can be placed in the milk room without any fear of creating dirt.

Appositeness—the heat can be applied where it is wanted; this is a valuable point in heating chicken brooders.

Controllability—automatic control on cither time or temperature basis is readily provided.

Labour saving—possibly the most important single factor.

Both radiant heat and dark heat methods are employed in the heating of chicken rearers. One type of brooder has a cluster of carbon-filament lamps to provide the heat; another type employs a dull red heat provided by a heating element of the usual type; while a third employs elements of the tubular dark-heating type.

Similar heater units may be employed for warming other kinds of livestock, but a more suitable method of applying heat in farrowing pens is to locate warming elements of very low voltage in the floor of the pen. One simple method of installation is to build the warmed floor section above the original floor of the pen. A layer of dry sand $1\frac{1}{2}$ in. thick is first spread on the section to be heated, and a sheet of corrugated asbestos shuffled into it until it is firmly bedded, with no air spaces beneath. The upper troughs of the asbestos are then half-filled with sand, the wires of the heater element laid in the troughs, and more sand added. Then a sheet of expanded metal is laid on top of the sand and the whole covered with 11 inches of concrete. The wires of the heater element are connecter] to a transformer which produces a very low voltage. In cold weather such an installation results in fewer piglet losses and an increase in growth rate. Water heating and steam raising are other important heating applications which have already been dealt with in Chapter Eighteen.

The heating effect of electricity is seen to great advantage in running such devices as incubators, which use very little power when once the correct temperature has been reached.

Where heat is required in very large amounts, such as in soil heating or crop drying, electricity can only compete with other sources of heat if power is obtained at a cheap rate. Improvements are, however, continually being made in such matters as the proper insulation of hotbeds and the design of grain driers for all-electric operation.

Soil warming is now largely effected by the low-voltage transformer-fed system, and this method may be successfully applied to the heating of hotbeds, propagating beds and benches, to cloches, and to beds and borders in glasshouses.

For these purposes the "dosage" method is generally employed in preference to methods involving thermostatic control. By the dosage method, whereby the soiled is warmed for a limited period each day, soil temperature may fluctuate considerably; but this does not appear to affect growth adversely unless the temperature falls to near freezing-point.

The whole subject of heating by electricity needs to be viewed with an open mind, remembering the special advantages already mentioned.

Power: Single-Phase and Three-Phase. It is in the Use of electricity to provide motive power that the farmer stands to gain most advantage. In his consideration of electric power, however, he is often confused by the appearance of the terms "single-phase" and "three-phase". They both refer, of course, to alternating current (A.C.). Single-phase is the simple form, already explained, analogous to direct current and requiring only two wires to provide the supply. Three-phase is an ingenious development from the basic single-phase and may be regarded as a super-imposing of three single-phase systems to form one system. Its main value consists in the fact that it is possible to carry these three systems on three conductors and not three pairs of conductors as might be expected. This results in a substantial economy in the cost of supply lines, and the system is used for all main transmission and much secondary transmission line. When, however, we come to small lines and comparatively small loads, such as are found in rural areas, it is often cheaper to use a pair of conductors supplying single-phase only. One fortunate fact is that single-phase, which is required for lighting, small power and domestic applications, can be taken from a three-phase system.

In the past there has been some prejudice against single-phase motors, but any justification for this has long disappeared. It is, however, true that the three-phase motor has a better starting torque, so that for given duties the single-phase has always to be of a more advanced type. This results, too, in the single-phase motor always being dearer for a given duty, although the extra is largely due to the more complicated starting equipment required to reduce the starting current on the single-phase motor.

A.C. electric motors offer the most simple and most robust type of motive power unit available to farmers. An A.C. motor of good make should last a generation without anything other than ordinary running attention. The simplest and most robust of all types is the three-phase squirrel cage motor which has no connections at all to the moving part. This type is permissible in practically all three-phase supplying areas up to 3 h.p.; but because of the high starting current some area boards ban it above this, and few allow it above 10 h.p. These bans are generally lifted if special types of starter are fitted. In any case, if the squirrel cage motor is banned, the slip-ring type can be fitted, but this costs more.

It should be clearly understood by farmers that these various restrictions are not imposed arbitrarily nor merely to make them spend money unnecessarily. They are devised for the common good, since the imposition of high starting loads would interfere with the general supply, possibly causing lights to dim, while in certain cases of sensitive protective gear other people's motors would stop. (This sort of thing happens in the rural supply systems of certain foreign countries where no such scheme for the common good has been enforced.)

The speed of the ordinary commercial type of A.C. motor depends on the frequency of the system (50 cycles per second in this country) and is related to it. Motors have, therefore, the following synchronous speeds: 3,000 r.p.m., 1,500 r.p.m., 1,000 r.p.m., 750 r.p.m., 500 r.p.m. The actual speed at full loads will be 2,800 r.p.m., 1,480 r.p.m., 960 r.p.m., 710 r.p.m., 450 r.p.m. respectively, being rather higher at part loads.

There have been attempts in the past to make a case for special "farm motors". Experience has shown, however, that there is no condition on the farm which cannot be met by a standard industrial motor of some type. Indeed, there is little need on the farm for any other than the ordinary protected type of motor, even for dusty situations. Farm dust is not usually abrasive or corrosive, and there is rarely any trouble from this source.

A special type of motor is, naturally, required for such specialized work as use in connection with the submersible centrifugal pump. Where, however, splashing by liquid or any acid is expected, then appropriate types must be used.

Electric Motors and the Cost of Power.

The relative costs of electric motors and internal combustion engines for power purposes vary from farm to farm. An estimate of comparative fuel costs for medium horse-powers may be made on the following basis:

Diesel engines use about 0.5 lb. of fuel per b.h.p.-hour.

Petrol and paraffin, about o. 75 lb. per b.h.p.-hour.

Electric motors use about 0.88 units of electricity per b. h.p.-hour.

But the first cost, depreciation and repairs on electric motors are much lower than those on petrol and paraffin engines, which are, in turn, lower than those on Diesel engines. Generally speaking, Diesel engines, with their high capital charges and low fuel costs are likely to be economical where a large amount of heavy work is done, while electric motors, with lower capital charges and higher fuel expenses, may be more economical where powers are small or the number of hours worked per annum is lower.

With electrical power at id. a Unit or less the electric motor can compete with the internal combustion engine on any count, and farmers with electric power available at such rates need not hesitate for a moment about choosing an electric motor in preference to an engine. The fact that power is available simply by pressing a button, that filling up with fuel, changing oil, etc., may be forgotten, is worth a great deal to most farmers. There is great scope for extension of the use of electric motors on British farms.

Sizes of Motors. It is uneconomical from the point of view of initial cost and of running cost to instal a larger motor than is necessary. The efficiency of most motors at half load is about 80 per cent., while at quarter load it is only about 66 per cent., and falls rapidly at loads less than this.

Typical sizes of motors required for common farm tasks are as follows:

Milking	1-2 h.p.
Refrigerated Cooler	$\frac{1}{4}$-$\frac{1}{2}$ h.p.
Grinding—Plate Mills	5-12 h.p.
Crushers	3-5 h.p.

Hammer Mills	8-20 h.p.
Automatic Hammer Mills	3 h.p.
Half-ton High-speed Mixer	3-5 h.p.
Grain drying	12-20 h.p.
„ cleaning	2-5 h.p.
Threshing	12-20 h.p.
Cake breaking	2-5 h.p.
Chaff-cutting	3-6 h.p.
Water-pumping	1-3 h.p.
Incubators	$\frac{1}{4}$-1 h.p.
Silage cutting and blowing	15-20 h.p.
Hoist	$\frac{1}{2}$-2 h.p.

Automatic Controls. One of the most useful characteristics of electric power is the ease with which automatic control can be contrived. This is generally on a quantity basis or on a time basis, and if required could be on a rate basis, a humidity basis or on a temperature basis. It is this quality which has allowed the development of the low-loaded hammer mill, controlled automatically by the amount of grain which has passed through and by the time of using to grind a required quantity. Pressure control is used for automatic water supply in the pressure-tank system, and height control when the gravity-tank system is used. Humidity is used to control grain drying in some ventilated silos and, of course, temperature combined with pressure is used to control coolers.

Load Characteristics.

Farmers are sometimes puzzled by the insistence of supply companies on keeping down the size of motors. This arises from the fact that an electricity supply system has to be built to supply the largest simultaneous demand which may arise, although this may be in effect for perhaps only 30 minutes in the whole year. This maximum demand is, of course, governed by the size of the individual motors, etc., which may happen to be on simultaneously. If the number of large motors on the system is small, there is less chance of their simultaneous switching on being serious.

Regarded on a wide scale this may not, to the farmer, appear to affect him. It will do so in the long run, however, since the only hope of reduced tariffs is an improvement in the ratio of Units consumed to maximum demand. Installation of a few big motors may result in additions having to be made to the supply system, and will in the aggregate prevent reductions in tariffs. A farm might be supplied through a single transformer of, say, 10 kW.; but using, say, a 9 h.p. and an 8 h.p. motor simultaneously would call for a larger and more expensive transformer. Looking at the matter nationally, it is to the common good to keep down the maximum demand.

As an example of what can be done in this direction there may be cited the case of the increasing popularity of hammer mills. As normally installed these call for motors of anything up to 20 h.p. or more, and the supply authority is sometimes unable to provide supply to take them. It is better from many points of view to have a 3 h.p. motor working for ro hours rather than a 30 h.p. one working for one hour. Practical experience has proved that an automatic feed, delivery and control are eminently practicable. The fact that a 3 h.p. mill can grind grain at only a fraction of the speed of the large mill is met by making arrangements for the small mill to run quite unattended and to stop when the required amount of grinding is done.

Special Uses of Electricity.

Among the special uses of electricity, stimulation of the growth of plants by a process known as electro-culture should be mentioned, though the results so far obtained have given little hope of appreciable increases in growth rates or yields. The treatment of stock by means of ultraviolet and infrared radiation is gradually developing, the latter being regularly used on a few farms in the United States for the treatment of calves with complaints such as pneumonia. Electric insect traps, which electrocute insects attracted to them are finding many uses, and research in America suggests that there is a wide field for the employment of this type of device.

Electric Fences. One of the most valuable special uses of electricity is in connection with the electric fence. British electric fencer units incorporate design features covered by a British Standard Specification. They invariably employ a 6-volt accumulator to supply the energy. This provides current in a primary coil, which induces a small current of high voltage in a secondary coil. As in the magneto, a contact-breaker continually makes and breaks the primary circuit. The current flows approximately once a second, and lasts for about one-hundredth of a second. The secondary coil is connected to the insulated fence wire, and animals touching the wire feel a sharp and temporarily painful sting.

An accumulator of 20 amp.-hours capacity will last for 20 days of continuous use, and will electrify up to 15 miles of fencing. One wire at 24 to 30 in. is adequate for cattle and horses. Sheep and pigs need two wires at about 12 and 18 in., and goats three at 10, 20 and 30 in. Light stakes at 20 to 30 ft. spacing provide adequate support for the fence wire.

Proper insulation of the wire is one of the essentials, and is secured by nailing or wiring solid porcelain or plastic insulators to the fencing post at the required height. The fence wire must be kept clear of weeds, grass and branches.

The fencer unit itself must have a good earth connection, and this is conveniently achieved by mounting the unit on a metal post which is driven well into the ground (Fig. 279). It is important to train stock to respect the fence, since it is not the wire itself which stops them, but only fear which keeps them away from it. To train stock, tempting food should be placed on the far side of the wire and the animals made to approach it slowly. They will soon learn to avoid the wire which gives them such an unpleasant sting.

Electric fences can contribute greatly to the improvement of efficiency on many farms, saving labour in the carting of crops such as grass and kale, and also indirectly often reducing the amount of manure hauling.

FIG. 279.—ELECTRIC FENCER UNIT AND SPOOL MOUNTED ON METAl. CORNER-POST. (WOLSELEY.)

Wiring.

The wiring of farm buildings provides problems distinct from those arising in wiring houses, owing to the special conditions such as damp, fumes, shocks and extremes of temperature to which the installation may be subject. Many types of wiring are available, including conduit tubing, metal-sheathed cable, rubber-sheathed cable, etc. The system most generally suitable for farm wiring is the tough rubber-sheath cable system. Whatever the system, however, care must be taken in installing it, for badly executed work may be dangerous as well as inefficient. All electric circuits should be protected from the effects of overloading by the use of fuses. A fuse is a simple safety device generally consisting of a short length of wire which forms part of the electric circuit and is placed in a readily accessible position. The passage of a current through the wire causes the temperature to rise, the rise in temperature being proportional to the resistance and to the current flowing. The fuse wire must be so chosen that when too heavy a current passes through, it melts and automatically breaks the circuit. This type of fuse, with its rather clumsy method of resetting, is gradually being superseded by small automatic circuit-breakers, which can be easily reset.

Electric Generating Sets.

Efficient small Diesel-electric generating sets are available for the production of a private electricity supply in isolated districts that cannot be reached by public supplies. Where such outfits are installed the main reason is usually a demand for lighting for domestic purposes. Demands for power are, of course, more efficiently supplied directly by engines rather than through the medium of a generator and electric motors.

REFERENCES

(33) Cameron Brown, C. A., "A Critical Study of the Application of Electricity to Agriculture and Horticulture." Technical Report W.T2 of British Electrical and Allied Industries Research Association.

(34) "Electricity on the Farm." British Electrical Development Association. London.

CHAPTER TWENTY-ONE

FARM TRANSPORT

"That carts possess many advantages over wagons is now pretty generally believed, and is proved by the fact, that whenever any large amount of work has to be done, carts only are employed, as a horse, when drawing singly, will do half as much more work than when acting with another."

Rudimentary Treatise on Agricultural Engineering. G. H. Andrews, 1852.

Improvements in transportation, with associated advances in bulk handling and storage, have in the past century, and especially in the last half of it, profoundly affected the life and industry of the more advanced countries, and are beginning now to exercise a great influence on less developed territories as well. In agriculture they have relayed a great and ever-increasing part. Their first effects were indirect, but, nevertheless, deep-seated. Importation of wheat from America and Australia, made possible by steamship improvement and bulk handling, has notably influenced production policy in Britain. Our livestock policy has similarly been determined by importation, in which refrigeration has played an important part. These broad considerations in themselves are a warning that agriculture must be profoundly affected by transport improvement, and that therefore British farmers must be alive to every possibility of exploiting these improvements in their own work.

It is not always realized what a high proportion of the work on farms consists of the transport of products and supplies such as feeding stuffs, manure and other aids to production. Transport is an important consideration on almost every farm, and on some it accounts for quite half of the total of man- and horse-labour. It is therefore important that the work should be done with high efficiency. Whereas a man can easily carry 1 cwt. on his back, he can as easily deal with twice that weight if equipped with a wheelbarrow, and ten times as much with the help of an overhead run-way. Vehicles in general have the function of reducing the power, labour and discomfort of transporting material from one place to another. On some farms, transportation is largely undertaken by horses and carts, which are immediately adaptable to all field and road conditions; but tractors are rapidly replacing them, while motor lorries are part of the regular equipment of many holdings.

FIG. 280.—OVERHEAD RUN-WAY FOR TRANSPORT-
ING FOOD AND MANURE

Some of the important aids to farm transport—e.g. front-mounted hydraulic tractor loaders, have already been dealt with in previous chapters and are therefore not discussed here. Others, such as universal elevators, are referred to in several places in connection with the wide range of jobs that they will perform. This chapter deals mainly with vehicles, but mention is made of hoists, etc., not discussed earlier.

Factors affecting the Draught of Vehicles.

From the earliest days of wheeled transport the object has been to reduce the draught of vehicles, and the wheels have been designed largely with this end in view. The draught is made up of axle friction, rolling resistance and slope resistance.

Axle Friction. The wheels of an ordinary farm cart are carried on plain metal bearings which are usually slightly tapered. The bearing surfaces should be kept well lubricated with a heavy grease, and if this is done axle friction is only a small item in the total draught. For instance, the axle friction with a load of one ton on an ordinary cart with spoked wheels and plain bearings only amounts to about 12 lb. (coefficient of friction about 0.08). The axle friction may be reduced to about one-tenth of this amount by the use of high-grade ball or roller bearings (coefficient of friction 0.005 to 0.01), but this reduction is quite insignificant compared with the other quantities making up the total draught.

Rolling Resistance. Points which favour a low rolling resistance are wheels of large diameter, lightness of the vehicle, and tyres and surfaces such that the wheel sinks in as little as possible. The steel-tyred cart is good on hard surfaces, but the loaded cart cuts into soft ground and forms ruts which, in addition to being undesirable in themselves, also increase the draught. When steel tyres have rounded edges they do not cut in as badly as if the edges are square, but the slight improvements possible in this direction are insignificant compared with the great reduction in draught that may be achieved in many circumstances by the use of pneumatic tyres. The main reason for the success of pneumatic-tyred wheels for farm vehicles is their low rolling resistance on soft ground.

Slope Resistance. When a load is drawn up an incline, the resistance is proportional to the slope *(vide* Appendix One). For example, if the incline is 1 in 12, the force required to overcome the slope resistance is one-twelfth of the weight of the vehicle and the load.

Pneumatic Tyres for Farm Implements and Vehicles.

When pneumatic tyres for farm carts were introduced in 1933, an official test showed that the draught of the rubber-tyred cart was less in all conditions than that of a similar cart fitted with iron-tyred wheels. The reduction in draught at a given loading ranged from 13 to 41 per cent., and increased pay loads ranging from 35 to 108 per cent, could be carried. Subsequent tests have shown the importance of having wheels of fairly large size, e.g. Swedish tests show up to 10 per cent, difference in draught between tyres of 16 in. and

20 in. diameter, and a similar difference between tyre widths of 6 and 7 inches.

When pneumatic wheels are fitted to the old type of box cart, some adaptations, such as the blocking up of the body of the cart above the axle, and the fitting of suitably curved shafts, are required to preserve the balance and facilitate tipping; but carts specially designed for the wheels may be made to tip easily and also to give a fairly low-loading line. Long low-loading carts with very large bodies (Fig. 281) are suitable for many purposes such as harvesting, though they are not as generally useful as the higher loading box cart owing to the difficulty of providing a simple and effective tipping mechanism.

Pneumatic-tyred wheels are now used on a wide range of implements and machines. One of the drawbacks to their increased use is the relatively short working season which some machines have. It has already been pointed out that pneumatic tyres are ideal equipment for the power-driven binder and mower, and the same applies to many other farm machines; but many of the machines work for only a few days in the year, and one of the unfortunate disadvantages of pneumatic tyres is that in such circumstances they may depreciate more during the idle period, owing to exposure, than they do during use, owing to friction and cuts. Care of the tyres by removal of grease and other foreign material before storage may minimize the damage during the idle period, but a much more satisfactory solution of the problem is the adaptation of the same tyres for use on many different machines.

The Tractor as a Transport Agent.

Since the advent of pneumatic tyres for tractors, the performance of the tractor as a transport agent has been much enhanced, for, like the horse, it becomes suitable for use both on the fields and on hard roads. A tractor fitted with steel wheels is usually capable of dealing with a heavily laden trailer on soft ground, but when it reaches the highway, road bands must be fitted; and this causes delay serious enough to render transport by such means uneconomic. Onalargefarm, having several tractors, the difficulty may be surmounted by using one tractor for the field and a differently equipped one for the road, but usually the only satisfactory solution is the fitting of pneumatic, tyres. A drawback to the use of pneumatic tyres for transport on fields is that the tractor often cannot exert as high a drawbar pull as it could if it were equipped with steel wheels or tracks. In difficult conditions this is a serious disadvantage, but the fitting of adjustable strakes may often effect a considerable increase in the maximum drawbar pull. Once on the road, the rubber-tyred tractor is an excellent haulage agent, for it can travel at high speeds if fitted with a suitable top gear.

FIG. 281.—LONG CART SHOWING ANGLE OF TIP

Tractor Trailers. Two-wheeled.

The usual type of 2-wheeled tractor trailer has a wooden platform 10-12 ft. long by 6 ft. wide, with the axle placed centrally or about 1 ft. to rear of the centre, so that the trailer is slightly heavy at the front. The trailer has easily detachable hinged sides and tail-board. A useful fitting at the drawbar is a combined jack and castor wheel, to facilitate hitching and permit moving of the trailer on level land without the help of the tractor. This type of trailer is frequently fitted with a screw tipping device.

FIG. 282.—TWO-WHEELED TIPPING TRAILER.
(STANHAY.)

Now that the use of power lifts on tractors is widespread, the common method of tipping 2-wheeled trailers has become by a hydraulic device. Cheap, self-sealing hydraulic couplings are fitted, and use is made of the tractor's hydraulic lift pump to operate a ram which is built in on the trailer chassis.

One type of tractor trailer (Fig. 283) is completely unbalanced, having the axle right at the rear, and is designed for use with a light tractor which benefits from the additional weight that is thrown on to it when the trailer is loaded. Experience has amply demonstrated the great advantages of this system, particularly where the trailer hitch is well designed to ensure ease of coupling.

The Sutherland power link is a mechanism which is designed to transfer a limited amount of weight, at will, to the rear wheels of a tractor which is equipped with a hydraulic lift and pulling a trailed implement or vehicle. The mechanism consists essentially of stiff springs which connect the lift arms with the trailer drawbar at a point behind the hitch. When the lift arms are raised, the springs pull upwards on the trailer drawbar, with the result that weight is transferred to the tractor's drive wheels from the trailer itself and also from the tractor's front wheels. Such a device, which can be employed on trailed implements as well as on trailers, may prove extremely valuable in lessening wheel slip where adhesion is difficult.

*FIG. 283.—UNBALANCED TWO-WHEELED TRAILER
FOR LIGHT TRACTOR. (FERGUSON.)*

Two-wheeled hydraulic tippers are suitable equipment
for many farms which need to haul sugar beet, potatoes and
corn in bulk, and are also often preferable to 4-wheelcrs
for manure hauling and many other jobs. Nevertheless,
4-wheeled trailers have many advantages, and it is often
best on medium-sized and large farms to rely mainly on
2-wheelcrs but to have also one or two 4-wheelers for special
uses.

Trailer Hitches.

The importance of an efficient hitch for tractor trailers has been emphasized by the development of a quickly coupled hitch for unbalanced 2-wheeled trailers, which enables a tractor equipped with hydraulic lift to pick up the front-heavy trailer without the driver having to step down from the tractor. The tractor (Ferguson) carries a hook which is coupled to the hydraulic lift linkage. The tractor backs up to the trailer until the hook is beneath an eye at the front of the trailer drawbar. On raising the hydraulic lift, the tractor hook engages with the eye on the trailer, and the hitch is securely locked as soon as the drawbar is fully raised.

Several other interesting and useful types of trailer hitches have been developed, both in Britain and abroad. On 4-wheeled vehicles it is often convenient to have at the front of the drawbar a short swinging arm which is telescopic until locked.

Trailer Brakes.

All types of tractor trailers need brakes, and legislation has been passed to make fitting efficient brakes compulsory for those which use public roads. Brakes may be of the over-run self-acting type or hand-operated. If hand-operated brakes are fitted, the lever must be within easy reach of the tractor driver. All trailers should have a ratchet device to keep the brakes on when required.

With automatic brakes it is important to ensure correct adjustment, so that the trailer brakes are applied smoothly as soon as the tractor slows down and the drawbar spring is slightly compressed.

Four-wheeled Tractor Trailers.

Four-wheeled trailers are advantageous where bulky loads have to be carried and a large platform is needed. They are easy to hitch, can be safely left loaded, and are particularly useful for long road journeys. Their chief disadvantages are the extreme difficulty of backing them, and tractor wheel slip in difficult conditions. Many 4-wheelers have a turn-table which necessitates the platform being high enough to enable the front wheels to pass beneath it, but automobile-type steering is now widely used, and if well designed it combines the advantages of a good lock with stability. Hydraulic side-tipping 4-wheel trailers are available.

FIG. 284.—TRAILER CHASSIS WITH FLEXIBLE AND EXTENDABLE CONNECTION OF REAR WHEELS TO POLE. (TASKERS.)

Fig. 284 illustrates a chassis of a type common in North America and now manufactured in Britain, to which various types of body can be attached. On one make the rear axle is attached to the tubular pole by means of a slotted connection which provides for swivel action when the trailer passes over undulating ground. Another make employs spring-loaded bolster plates to give flexibility. A choice of wheel-bases to suit the type of body fitted is provided for by attaching the rear axle to the pole in alternative positions. With the type of chassis which has a flexible connection between front and rear axles, the trailer body is bolted to the front stanchions only, the rear being allowed complete freedom of movement. The body itself must be flexible, in order to allow it to follow to a certain extent the swivel action of the chassis.

Self-Emptying Trailers for Green Crops.

The use of self-emptying trailers for the transport of green crops can save labour at the silo or drier, and sometimes also in the field. The need for saving labour at particular times will determine whether the purchase of such specialized trailers can be justified. Two main types are in use, viz. those with P.T.O.-driven chains and slats moving over a solid floor (or an endless belt forming the floor), and those which have hinged doors which open at the bottom. In the P.T.O.-driven type, loading may be assisted by depositing the green crop from the loader on to the front of the trailer, and then moving the load back occasionally by engaging the power drive as the work of loading proceeds. This method of loading without a man has the advantage of making unloading and spreading easy. In the bottom-opening type of trailer, the sides are very high, and the bottom is V-shaped. The method of emptying is to release a catch which allows the rear part of the V to swing to a vertical position, thus depositing part of the load. If the trailer is then moved forward, the vibration will cause the rest of the load to fall to the ground. The tractor driver pulls a cord to re-fasten the door.

Some self-emptying trailers can have a variety of uses. For example, they may be used for carting and distributing kale to cattle, for carting and unloading potatoes and sugar beet, or sometimes for carting corn in bulk. One manufacturer provides a manure spreading attachment which can be fitted to the rear.

Tractor and Implement Transporters.

The transport of tractors and other heavy equipment is often a serious problem, and various types of pneumatic-tyred trailers are employed for this purpose. Such trailers need to have very low platforms so that tractors can be driven straight on, and heavy implements such as disc harrows may be pulled on by means of a hand winch attached to the trailer.

The simplest type of tractor transporter is a 2-wheeled trailer with the bed mounted directly on the axle, and consisting merely of two members which are correctly spaced and are just wide enough to accommodate the tractor wheels or tracks. A hinged ramp is provided at the rear, and this is raised for travelling. Jacks are provided at the corners to steady the trailer during loading. Such 2-wheeled transporters can only accommodate tractors of moderate weight. For heavier tractors (up to 4 tons) a somewhat similar 4-wheeled trailer with the two axles set close together may be used. Such a vehicle does not need any special arrangement for steering, and is used just as if it were a 2-wheeler.

For transporting implements, 2-wheeled trailers fitted with a flat platform and a hand winch may be used. An alternative method for heavy implements is to employ a long 4-wheeled vehicle with a low platform which is loaded from the side.

A specialized type of implement transporter suitable for use by contractors consists of a 3- or 4-wheeled chassis having a frame consisting of a high arch. The transporter is manoeuvred until the implement is beneath the arch. The implement is then lifted off the ground by a number of chains and a winch. When well clear of the ground the implement may be prevented from swinging by the use of check chains.

Specialized Self-propelled Transport Vehicles.

Many types of self-contained transport vehicles including dumpers, engine-driven carts and motorized trucks are available, and find economic employment on some farms where there is a great deal of regular specialized transport work to be done. Dumpers are tractors which carry a transport box over the large drive wheels and have their gears and controls arranged so that the vehicle normally travels with the box at the front, and the steering wheels at the rear. They are used on some large farms for carting sugar beet and potatoes. Advantages are good drive wheel adhesion and easy tipping, but with good hydraulic-tipping tractor trailers now available, the continued use of special-purpose dumpers seems unlikely.

One type of mechanical cart is like a dumper except that it normally travels with the steering wheels ahead. Another type (Motocart) is a 3-wheeled vehicle having a single-cylinder air-cooled engine mounted alongside the large-diameter front drive wheel. It is normally a 30-cwt. vehicle (Fig. 285). Advantages include a low fuel consumption compared with a tractor and trailer, and easy manoeuvrability. The vehicle can be "led" by a man walking alongside.

Smaller engine-driven 3-wheeled trucks, designed to carry loads of about 10 cwt., are useful for such regular work as transporting food and water on poultry farms (Fig. 286). Air-cooled engines of 1 to 3 h.p. are employed to drive the single front wheel. These vehicles are easily "led" from fold to fold, and use very little fuel.

Motor Lorries.

Where there is sufficient regular road work, a motor lorry may be a useful part of the farm transport equipment. This applies especially to farms having such regular work as transporting milk to a town or station. A motor lorry also makes it possible to grow bulky and perishable market-garden products and fruit at great distances from a town or railway. The motor lorry can be used for a great variety of work, especially if equipped with a general-purpose body which can be adapted for the transport of livestock. In some instances farmers find it convenient to do contract work for their neighbours when the lorry is not fully employed at home, while others find it more satisfactory to rely on regular road haulage contractors for journeys off the farm.

For most farms the most useful type of lorry is a 3-5 ton vehicle with a long wheel-base, and fitted with drop sides. For jobs such as carting sugar beet, green crops, etc., extension sides can be fitted. For carrying livestock it is usually necessary to provide, in addition, a canvas cover and a loading ramp fitted with side-rails.

Ex-W.D. four-wheel-drive 3-ton trucks have proved valuable for certain types of transport work, being able, when loaded, to tow a heavy trailer over fields and rough farm roads. Their disadvantage is a heavy fuel consumption. The high manufacturing cost makes it unlikely that such vehicles can be employed in agriculture except on a second-hand basis, after use for other purposes where a high purchase price is of less importance.

A less powerful type of 4-wheel-drive general-purpose vehicle (e.g. "Jeep" and "Land-Rover") weighs a little over 1 ton and is designed to carry $\frac{1}{2}$ ton. Such vehicles can traverse rough land and have a good capacity for keeping going in difficult conditions. They can be used for light field work— e.g. light rolling and harrowing, mowing, spraying, etc., and also for light belt work. Their primary function is, however, the transport of men and of small loads over difficult roads or over fields.

(N.I.A.E. Photo)

FIG. 285.—MECHANICAL CART DRIVEN BY SINGLE FRONT WHEEL WITH SIDE-MOUNTED ENGINE. (OPPERMAN.)

("Poultry Farmer" Photo)

*FIG. 286.—MOTORIZED TRUCK IN USE ON POUL-
TRY FARM. (WRIGLEY.)*

Hoists and Lorry Loaders.

Friction hoists are employed for raising corn in sacks
to the first floors of granary buildings. They may be driven
by an engine, a line of shafting, or by electric motor. Where
an electric motor drive is to be employed, the most suitable
type is usually a machine in which the motor, gearing and
winding drum are assembled as a single compact unit, which

is suspended from an overhead hook or rail. Such units can be mounted on wheels and arranged to run along an overhead steel joist, so as to transport the sacks well inside the building. The usual type of hoist is a 2-way unit capable of lifting $2\frac{1}{2}$ cwt. at a time.

Portable self-contained engine-driven hoists, complete with a light, swinging jib, are available for either stationary work, or for mounting on a trailer or lorry platform to lift sacks and bales in the field. In one example, the whole of the mechanism is mounted on the jib, and the device is readily set up for work by dropping the lower end of the jib into a suitable socket-plate, which may be left in position in the granary or on the vehicle.

Another type of lorry loader consists of an engine-operated 2-wheeled hoist which is hitched behind the trailer to be loaded. It can pull sacks from 25 ft. away, and can lift up to 10 ft.

Some lorry-loaders consist of a lifting platform which is attached to the rear of the lorry or carried on its own two wheels. The platform is usually engine-driven, and moves vertically from floor level to lorry height.

Universal Elevators and Loaders.

Reference has already been made to the wide use of elevators for loading sacks, bales and sugar beet. They usually consist of a simple conveyor of chains and slats, running on a bed which may be solid or in the form of a grille (Fig. 287). A hopper can be fitted at the base for handling crops such as sugar beet. The angle of the bed is usually adjustable: alternatively, the end section of a long elevator may be hinged and adjustable to suit various loading heights. The smaller machines are usually carried on two pneumatic-tyred wheels, but some machines with a high lift have four wheels. The engine, usually of $1\frac{1}{2}$-2 h.p., is mounted on the chassis beneath the elevator. A typical elevator can deliver up to 12 2-cwt. sacks at a height of 9 to 12 ft. in a minute, but such a high rate of work is seldom needed. On some machines it is possible for one man to use the loader by putting two or three sacks on the conveyor, declutching, and then engaging the clutch by means of a rope when he is ready to receive the sacks. The clutch mechanism usually consists of fast and loose belt pulleys. Where sacks or bales are to be loaded on to vehicles from the field, it pays to dump them in heaps so that several can be loaded without moving the elevator.

Long bale-elevators suitable for raising bales to the top of a Dutch barn are now available (Fig. 288), but as mentioned elsewhere, there is no reason why a hay and straw elevator should not be equipped to do this work.

Bale Pick-up-Loaders.

Though many farmers find it best to load bales with the help of a universal elevator towed behind the trailer or truck, some prefer to employ a pick-up bale loader. Such a machine may be an engine-driven endless conveyor type loader which is attached to the tractor, or to the side of the vehicle. It is fitted with V-shaped guides to push the bales into position. A simpler type of loader has no engine and is driven by its own two land wheels (e.g. International).

Tractor-Mounted Cranes and Lifts.

Many kinds of light, tractor-mounted cranes and lifts are available for such work as lifting sacks and bales. The simplest mechanism is a sack-lifter which is attached directly to the 3-link hydraulic lift system of a tractor, and will raise loads to lorry height. Among other less expensive types of lifter are light jib-type cranes operated from the tractor's normal hydraulic lift pump. Other more powerful types, suitable for lifting heavy loads, are manufactured; but they are usually more suitable for specialized industrial lifting jobs than for farm work.

FIG. 287.—HYDRAULIC TIPPING TRAILER DELIVER-ING SUGAR BEET DIRECT INTO HOPPER OF OPEN-GRID TYPE UNIVERSAL ELEVATOR. (WOLSELEY.)

FIG. 288.—LOW-LOADING HIGH-LIFT BALE ELEVA-TOR. (BLANCH).

784

Transport in the Farmyard.

The aggregate importance to British agriculture of mechanization of transport about the farm buildings can hardly be over-emphasized. Efficient mechanization usually involves planning to eliminate unnecessary movement of materials, and re-arrangements designed to cut down the distances travelled and the difficulty of the work. Many buildings were not designed for their present uses, and there are often economic limitations to the amount of re-modelling of farmsteads that can be undertaken. Nevertheless, it is frequently better to tackle the problems by drastic re-organization rather than by using mechanical gadgets to overcome the difficulties of layouts which are fundamentally unsound for present needs.

Much can often be done by opening up gateways and passages so that tractors and trailers can be used, and there are many farms where it is possible to achieve substantial economies in transportation by making smooth, concrete feeding passages, and introducing such simple aids as rubber-tyred feed barrows. The value of a piped water supply in saving transportation is becoming widely appreciated.

Many farms demonstrate what may be done to facilitate transport by the use of overhead carriers for food and manure in cowsheds and piggeries, but in recent years there has been little tendency for the use of such carriers to increase, most farmers preferring to have smooth roadways and rubber-tyred transport vehicles whose movements are not restricted to a fixed line.

CHAPTER TWENTY-TWO

HORTICULTURAL MACHINERY

Many of the machines used by market gardeners and fruit growers are also used on farms, and have been dealt with in previous chapters. In this way, mention has been made of horticultural tractors, rotary hoes, drills and sprayers. This chapter deals very briefly with equipment of vital interest to some horticulturists, which does not fit into other parts of the book.

FIG. 289.—WALKING TRACTOR WITH ADJUSTABLE WHEELS AND FRONT TOOL-BAR. (AUTO-CULTO.)

Special Power Units and Cultivation Implements.

On open land, most kinds of market garden crops can be successfully grown by using a light-weight medium powered 4-wheeled tractor equipped with hydraulic lift, a good range of standard agricultural equipment, and wheels which can be set out to a maximum track width of over 6 ft. The crops may be drilled or planted in beds, with row-widths down to 11-12 inches, provided that wider paths are left for the tractor wheels at the outsides of each bed. Thus, each bed may comprise, according to the nature of the crop grown, 6 rows 11 inches apart, 5 rows 14 inches apart, 4 rows 17 inches apart, and so on, with pathways of 20 to 25 inches between beds. For work on such crops, self-propelled tool-bars for drilling, hoeing and spraying are helpful, but by no means essential.

The need for special types of power units and/or cultivation implements arises when it is necessary to work between fruit trees or in hop-yards, where the planting distance has already been decided, and the equipment has to conform to it. Even here, there is often a case for planning planting distances to suit machinery, when the opportunity for re-planting occurs.

Narrow rows of fruit and hops sometimes necessitate the use of extremely narrow tractors, with overall widths of about 4 ft. Such tractors are now available for this work, and for use in vineyards abroad. They require specially adapted mounted equipment, and use, among other implements, types of plough which are designed for ploughing up to or away from both sides of rows of hops or fruit. One type of hop plough suitable for very narrow work turns two furrows to the right and two to the left simultaneously, covering all the ploughable land in one bout. Cut-down tined implements and disc harrows are also used.

Another type of special-purpose implement is needed for rather wider rows, where it is desired to work as close to one side as possible. For this work, offset ploughs which will work right up to the outside of the tractor wheel mark can be obtained.

Cutter-Bars for Market-Garden Tractors.

Most modern single-wheeled and two-wheeled tractors can be equipped with a cutter-bar which is useful for tidying up small areas of land where weeds grow, as well as for more straightforward work on long grass. A useful feature on some machines is an independent clutch which enables the cutter-bar to continue working when the tractor is out of gear. Fig. 291 shows a cutter-bar attached to the front of a machine which is primarily a small rotary cultivator, but which, like many other light tractors, can also operate a cylinder-type lawn mower. The cutter-bar is free to follow ground undulations, and such small machines, with closely-spaced fingers, can do excellent work. Fig. 292 illustrates a larger two-wheeled cutter of a type which is widely used for all kinds of work where access is difficult. Such machines, like most other modern two-wheeled tractors, are extremely versatile, being capable of handling a fair-sized trailer and pulling a small roll or disc harrow, as well as driving a large-capacity pump and driving the generator for a fair-sized hand-supported hedge trimmer.

FIG. 290.—OFF-SET DISC HARROW WORKING BE-TWEEN FRUIT TREES. (FERGUSON.)

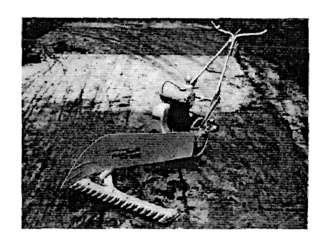

*FIG. 291.—CUTTER-BAR ATTACHED TO LIGHT RO-
TARY CULTIVATOR. (ROTARY HOES.)*

(N.I.A.E. Photo)

*FIG. 292.—TWO-WHEELED WALKING TYPE MOTOR
SCYTHE. (ALLEN.)*

Hand-type Hedge Trimmers.

Most horticultural trimmers have a short reciprocating knife which is driven between closely-spaced guards by an electric motor mounted at the inner end of the bar. Electricity for driving the trimmer is generated by a dynamo which may be mounted on and driven by various kinds of walking-type tractors. A typical small unit employs a 110-volt generator and a $\frac{1}{4}$h.p. electric motor. A long rubber-covered lead is provided, so that the tractor need only be moved infrequently. These trimmers can do good work on one-year-old wood, and the common type is capable of dealing with single stems of up to $\frac{1}{2}$in. diameter. For heavier work, a larger trimmer driven by a $\frac{5}{8}$h.p. motor may be obtained for use with some tractors. Such trimmers can deal with single stems $\frac{3}{4}$in. thick. One type of trimmer is mechanically driven through a flexible drive, and another employs compressed air. The rate of work possible with one of these trimmers is very much higher than is possible with shears or hedging hooks.

One type of trimmer employs a circular cutting head.

Soil Sterilizing Equipment.

Soil may be sterilized either by chemical means or by heating. Chemical methods do not involve the use of expensive equipment, but they are rather limited in their action. Heating the soil to a sufficient temperature, on the other hand, not only kills all pests and diseases, but increases fertility at the same time.

Small quantities of soil may be heated in various ways, but the normal method used for the soil in commercial glasshouses is steaming. Boilers commonly used for this purpose are portable loco-types, preferably with forced draught and a high steam-raising capacity of 1,000 to 2,000 lb. steam per hour; but much cheaper smaller types, with a maximum output of about 600 lb. per hour, can be successfully used provided that the operating principles are properly understood.

The steam is delivered to the soil by means of perforated iron pipes. The system generally preferred is, briefly, to excavate a trench to a depth of 12 to 18 inches; lay in one or more perforated pipes (Hoddesdon pipes) 18 to 20 inches apart; cover with the soil from the next trench, then cover with a tarpaulin, and pass steam until the temperature of all the soil reaches boiling point. When steam is passed into soil that has been well broken up, it at first condenses on the soil particles surrounding the pipe. As more steam is passed in, the soil near the pipe is raised to boiling point, and the steam then passes through, and condenses farther away from the pipe. For efficient steaming, the flow of steam must be just sufficient to cause the heated layer to advance upwards at a steady rate. An excessive rate of steaming is wasteful, because the surplus which cannot be condensed rapidly enough finds its way to the surface as steam, instead of delivering up its heat to the soil. In general, the steaming rate for economical operation should be such that it takes 20 minutes or more for the surface soil to reach boiling point. If then the steam is turned off, and the cover left on for a further 5 minutes, sterilization will be effectively achieved. Near concrete and wet patches, the steam should be passed slowly for a further five minutes after the surface is cooked. As with dairy boilers, there is no virtue in high pressures as such. The loco-type boiler, working at a pressure of 80 lb. per sq. in. or more, should be used with a water trap and pressure reducing valve close to the Hoddesdon pipes. The area that can be sterilized in a given time is not reduced by working at a low pressure. The lower pressure results in more time being taken to cook a particular part of the soil, but more pipes can be employed on the same boiler. Steam mains should be well lagged and as short as possible, with no steam leaks. A small boiler with an output of 600 lb. per hour can work quite economically at a pressure of 10 lb. per sq. in., with a 2-inch main up to 100 ft. long and a grid of two pipes only.

For fuel economy, the soil should be as dry as possible, since much of the steam's heat is used in raising the temperature of the soil moisture, which has a high specific heat. A really wet soil needs much more heat than a dry one to achieve effective sterilization—a point of particular importance in heavy soils, since these can contain so much more water than light. Detailed information on the subject of sterilizing finely divided glasshouse soils is contained in a report by the N.I.A.E.35

Compost Shredders and Mixers.

The preparation on a large scale of uniform potting compost is greatly assisted by use of a compost shredding machine such as the Pneulec-Royer. This machine has an inclined hopper, the lower side of which consists of a high-speed combing belt. The endless rubber belt is covered with steel teeth and driven by a built-in electric motor or petrol engine. As the belt moves upwards through the hopper its surface combs and shreds the material until it is fine enough to pass out in a thin layer through the opening between the belt and the upper end of the hopper. The shredded material is thrown from the machine by the speed of the belt, and may be delivered into a large heap or deflected into a container.

Mixing may be achieved by putting the various constituents through the machine together.

Where no compost shredding machine is employed, the mixing of previously shredded composting materials may be assisted by use of an ordinary concrete mixer.

When manure compost for application to the land or for making mushroom beds is being prepared, some of the ordinary farm manure handling equipment may be used with advantage. For example, a tractor-mounted hydraulic loader with manure bucket may be employed to put the long compost into a stationary P.T.O.-driven farmyard manure spreader, the spreader being arranged to deliver the compost into a heap after shredding it. If the need to do this job occurs frequently, it may be worth while to equip the spreader with electric motor drive, or to invest in a compost shredder expressly designed for such work as preparing mushroom beds (e.g. the De la Pré).

Soil Block-making Machines.

Soil block-making machines consist essentially of a mould and a press, designed to make moist potting compost into firm blocks in which seedlings are planted for propagation, no pot being required. The blocks are usually hexagonal in shape, as this allows them to be stacked close together. The dimensions normally used for tomato seedlings are 3 in. across the parallel sides of the hexagon, and about 4 in. high; but much smaller blocks (e.g. 2 in. by 2 in.) may be employed for other crops. John Innes compost makes good soil blocks if the loam is sufficiently fibrous, and the mixture just wet enough. The moist compost, which should be at a temperature of about 60° F., is vibrated into the moulds until they are full. Pressure is then applied by a plunger which compresses the blocks and sometimes makes a small planting cup in the top of each. In other models, the seedling is held in position whilst the block is being made. The mould has a removable bottom, and the blocks are pushed out ready for use. The range of equipment now available includes machines which make blocks of various shapes and sizes, in batches of one, four, six, eight or ten at a time.

There can be no doubt that soil block-making machines have come to stay. They contain more compost than the pots which they replace, and results show that tomatoes propagated in them often grow away better than those from pots. Block-making machines can eliminate all the troubles of providing, carrying, cleaning up and washing pots, and a great extension of their use is likely.

Fruit Grading Machinery.

Grading for size and quality is an essential preliminary to marketing of fruit grown on commercial holdings, and many types of machine are available to assist this work. Some are cheap and simple sizing or weighing machines for grading fruit which has been sorted for quality beforehand. Others are complex and expensive machines which incorporate devices to facilitate sorting for quality, accurate sizing mechanisms, and bins to facilitate packing.

Mechanisms employed to assist sorting for quality include roller feed conveyors, which turn the fruits over as they pass in front of operators, who pick out inferior ones. This method is advantageous where a large throughput is an important objective, and a small proportion of blemished fruit in the pack is not serious. An alternative method, by which the operators pick up each fruit individually and put them according to quality grading into appropriate sections of the size grading mechanism, results in a more perfect quality grading but a lower throughput. On some machines employing the second method (e.g. the Jansen) the operators sort direct from the boxes, while on others (e.g. the K.E.F.) the fruit continues to pass in front of the operators on a sorting belt until it is picked off and placed in the appropriate section of the sizing mechanism.

Sizing mechanisms may be broadly grouped into those which grade by diameter and those which grade by weight. Machines which grade by diameter vary greatly. Some size the fruit in only one or two directions, while others rotate it and thereby achieve a more accurate sizing. In general, machines that grade by diameter tend to have higher throughputs, or to be rather less expensive for a given throughput, than those which grade by weight. On the other hand, a good weight grader has many advantages. It can be used for fruit of any shape, is easily adjustable, and can be used for fruit which is easily blemished. Grading by weight can also be more accurate than grading by diameter, though good diameter type graders are usually accurate enough for practical purposes.

Good graders do little damage to apples, most of the bruising that does occur being caused when the fruit rolls into the receiving bin. This is usually negligible compared with damage incurred in picking, and getting fruit to the grader.

Throughputs of graders vary widely according to the number and ability of the operators, as well as according to the machines and auxiliary equipment employed. Some graders are equipped with automatic box-tipping devices for loading, but throughputs on such high-speed machines are frequently limited by the speed of packing.

No one grading machine has all the advantages, so choice of a grader requires careful consideration of the most important needs for the conditions in which it is to be used. In the brief description of types of mechanisms which follows, it is not possible to give an adequate description of any individual machine; but further information may be obtained from a comprehensive Report36. Prices of graders range from under £100 to over £2,000.

Size Graders. (a) *Fruit falls through a circular hole of fixed size.* Machines of this type include the Notenboom tomato grader, in which the sizing board has a number of rows of holes of different sizes, and the fruit is carried along by means of a series of wooden laths operated by a crank mechanism. The effect is to move the fruit, without pushing or rolling it, to successively larger holes, until it falls through.

The Drake and Fletcher machine has a series of rubber conveyor belts with circular holes. Fruit which does not fall through the holes in the first belt passes to a second, in which the holes are larger.

(b) *As fruit moves along conveyor, size of opening increases.* Graders with mechanisms of this type are common, and include the following: the Helix tomato grader, which has a pair of augers of gradually increasing pitch, between which the tomatoes are conveyed until they fall through; the Jansen apple grader, in which the fruit is rapidly rolled along by a soft rubber belt, while leaning against a rubber barrier set at a gradually increasing distance from the belt; the West Friesland grader, with diverging rubber belts; the Ardee grader, in which the fruit is placed on a rubber iris diaphragm which gradually opens as the two belts which form its sides progressively diverge; and the Lightning grader, in which the apples carried on an endless conveyor lean against rubber rollers as they move along. The larger fruit is pushed away by the rollers, while the smaller falls through to the receiving bins.

(c) *Fruit rests or revolves on conveyor belt, and is brushed aside by revolving paddles, wheels or belts situated above.* An example of this type is the Tod, a small, low priced machine consisting simply of a sorting table and a V-belt conveyor above which are mounted rotating rubber vanes. The distance of the vanes from the conveyor is adjusted so that the fruit is removed, according to its height, by being brushed sideways. This machine is mostly used for tomatoes, but can also be used for apples. Others working on a similar principle include the Crossland, Galbraith and the Canadian Bartlett.

Weight Graders. In weight graders, each fruit rests in an individual hinged cup, and for much of the journey along the conveyor the cup and fruit are supported in an upright position by a fixed guide rail. At suitable intervals, however, the support is provided by an adjustable lever type weighing device. When the moment exerted by fruit and cup is sufficient, the cup tips, and deposits the fruit into its appropriate bin.

In the Oxlo grader, the bakelite cups tilt an adjustable balance arm which has weights attached. This is a simple machine which is very accurate. Larger machines employing a similar principle and many auxiliary devices include the K.E.F. and the Mather and Platt.

Vegetable Washing Machines.

It has long been realized by progressive growers that financial returns can depend on the manner in which produce is prepared and packed for market. Even first-grade vegetables may realize an unsatisfactory price unless they are trimmed, washed, and properly packed. Manual methods of washing, in a tank of water or with a hose-pipe, may be economic on small holdings; but growers on a large scale need labour-saving washing machines. Vegetable washers of many types have been developed to suit various crops, and the methods employed are briefly discussed below. They must depend primarily on the type of crop to be handled, but of equal importance is the quantity of water available, which in turn influences the means adopted to deal with the dirty water. The principal methods adopted in washing machines are (1) spray jets, (2) mechanisms for "tumbling" the vegetables in the presence of water, and (3) scrubbing. Individual machines employ any one of these methods or a combination of them. In addition, many machines employ a soak tank, either before or during the washing process, and most continuous-process machines have a final "picking-over" conveyor which enables any blemished vegetables to be removed before packing.

Jet Washers. Where plenty of water is available, or easily-damaged crops are to be handled, the vegetables may be moved on an endless conveyor through a spray chamber, where they are exposed to jets of water. One of the simplest types of conveyor is a rotating circular table. Many home-made washers of this type are hand-turned, while a machine devised by the N.I.A.E. is driven by a hydraulic two-piston oscillator of the type used on irrigation spraylines. The type of conveyor usually employed on commercial machines consists of a roughly horizontal endless belt (e.g. the Esher and the Wedco-Allsebrook).

The circular machine cannot easily utilize a pre-soak tank, but use of such a device greatly assists the action of spray jets, and can fairly easily be incorporated in some types of endless belt machines. A further refinement is found in one machine, which has a roller conveyor that gently turns the crop as it passes through the chamber (Cooch).

"Tumbler" Machines. Tumbler washers fall into two main types, viz. rockers and rotary machines. The rocker is

a very simple device which has long been in use for washing carrots. It consists essentially of a large tub with a flat, slatted bottom, which is rocked to and fro on semi-circular hoops by means of an engine-driven crank (e.g. the Cole). The top of the box has a grill lid which can be lifted for filling, and is closed to prevent the crop being thrown out while undergoing treatment. As the tub is rocked, water is poured in at the top, and runs out at the bottom; and washing is continued until the crop is judged to be clean enough. About 4 cwt. of carrots are dealt with in a batch, and washing can be fairly effective.

Rotary tumbler machines with spray jets (e.g. Edlington) are also used mainly for carrots. The tumbler mechanism consists of a nearly horizontal rotating cylinder of wire mesh, and spray jets are directed on to the crop as the cylinder revolves.

One type of rotary tumbler machine (Pyleboro) has the rotary drum partially submerged in a soak-tank. The water has a cushioning effect, and such machines, with a mechanism designed to handle crops gently, can be satisfactorily used for many types of vegetables.

Continuous-process rotary tumbler machines are normally equipped with a short belt conveyor for picking over and bagging-off."

One very simple type of rotary tumbler is designed for potato washing (e.g. Penney and Porter).

Scrubbers. Some washers expressly designed for carrots employ a mechanical scrubbing process. One type scrubs the carrots under water as they pass along a horizontal conveyor trough. The mechanism employed in this case consists of a reciprocating brush which resembles an inverted doormat. Other types of scrubbing machines use circular revolving brushes. The scrubbing mechanism is sometimes employed between a pre-soak tank and spray jets.

Water Supply, Drainage and Installation Problems. Whatever type of vegetable washer is chosen, problems of water supply and drainage need consideration. Machines which do not use high-pressure jets require a minimum of water, and are therefore suitable for use in the field where water has to be carted. Machines requiring a large volume of water tend to create a drainage problem. Where the water supply is drawn from a nearby stream, it may be returned by an open furrow of gentle slope, which gives the soil opportunity to settle out.

With fixed washing machine installations of types that need to use a large volume of water, limitations of water supply, or a high cost, may necessitate some economy in the total amount of water used. Such considerations, together with the need to keep mud out of the drains, may make it essential to construct a settling tank and filter system which enable the water to be used repeatedly.

If a washing machine is to be installed in a packing shed, it is necessary to plan the layout so that the dirty side of the process is well separated from the clean, and to allow plenty of light and room for sorting and packing. The packing and despatch floor should be at lorry platform level, in order to facilitate loading.

REFERENCES

(35) Morris, L. G., "The Steam Sterilizing of Soil." N.I.A.E. Case Study No. 14 (1951).

(36) Courshee, R. J., "Fruit Grading Machinery." N.I.A.E. Report (1952).

CHAPTER TWENTY-THREE

CARE AND MAINTENANCE OF FARM MACHINERY

"What husbandly husbands, except they be fools, but handsome have store-house, for trinkets and tools? And all in good order, fast locked to lie, whatever is needful, to find by and by."

Five Hundred Pointes of Good Husbandrie. Thomas Tusser, 1573.

Farm machinery must be cared for intelligently and kept in good repair if it is to render efficient service. Investigations into the life of implements and machines have shown that there is little correlation between the life of the machine and the number of days' use per annum, and this is partly due to neglect of the machinery during the long idle periods. Many machines are discarded because of a small breakage when only a minor repair would put them back into service, and others need but a small alteration to rescue them from obsolescence and make them equal in performance to their costly new counterparts. Few farm implements actually wear out; most, such as the plough, have wearing parts which may be easily and cheaply replaced, and only a few, such as the disc harrow, in which many parts are subject to wear, become so generally worn out that it is cheaper to buy a new implement than to undertake a general replacement of the worn parts. Machinery costs are an important item in the total farm costs, and one method of reducing them lies in proper care and repair of the equipment. The maintenance in good order of the equipment includes protection from the weather by housing, protection from corrosion and decay by painting, regular attention to lubrication, and the adjustment and replacement of parts.

Housing of Farm Machinery.

The need for housing farm machinery varies with the type of equipment concerned. The binder that stands outside, or forms a perch for poultry for eleven months of the year, is not likely to give satisfactory service in the twelfth month; and where implement sheds are available it obviously pays to use them. But the capital and maintenance costs of farm buildings are so high that it is uneconomic to put up implement sheds for ploughs, unit cultivators and similar implements which consist almost entirely of painted steel bars of generous proportions. The best and cheapest form of protection for such implements is to re-paint the frame-work when necessary, use the grease gun on any bearings, and cover the bright parts with one of the many rust-preventive compounds which are now available. This takes a little longer than pushing the implement under an open shed, but if done thoroughly, there is no reason why the simple implements should come to any harm when left outside for long periods. Complicated machines and tractors, on the other hand, need to be properly housed. Tractors need a closed garage, if only to make starting easier on cold and damp mornings.

Investigations in America into the influence of housing indicate that in that country, losses from exposure are not so great as is often contended or imagined; but these investigations showed that housing a machine such as a binder greatly prolonged its life, and that the cost of repair of housed machines was lower. Moreover, when machines are housed they are always more readily available for use, and can more quickly be put in good working order.

The requirements for implement storage are simple, for the building need only provide protection from moisture and the sun. The exact type of shed and its orientation must depend upon local circumstances, but an open building facing approximately north-east is suitable for most districts in Britain, since little rain usually enters, and the sun, which causes wood to shrink and warp, is excluded. Unfortunately for the implements, many implement and cart sheds have been converted to other purposes. These old sheds, with bays about 9 ft. wide and very deep, were ideal for implement storage; but when modern sheds are built, added adaptability is secured if some of the bays are wider.

If open sheds are constructed for housing simple tractor-mounted implements, they can be low and shallow. A lean-to shed with 6 ft. head-room and 9 ft. deep is adequate for the housing of unit ploughs, cultivators and harrows.

Care, Repair and Adjustment of Equipment.

When implements are put away after use, any bright parts, such as binder knotters, plough breasts, etc., should be given a coat of waste oil or grease in order to protect them from rust.

With a machine such as a binder, which will not be used again for a very long period, it is best to use a proprietary rust-preventive compound on parts such as the knotter bills. A simple precaution like this, which occupies only a few minutes, is well repaid in the reduced depreciation and the easier operation of the implements when next they are used. For example, two minutes spent on protecting knotter bills at the end of one harvest may save a difficult hour or more caused by rusty bills at the beginning of the next.

Some machines deteriorate rapidly if they are not thoroughly cleaned before being stored. Fertilizer distributors and spraying machines may become so completely corroded as to be quite unworkable if chemicals are left in contact with the metal parts. It is good practice to give a fertilizer distributor a thorough wash before oiling and storing it for a long time; for even the dust left after sweeping it out may cause serious corrosion.

During slack periods it pays to carry out any repairs or adjustments necessary to get machines ready for service. Time spent on jobs like filling oil baths, tightening nuts and bolts, and lubricating and adjusting chains before an implement is put to work, saves breakages and the disaster of equipment being out of action when it is urgently needed.

Farm Workshops.

Practically every farm to-day needs a workshop of some kind, though only the larger farms can justify keeping a full-time mechanic. The type of workshop needed depends partly on the capabilities of the farmer and of the men who will use it. At one extreme a solid bench and vice, and a few hand tools will suffice; but many farmers who are especially dependent upon their machinery find it economic to maintain a workshop fitted with a forge and a few power-driven tools, such as a drilling machine, a hack-saw and a "buffing" machine. A differential pulley to assist in raising heavy parts, and an oxy-acetylene welding outfit are welcome additions to the equipment on large farms where thorough overhauls and repairs are undertaken.

Oxy-Acetylene and Electric Welding.

Nearly all worn or broken iron or steel parts of farm machines can be quickly and economically repaired by welding. With a good job of welding the part repaired is usually quite as good as new.

While farmers need not necessarily know how to weld, or themselves employ a man who knows, all farmers should understand just what can be achieved by welding, so that parts which can be economically repaired are not scrapped. Welding may be used in farm repair work for two main types of job, namely (1) to rebuild and hard-surface worn parts; and (2) to join together broken parts.

(1)Rebuilding And Hard-Surfaging. Worn steel or cast iron parts can be rebuilt to their original shape by depositing ordinary weld metal; and they may be finished with a deposit of tough glass-hard and abrasion-resisting metal. Axles, shafts and journals may be thus built up, and ground or turned to a smooth finish. Similarly, all such parts as steel tractor lugs or strakes, tracklayer grousers, track links and pins, gear teeth and sprockets can be built up, and turned to size where necessary. Welding is particularly useful for all cultivation implements, especially ploughs. Thus, worn plough breasts can be patched; and steel shares can be built up and hard-faced. A properly rebuilt and hard-surfaced share will often last twice as long as the original one; and it has been found that the process can be repeated up to five times.

(2)Joining Broken Parts. In general, all broken steel or cast iron parts can be repaired quickly and efficiently by welding. Before the part is welded and put back to work, it is advisable to consider the cause of the breakage. If the metal of the broken part was faulty, welding may not be worth while. In some circumstances reinforcement of the part may be advisable; while in others, adjustment of the machine to ensure that the trouble does not recur may be necessary.

The Rural Industries Bureau has assisted many country smiths to become expert welders, and it is to be hoped that village blacksmiths of the future will have an oxy-acetylene or electric welding outfit as a regular part of their equipment. For large workshops it is desirable to have both oxy-acetylene and electric welding outfits. Most straightforward welding work can be done more rapidly and efficiently by the electric outfit; but some welding jobs and cutting are more efficiently carried out by an oxy-acetylene outfit.

Fuel and Spares Trailer.

Servicing the tractor and. all equipment used with it is often facilitated by the provision of a two-wheeled trailer suitable for carrying fuel, tools and spares. When constructing or purchasing such a vehicle it is advisable to choose one capable of carrying just over a week's supply of fuel. The tank should be fitted with a semi-rotary pump, and a length of delivery hose long enough to reach the tractor tank. Such a trailer saves a great deal of time being wasted in running about after fuel. It also ensures a clean fuel supply, free from dirt and water.

The trailer should be fitted with a lock-up box for carrying supplies of petrol, oil and grease; shackles, tow-chain, funnel and tools; and such spares as sparking plugs, plough shares and a few assorted nuts and bolts. There should generally be accommodation on the trailer for the tractor driver's bicycle or motor cycle.

REFERENCE

(37) Hine, H. J., *The Farm Workshop*. Crosby Lockwood & Son.

CHAPTER TWENTY-FOUR

DEVELOPMENT AND ECONOMIC ASPECTS OF THE MECHANIZATION OF AGRICULTURE

"The implements which mankind have employed in the cultivation of the earth, and their gradual improvement, is a theme closely interwoven with the history of agriculture."

The Implements of Agriculture. J. A. Ransome, 1843.

There is good evidence that the ground was cultivated as early as 1000 B.C, and when the building of the pyramids was in progress the first primitive mechanized farming was in full swing. The ox, the ass and the goat were used for pulling primitive cultivating tools, and it became possible for part of the population to produce sufficient food for the community. From these early historical times agriculture has developed as an art, for 3,000 years to our knowledge, and probably much longer. For nearly the whole of this period the chief tools were such as the simple plough, the hoe, the sickle and the flail. The operation of these was mostly by human muscular effort, but increasing use was made of the ox and the horse.

It may be of interest to review briefly, as an example of the gradual course of mechanical evolution, the progress of mechanization in Great Britain. Here, the development of the plough can be roughly traced from Saxon times. In the Manorial period, seeds were generally sown broadcast, and a hawthorn tree, weighted by logs, was used as a harrow. Weeding was done by hand or with a forked stick, and reaping was carried out, as in Roman times, in two operations, the ears being first removed and then part of the straw gathered.

During the Dark and Middle Ages and up to the end of the sixteenth century little advance was made. Ploughs were extremely clumsy, and wheat was still harvested with the sickle. The writings of Fitzherbert and Tusser show that little progress had been made during the past thousand years. During the seventeenth century there were signs of the developments to follow, for towards the end of the century the ancestors of many of our modern machines, such as the seed drill and the reaper, appeared in their first primitive forms. The eighteenth century witnessed a. gradual improvement in all kinds of machinery, though progress was slow and uneven. In 1733, Jethro Tull published his *Horse-hoeing Husbandry*, and the maxims of clean farming and economy of seed by the use of drills and horse hoes were gradually put into practice.

In the early part of the nineteenth century many successful new mowers, reapers, horse rakes, chaff cutters, root cutters, threshing-machines and so on were invented, but, generally speaking, comparatively little progress was made in the evolution of the tools used by the farmer during the hundred years from 1750 to 1850, when machinery was transforming the industrial world. The introduction of machine methods into agriculture, making possible a general substitution of horse or mechanical power for human muscular effort, may be considered to have begun in earnest about the middle of the nineteenth century.

With the introduction of the internal combustion engine the movement gained momentum, and in the present century, especially during the past twenty-five years, the use of mechanical power and appliances in agriculture has increased at a rate unparalleled in history. In some parts of the world to-day agriculture is mechanized to a degree which, a few decades ago, would have seemed revolutionary and impossible. On the other hand, there are countries where the existing farming methods and machines differ little from those of a thousand years ago.

Mechanization of Agriculture in Foreign Countries.

Though the mechanization of industrial processes is almost universal, the majority of farmers throughout the world are still content with simple implements and machines which differ little from those used by their forefathers. The agriculture of most of Asia and Africa, and also of some parts of Europe, is characterized by an innate conservatism. The widespread and scattered nature of farming, the low standards of living, the relative cheapness of human labour, and limited resources in land and capital, prevent any rapid adoption of new machines in these regions; but in the newer countries, shortage of labour and almost unlimited amounts of land at an early stage in their development provided the incentive for important advances in agricultural engineering. During the years between the two wars there occurred a tremendous development in the application of mechanical aids to agricultural production, especially in the United States, Canada and Russia. Countries like our own, which could not readily revolutionize the traditional methods even if that were desirable, adopt or develop such machines and mechanical methods as can be fitted into the existing farming systems.

FIG. 293.—TRACKLAYER AND WIDE DISC

HARROW COVERING 6½ ACRES PER HOUR ON A
LARGE GRAIN FARM. (U.S.A.)

In the United States, under the stimulus provided
by high labour costs, the mechanization of agriculture has
progressed at an extremely rapid rate. The aim in agriculture,
as in industry, has been to remove drudgery and to secure the
maximum output per unit of labour. The U.S. Department
of Agriculture has estimated that the amount of manual
labour required to produce 20 bushels of wheat per acre was
57 • 7 man hours by the rudimentary methods of 1830,
when the sickle and flail were used; in 1896, with the binder

and stationary thresher, it was reduced to 8 • 8 man hours; and only 3 • 3 man hours were required when using the combine harvester in 1930.

Farmers have equipped themselves with tractors, motor lorries, combine harvesters and a whole range of other power-driven machinery in order to raise the efficiency of food production. Tractors, motor lorries and cars now do the work that used to be done by horses, and there has been a great decline in the number of working horses and mules used.

A feature of the last quarter-century has been the rapid adoption of the combine for harvesting cereals, and to a lesser extent crops such as soya beans.

Yet the common impression that American agriculture consists mainly of large-scale wheat farms is erroneous. There is a great variety in systems of agriculture, and while wheat farms of over 1,000 acres are common in the Great Plains, small mixed holdings are more typical of American agriculture generally. Eighty-eight per cent, of farms range from 3 to 260 acres, and 60 per cent, of the farming population is on holdings of less than 100 acres. There is a marked tendency in favour of the family size of farm, whether it be a 640-acre farm growing only cereals, a 160-acre farm growing several crops and producing milk and pigs, or a much smaller holding where fruit and vegetables are grown on an intensive system. Dairying is the most important branch of agriculture. In Alabama, which is fairly typical of the Southern States, 84,000 of the 231,000 farms are "one-mule" units.

Rapid developments in rural electrification have occurred since the setting up of the Rural Electrification Administration.

One of America's greatest farming problems is soil erosion, and engineering methods are employed in dealing with both water and wind erosion. For example, where water erosion is serious, terracing machines are used to set up the land in a series of terraces so that steep cultivated slopes are avoided; and in the dry-farming areas, where wind erosion is severe, special cultivation implements which leave the rubbish on the surface and create a "stubble mulch" have been developed.

It is impossible in a few words to provide an adequate picture of the mechanization of American agriculture. Reference should be made to American books and periodicals for further information on this subject.

The results achieved in terms of labour productivity are striking when compared with the British average 38, 39.

As in the United States, the agriculture of *Canada* is in large areas well suited to mechanization, and the use of tractors and combine harvesters has spread with rapidity in the Prairies. The winters are long, the period of vegetation short, and the work of cultivation and harvesting must be carried out in a short time with scarce and expensive labour 40.

In *Australia,* the chief farming enterprises are wheat-growing and sheep-feeding. Conditions are somewhat similar to those prevailing in Argentina in that inexpensive grazing for horses is available, reducing their cost of upkeep. Tractors are used on most of the larger farms, but horses remain an important source of power. In many directions the construction of machinery has proceeded along lines independent of developments in North America or Europe. An example is the invention of "stump-jump" ploughs and cultivators for the tilling of fields containing roots or stones. These are so constructed that when the implement strikes an obstacle it is lifted over it in such a way that nothing is broken. Special types of harvesting machines including "headers", "strippers" and combines, have also been developed along independent lines.

In *Russia,* mechanization has taken place by systematic planning rather than by spontaneous development. The soil and climate, the vast areas, and the existence of home supplies of fuel have provided an incentive to mechanical advances. Russian agriculture to-day consists chiefly of collective farms, a typical unit comprising 100 to 200 peasant families tilling 1,000 to 2,000 acres of socialized land and having a communal existence.

The modern tractors are mainly tracklayers, these being considered more dependable and effective than wheeled machines for Russian conditions. The engines are mainly of the Diesel type on account of their high efficiency in terms of h.p.-hours per gallon of fuel. The combine is considered indispensable for cereal production. The medium and small collective farms do not generally possess many machines of their own; these are provided by the state tractor stations, and work done is paid for by part of the harvest.

In 1939, 6,647 state tractor stations were operating over 400,000 tractors and 130,000 combine harvesters. The productivity of labour was doubled in the twelve-year period 1925-7 to 1937-9.

In addition to the collective farms there are a few enormous State farms. An example of these is one on prairie land near Rostov, comprising about 400,000 acres, where sometimes the only operations performed are ploughing in the grain and harvesting the crop by means of combines.

The situation in most of *Europe* differs from that in the newer grain-exporting countries. The size of farms is small, and the distribution of fields often unfavourable to the use of large machines. There is a surplus rather than a deficit of labour, and the wide variety of crops grown in the mixed farming systems makes the adoption of mechanical aids a more gradual process. Moreover, many countries are much concerned with the maintenance of a "balance" between agriculture and industry, and state effort sometimes aims at preserving a large agricultural population. There is no desire to obtain the last ounce per man by means of machinery; rather, the object of the farm is to provide work and a living for the family. The objective is to increase yield per acre and to lower the costs of production rather than to cultivate large farms cheaply. The fiscal policy is generally bound up with social, political and military considerations, and the general effect in these circumstances militates against any startling changes in farming practice. But there is inevitably gradual progress in the mechanization of the traditional farming operations.

Over a large part of the world, notably *Tropical Africa, India* and the *East,* very primitive tools are still used. In these countries agriculture is mainly in the hands of the peasants. The holdings are small, and the land of individual holdings is frequently widely scattered in very small plots. Hand tools are, and for some time are likely to remain, the prime implements of the peasant farmers in these countries.

Economic Aspects of Mechanization.

The importance of efficiency of mechanization in British agriculture is indicated by the fact that labour now represents anything from 30 per cent, to over 50 per cent, of total farm costs. On average, machinery costs now account for a little over 20 per cent, of farmers' annual outgoings, while capital investment in machinery averages just over 20 per cent, of total tenant's capital. The increase in productivity of farm labour in Britain has not been as marked as in U.S.A. The increase in the net output of British agriculture between the 1930's and 1950 has been reliably estimated to be about 40 per cent. During the same period the total labour force increased by 16 per cent. The increased output per man during this period was therefore approximately 20 per cent. In assessing the importance of mechanization, however, the effects of shorter working hours and extra holidays with pay-must be considered. Due regard should also be paid to the fact that on many farms during this period there has been extra work on land reclamation, and on remedying the results of years of neglect of drainage, hedging, building repairs and maintenance work generally.

In comparing the productivity of labour in U.S.A. and in Britain due allowance must also be made for the advantages of typical Mid-Western farms in respect of their more uniform climate and easy-working soils. Among factors which we in Britain should study fully, and emulate where possible, are (1) sensible farm layouts, with a limited number of fields and a minimum of fencing; and (2) simple cropping and livestock husbandry systems, leading to individual enterprises which are big enough to justify mechanization. There is no reason why many British farms should not specialize in rather fewer enterprises, without in any way neglecting the principles of good husbandry.

Perhaps the most important difference between British and American farming is in the proportion of hired labour. In U.S.A., hired labour represents only one-fifth of the total engaged in agriculture, whereas in Britain the number of paid workers is about double the number of farmers.

A Farm Management Survey conducted by Reading University Department of Agricultural Economics has shown that a high level of mechanization tends to be associated with a high output per acre. The more highly mechanized farms in Southern England were found to spend more per acre on labour, as well as on many other things; but because of high output per acre, they achieved higher profits than less mechanized farms.

Mechanization and Management.

The purchase of machinery by itself achieves nothing. Machinery can never compensate for bad planning or bad farm management. The first requirement in planning mechanization is a detailed study of the management of the individual farm. When management policy has been suitably adjusted, successful mechanization necessitates *(a)* choosing a set of equipment suited to the particular needs of the farm, *(b)* learning the most efficient techniques for operating all the equipment, and *(c)* evolving a system of looking after the machinery which will ensure that it is always efficiently maintained. When a particular farming operation is to be undertaken, its satisfactory performance almost invariably involves the use of power and machinery; and success in farming depends more and more on the ability to choose and use machinery to greatest advantage.

In Britain, good farm management consists of getting the best out of limited resources in land, labour and capital. The relationship between mechanization and these three resources is briefly discussed in turn below.

Land. The total area of land in Britain is limited, but mechanization influences the use that can be made of it. For example, the displacement of some 300,000 working horses by tractors during the period 1939-51 released for the production of human food something like a million acres of land which was formerly needed for growing fodder for the horses. There arc also vast areas of land such as the heaviest clays and the chalk downs, which could only produce poor grass in the absence of efficient cultivations, but which are now enabled, by the use of such devices as powerful tractors and milking bails, to produce heavy crops of corn, and milk. Efficient mechanization also enables more to be produced on the best land, owing to such factors as *(a)* deeper and more thorough cultivations, carried out at exactly the right time; *(b)* crop protection techniques such as spraying for weed control; and *(c)* improved harvesting processes which reduce the losses caused by unfavourable weather.

Labour. The importance of mechanization with respect to saving of labour has already been referred to. On individual British farms, however, the primary object of mechanization of a particular operation is often not to reduce the total labour cost of the farm. Sometimes the object is increased farm output from the same labour force, while often the existing labour force is insufficient to get all essential work done at the right time, and mechanization is resorted to in order to overcome this difficulty.

A common reason for mechanizing a particular operation is the need to reduce the peak demand for labour during certain seasons. For example, the supply of casual labour for assisting with spring work in sugar beet, the hay harvest, the corn harvest and the harvesting of root crops steadily diminishes, and any machine which will help to reduce labour needs at these times is of interest. Whether it will pay to mechanize one or more of these operations in a particular way may, however, depend partly on the labour needs at other seasons. For example, it may be clear that the use of a green-crop loader will substantially reduce the labour needed for making silage; but if capital is limited, if the silage difficulty can be overcome by paying men to work overtime, and the most difficult season for labour is during beet harvesting, it is clear that any capital that can be spared for mechanization should be used to buy a beet harvester rather than a green-crop loader.

On many farms there is a high peak in man-labour requirements when work in root crops clashes with the hay harvest. The equipment of the workers with suitable mechanical aids such as hay-loaders, sweeps, elevators, tractor root hoes and so on, may increase the efficiency of man-labour and also of the complete farming organization. The farmer's problem must always be viewed as a whole. Adequate power and machinery must be available to do the work at the right time; but expenditure on any equipment that is unnecessary must be avoided. The criterion is whether additional expenditure on equipment will improve the balance of total farm income over total expenditure.

CAPITAL. The level of capital investment in machinery that can be justified in particular circumstances cannot be defined without reference to all relevant details. In general, capital investment has to be justified by reduced running costs or increased output, or by preventing a deterioration in either of these factors in times of increasing difficulty. On average, some £50 million was invested annually in machinery by British farmers during the years 1949-52. On typical farms this represents an investment of about £2 per acre each year. The capital invested in machinery on individual farms has steadily increased. When a farmer has to decide whether a particular piece of equipment should be purchased, the problem should be tackled in two stages, viz. (1) preparation of a budget in which the effect of introduction of the machine is studied in isolation, and (2) consideration of the repercussions of introducing the machine on farm management as a whole.

For example, suppose that there is a technically efficient potato harvester which is quite suitable for the soil conditions of the farm. The first step is to calculate the cost of lifting the crop with and without the machine. The next is to consider such problems as the effect of introducing the harvester on labour needs; whether the acreage of potatoes can profitably be increased; and the repercussions that such a step would have on other enterprises. Introduction of a milking machine sometimes facilitates an increase in size of herd, but the repercussions of such a step are likely to include the need to grow more cattle food, and perhaps to provide a larger cowshed.

Methods of Calculating Operating Costs.

The factors involved in operating costs of machinery are *(a)* depreciation of the machine, *(b)* running cost (fuel, oil, etc), *(c)* repairs and maintenance, *(d)* interest on capital, and *(e)* miscellaneous charges like housing, licence, insurance, etc. These factors are fully discussed in another book by the present writer.41

As a simple example of the methods that may be employed, consider the introduction of a sugar beet harvester costing £500 on a farm which grows 50 acres of beet annually. Assume that the machine will do 2 acres in an 8-hour day, and because development of beet harvesters is still progressing, assume that the machine must be written off in 5 years. The cost per acre of harvesting by the machine may then be reckoned as follows:

Item	*Cost per Acre*		
	£	s.	d.
Depreciation ($\frac{1}{50}$ x 20 per cent, of £500)	2	0	0
Repairs and Maintenance, say £40 per annum		16	0
Interest on Capital ($\frac{1}{50}$ x 4 per cent, of £250)		4	0
Tractor at 4s. per hour (4 hours)		16	0
Labour, 2 men for 4 hours at 3s	1	4	0
Cost per acre	£5	0	0

The cost arrived at must be compared with the cost of hand work, which may be higher. The effects of the machine on "quality" of the work will also need to be considered, both as regards the dirt and top tares on beet despatched to the factory, and also as regards the effects on the tops if these are needed for feeding. As already indicated, the cost comparison may be of secondary importance compared with a serious scarcity of labour. The reader should note the proportion of the cost accounted for by depreciation and repairs in the above example, and should understand that different methods of reckoning could give quite different answers.

Factors affecting the Choice of Equipment.

Factors which must be considered when deciding whether an increase in mechanical plant will justify itself are the initial cost and life of the machine, its annual use, running costs, adaptability and technical efficiency, and the effect that its introduction will have on farm output.

The life of different farm machines varies between wide limits; some are worn out in a year or two and some last for half a century. The useful life of farm machinery is frequently ended by obsolescence; for many are discarded long before they are worn out, owing to rapid improvements in design.

Farm operations are mostly confined to limited seasons or periods, and the annual use is generally low compared with that of machines in factories. Many harvesting machines are used for only short periods, while certain special machines, such as potato sprayers, may be used for only a day or two in the year. On the other hand, equipment such as the milking machine is used twice a day on every day of the year. The desirability of the purchase of any machine depends, in part, on the number of days per year on which it can be productively employed. Ease of adaptability to a variety of operations leads to a full annual use and thus Lo more efficient operation.

Running costs of various items of power and machinery must be carefully considered, for while the initial cost of most forms of equipment is fairly constant throughout the country, the running costs vary from farm to farm. In considering the advantages and disadvantages of using various items of labour-saving machinery it is important to consider the effect of the use of the machine on the *total,* running cost of the farm. Though a certain machine may in theory economize in labour, its use may actually increase the total farm costs unless it can be of service in the periods when the demand for labour is greater than the supply.

The production of a new farm machine involves a considerable expenditure of time and money in design, testing, production and trial before a satisfactory model can be put on the market. Machines that are still in the experimental stage are sometimes offered for sale, but the purchase of such untried machines may generally be avoided by dealing with well-established manufacturers.

In addition to the considerations outlined above, the effect of the use of mechanized methods on crop yields and quality must be considered. If work can be done more quickly, an increase in the actual cost of working may be more than repaid. Examples of operations to which this consideration frequently applies are the harvesting in good order of clover seed, hay, and barley, the autumn drilling of corn, particularly on heavy land, and the preparation of spring seed beds.

Mounted Implements. Many of the most efficient modern implements and machines for use with all-purpose tractors are designed for mounting directly on the tractor's hydraulic lift and linkage system. A measure of standardization of linkages and hitch-points has been agreed, but most mounted implements can only be effectively used with the make or makes of tractor for which they were designed. This creates a problem for farmers who operate only two or three tractors, a matter which has been discussed in Chapter One.

Social Factors. A factor which has an important bearing on the choice of equipment, but which cannot be measured quantitatively, is the effect of mechanization on the nature of the life and work of farm workers, including farmers themselves. A machine which eliminates drudgery from farm work may achieve something that is more important than the overcoming of a temporary labour shortage. Such mechanization can alter the whole character of farm work. In a country such as Britain, where farms have to compete with factories for a supply of labour, attracting intelligent and well educated boys to farm work is lilely to depend to a great extent on making the work reasonably attractive. This sometimes means eliminating continuous drudgery and uncongenial jobs with the help of mechanization, even though use of the machine may not result in an immediate economic advantage.

Mechanization of Small Farms.

Many surveys have shown that expenditure per acre on equipment tends to decrease progressively with increasing size of farm. The decrease is rapid up to a size of 150 to 200 acres and becomes small above 300 acres. Intensity of production is often greater on small farms, but the tendency to higher capital investment per acre on small farms is mainly accounted for by the fact that a greater acreage is usually operated by a single implement on large farms than on small ones. For example, many farms with less than 20 acres of corn possess a binder, yet one machine could harvest five times that area.

Nevertheless, mechanization is just as important on many small farms as on large ones. When one man has to do all the work on a 60-acre dairy farm which has 15-20 acres of tillage, he can only do it effectively if he has a milking machine to make the best use of his time in the cowshed, and a medium-powered tractor and equipment, that will enable him to get on with field work quickly and efficiently. Much of the new equipment such as buckrakes, hydraulic manure loaders, etc., is very suitable for one-man operation; and the only serious problem is finding sufficient capital to allow the purchase of such up-to-date equipment.

Contract Work and Co-operation.

On many small and medium-sized farms it is clearly impossible for the farmer to justify purchasing all the mechanical equipment that is needed. In some districts agricultural contractors are well established, and there are many types of work which they can do efficiently for all kinds of farmers. Examples are deep ploughing, subsoiling, spraying, baling, threshing, manure and lime spreading, mole draining and hedge trimming, most of which require specialized equipment of a kind that few farmers can own. Work such as pick-up baling and combine, harvesting may sometimes be satisfactorily done by contract, but is more difficult owing to the limited time during which the best results arc obtainable. Some small farms can make good use of contractors for the more ordinary kinds of farm work, but on most farms the acquisition of a medium-powered tractor and a few implements is more satisfactory.

There is great scope throughout Britain for more co-operation between neighbouring farmers in regard to the hiring of little-used machines. Many machines such as special-purpose drills, farmyard manure spreaders and hedge trimmers are so little used on one small or medium-sized farm that there is every advantage in small groups of farmers arranging to hire out equipment to one another, or to do work for one another on a contract basis. This is a practice which, under the name of "custom" work, is common in many parts of U.S.A. There, it is not unusual for one of three neighbours to own a corn-picker, another a forage harvester, and the other a pick-up baler, each farmer working with his machine for the others, in addition to doing his own work.

Another method of co-operation which has been successful in Buckinghamshire is the joint ownership of "pools" of machinery which can be hired by the contributing members.

REFERENCES

(38) Report of the Anglo-U.S. Productivity Council on Productivity of Labour in Farming.

(39) Menzies-Kitchin, A. W., "Labour Use in Agriculture." Camb. Univ. Farm. Econ. Branch. Bull. No. 36.

(40) Report of the British Agricultural Machinery Mission to Canada. H.M.S.O. London. 1949.

The following American text-books give an account of equipment used in the United States:—

(41) Davidson, J. B. *Agricultural Machinery.* Wiley.

(42) Smith, H. P. *Farm Machinery and Equipment.* McGraw-Hill.

APPENDIX ONE

FUNDAMENTAL MECHANICAL PRINCIPLES AND SIMPLE MACHINES

"For now so large a part of their (farmers') business is performed by the aid of machinery, it is positively necessary that they should become acquainted with the principles and construction of mechanical contrivances; and this is by no means so difficult a matter as they would at first be likely to imagine."

Rudimentary Treatise on Agricultural Engineering. G. H. Andrews, 1852.

An intelligent study of machines cannot be made without some reference to certain fundamental mechanical principles or the use of certain technical terms. A few simple principles and definitions are therefore stated below.

A **force** may be defined as "any action on a body which produces or destroys, or tends to produce or destroy, motion". It is the equivalent of a push or a pull. Thus, a horse exerts a force upon the load which it pulls, and the gases which explode in the cylinder of an internal-combustion engine exert a force on the piston. A force has two properties, viz.

direction and size: both of these characteristics may be represented by a straight line, and for this reason, a force is said to be a **vector** quantity. The unit of force is the weight of one pound.

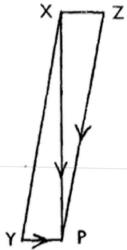

FIG. 294.—;COMPOSITION OF FORCES

XP, resultant of two forces, **ZP** and **YP**.

The Composition of Forces.

When more than one force acts on the same member of a machine, it is frequently necessary to find the combined or **resultant** effect of the forces. Thus, if the direction and magnitude of the forces acting upon the share, the breast, the landside and slade of the plough can be. determined, it is possible to find their combined effect. Forces may be combined graphically by use of a diagram called the **parallelogram of forces.** For example, suppose ZP and YP (Fig. 294) are the vectors for the forces acting on the breast and landside of a plough, respectively; then, if the parallelogram ZPYX be completed, XP is the vector for the resultant force. Similarly, any number of intersecting forces may be compounded into one resultant.

The Resolution of Forces.

When it is necessary to resolve a single force into two or more forces acting in given directions, the converse of the foregoing method may be employed. For example, if the force exerted by three horses on a plough working on horizontal ground is 600 lb. and the traces make an angle of 20 degrees with the horizontal, the force tending to lift the front of the plough in a vertical direction and the **component** of the pull in a horizontal direction may be found as follows:—Draw AB, the vector for the pull of the horses (Fig. 295). Through A draw AC at an angle of 70 degrees with AB. Complete the rectangle ABCD. Then AC represents the vector for the lifting force and AD that of the horizontal component. AC = 204 lb. AD = 564 lb.

FIG. 295.—*RESOLUTION OF FORCES* **AD, AC,** *TWO*
COMPONENTS OF FORCE **AB.**

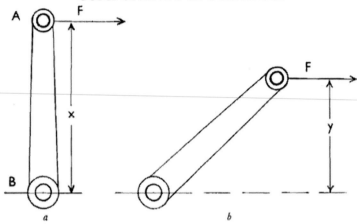

FIG. 296.—*DIAGRAM ILLUSTRATING MOMENT OF*
A FORCE **AB,** *LEVER;* **F,** *FORCE;* **FX, FY,** *MOMENTS*
OF FORCE **F** *ABOUT* **B.**

The problem may, of course, be more accurately and easily solved trigonometrically. Thus,

AC=AB Sin 20° = 600 x 0.34=204 lb.

AD=AB Cos 20° = 600 x 0.94=564 lb.

842

Moments.

The **moment** of a force about any point may be defined as "the tendency of the force to rotate the body upon which it acts about the point", The moment is measured by the product of the force and the perpendicular distance from the line of action of the force to the point of rotation. For example, if a force of F lb. (Fig. 296 *(a)*) is exerted through A, at right angles to the lever AB which pivots at B, the moment of the force about B is *Fx* lb.-ft. If the lever moves into the position shown in Fig. 296 *(b)* and the force continues to act in the same direction, the moment is given by the product *Fy*.

The Principle of Moments.

The principle of moments states "If a system of forces in one plane act upon a body and keep it in equilibrium, the algebraic sum of their moments about any point in the plane is zero." This principle is of value in the solution of many engineering problems.

Example.—A uniform beam, 12 ft. long and weighing 6 cwt., is supported on a wall at each end. A load of 30 cwt. is to be placed 3 ft. from one end. What will be the forces exerted by the beam on each support?

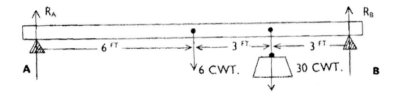

FIG. 297.—ILLUSTRATING PRINCIPLE OF MO-
MENTS

Fig. 297 shows the forces acting on the beam and supports. The weight of the beam acts through its centre of gravity. The sum of the vertical thrusts (Ra and Rb) exerted by the walls on the beam must equal the weight of the beam and its load. The thrust exerted by the beam on each support must be equal and opposite to the thrust exerted by the support on the beam. By the principle of moments, since the system is in equilibrium, the algebraic, sum of the moments about any point is zero.

Considering point A, and calling clockwise moments positive and anticlockwise ones negative,

$$(+ 6 \times 6 \text{ cwt.-ft.} + 30 \times 9 \text{ cwt.-ft.}) — (Rb \times 12 \text{ cwt.-ft.}) = 0.$$

$$\therefore 12\ Rb = 306 \text{ and } Rb = 25.5 \text{ cwt.}$$

i.e. a force of 25.5 cwt. is exerted upon the support B.
The force exerted upon the wall at A may be obtained by difference or by taking moments about B. $Ra = 30 + 6 - 25.5 = 10.5$ cwt.

Work, Power and Horse-Power.

In a technical sense, **work** has a restricted meaning. It is force acting through distance and representing a definite expenditure of energy. "When a force acts upon a body and causes it to move, it is said to do work on the. body." If the force is constant, the work done is measured by the product of the force and the distance through which the body moves in the direction of action of the force. The British Engineers' unit of work is the foot-pound (ft.-lb.).

Power

is the **rate** of doing work, or the amount of work done in a given time. The British Engineers' unit of power is the **horse-power** (h.p.) and represents the expenditure of 33,000 ft.-lb. of work per minute. The unit horse-power was fixed about 1780 by James Watt who, as a manufacturer of steam-engines, compared the power of his engines with that of horses. He established the unit by experiments with horses winding up weights. Draught horses weighing 1,400 lb. or more can develop 1 h.p. continuously for a working day. An example illustrates how horse-power may be calculated.

Example.—What is the least horse-power that a pump will require when lifting water at a rate of 400 gallons per minute through a vertical distance of 20 feet?

Weight of water lifted per minute=400 x 10 lb.

(1 gallon of water weighs 10 lb.)

Work done per minute=weight lifted per minute x height.

=4,000 lb. x 20 ft.=80,000 ft.-lb.

Since 33,000 ft.-lb. per minute represent 1 h.p.,

minimum* power required to work pump = $\dfrac{80000}{33000}$= 2·42 h.p

Energy.

Energy is defined as "the capacity to do work". It exists in several forms such as electrical energy, heat energy and the stored or "potential" energy of unburnt fuel or a coiled spring. "Kinetic" energy is the energy possessed by a moving body. Energy can be converted from one form to another, and one of the functions of the engineer is the conversion of natural energy into forms from which useful work may be obtained. Thus, in the internal-combustion engine, the energy of the fuel is converted into the energy of the exploded mixture, which may be applied to the performance of useful work.

The Conservation of Energy.

"Energy cannot be created or destroyed, but can be converted from one form into another." All the energy quantities are related; e.g. one British Thermal Unit (the amount of heat required to raise 1 lb. of water through 1° F.) is equivalent to 778 ft.-lb. of mechanical energy. This relationship is called "the mechanical equivalent of heat". The electric unit of power is the "watt", and 746 watts are equivalent to 1 h.p.

$$1 \text{ h.p.} = 33,000 \text{ ft.-lb. per minute.}$$
$$= 42 \cdot 4 \text{ B.Th.U. per minute.}$$
$$= 746 \text{ watts.}$$

Although energy cannot be destroyed, its conversion from one form to another always involves some wastage. In the conversion of the energy of coal into mechanical energy in the steam-engine, for example, about 85 per cent, of the heat energy in the coal is wasted and only 15 per cent, converted into work. Internal-combustion engines can generally convert only about 20-30 per cent, of the potential energy of the fuel into useful work.

Machines: Mechanical Advantage, Velocity Ratio and Efficiency.

In the broadest sense, a machine is a device which receives work from some outside source of supply, and modifies the forces and motion in such a way as to achieve some practical advantage by delivering the work in a form suitable for the purpose required. One of the simplest and most ancient forms of machine is the lever, which is referred to again below. In the form of a crowbar, it enables a man to move a very heavy weight which he could not move without the aid of some machine. In most simple machines such as hoisting tackle, the force delivering work to the machine (the **effort**) is smaller than the **resistance** overcome. The **Mechanical Advantage** of a machine is the ratio of these two forces. Thus, if a weight of 2 to lb. can be lifted by exerting a force of 60 lb. on a pulley tackle,

$$\text{Mechanical advantage} = \frac{\text{Resistance}}{\text{Effort}} = \frac{210}{60} = 3\cdot5.$$

The **Velocity Ratio** of a machine is the ratio of the speed of that part of the machine on which the working agent acts, to the speed of that part which acts on the resistance. In the case of the pulley tackle mentioned above, if the rope pulled by the operator moves 4 ft. while the weight is being raised 1 ft., the velocity ratio is 4.

If there were no waste of energy in machines, the mechanical advantage would be numerically equal to the velocity ratio. In practice, some of the forces overcome by the machine do not represent useful work. For example, energy is wasted in overcoming the frictional resistances of bearings. The ratio of the total quantity of useful work done by a machine to the total quantity of work delivered to it is called the **Mechanical Efficiency.** In the pulley tackle discussed above,

$$\text{Efficiency} = \frac{\text{Output}}{\text{Input}} = \frac{1 \times 210}{4 \times 60} = 87 \cdot 5\%.$$

The complete relationship between the quantities considered above is Mechanical advantage=Velocity ratio x Efficiency. The velocity ratio is always a fixed quantity, governed only by the dimensions of the elements of the machine. The mechanical advantage, on the other hand, depends upon the energy wasted in the machine, i.e. on the efficiency, as well as on the size of the elements.

Simple Machines.

Many simple, essential mechanisms are common to all machines. In the ultimate analysis, all machines may be reduced to two elements, viz. the lever and the inclined plane. Usually, however, six simple machines are recognized, viz. the lever, wheel and axle, inclined plane, screw, wedge and pulley.

The **Lever** is one of the simplest and oldest machines. In its most simple form it consists of a plain rigid bar that pivots about a point called the fulcrum. There are three "classes" of levers (Fig. 298) and the same law applies to all, viz. the product of the effort (F) and the length of the power arm (P) is equal to the product of the resistance (W) and the length of the weight arm (1).

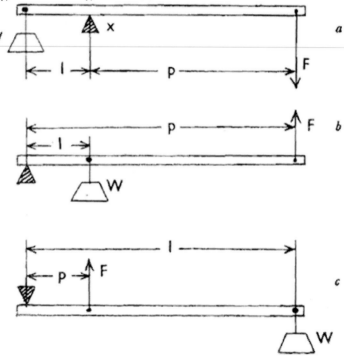

FIG. 298.—THE THREE CLASSES OF LEVER

a, first; b, second; c, third. **F**, force; **W,** resistance; **x,** fulcrum.

The **Wheel and Axle** (Fig. 299) may be regarded as a continuous lever. The velocity ratio is given by dividing the radius of the wheel by the radius of the axle, i.e. $\frac{R}{r}$. Familiar examples are vehicle wheels and pulleys.

The **Inclined Plane** (Fig. 300) enables a small force acting parallel to the plane *(a)* or parallel to the base of the plane *(b)* to raise a large weight vertically by rolling or pushing it up the plane. In *(a)*, neglecting the effects of friction, the work applied by the force, F, in moving the weight, W, from A to B is F x AB. In doing this, the weight, W, is raised a vertical height, BC, against the force of gravity, i.e. the work done is W x BC.

$$F \times AB = W \times BC, \therefore \frac{W}{F} = \frac{AB}{BC}$$

In *(b)*, where the effort is parallel to the base of the plane, the force F', in moving the weight W from A to B moves it a distance AC in the direction of action of the force, and

$$F' \times AC = W \times BC, \therefore \frac{W}{F'} = \frac{AC}{BC}$$

For example, the force acting parallel to the plane that is required to move a weight of 2 tons on frictionless wheels up a i in 10 incline would be $\frac{1}{10}$x 2 tons = 4 cwt. If the force acted parallel to the base of the plane, it would need to be very slightly greater. A frictionless inclined plane has been postulated for simplicity in the above considerations, but in practice, friction is a large and important quantity in most machines employing the principle of the inclined plane. Friction is dealt with in Appendix Three.

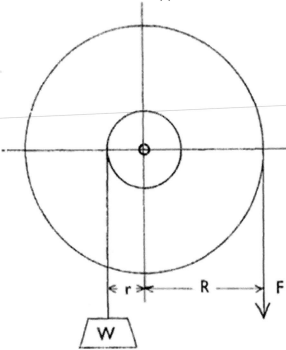

FIG. 299.—THE WHEEL AND AXLE

F, force; **W**, resistance.

852

The Screw.

The screw may be considered as an inclined plane wrapped around a cylinder, i.e. a continuous inclined plane. The **pitch** of a screw is the distance the thread advances along the axis in one revolution. In the screw-jack, the **driver** is the end of the handle, where the force is applied. (The term "driver" is given to that part of a machine to which the effort is applied; the follower is that part which applies the final force to the resistance.) In one revolution of the handle, the driver moves a distance $2\pi R$, where R is the radius of the circle through which it moves. At the same time the follower, i.e. the screw, moves upwards a distance equal to p, the pitch. The velocity ratio of the simple screw jack is

therefore $\dfrac{2\pi R}{p}$.

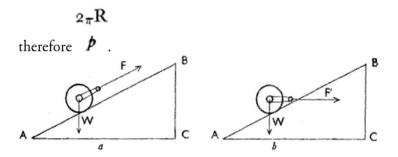

FIG. 300.—THE INCLINED PLANE

a, force parallel to plane; *b*, force parallel to base of plane.

The frictional resistances in screw-jacks and presses are very high, and this is necessarily so. If the frictional resistance were below a certain limiting value, a jack would "overhaul"—i.e. the weight would lower itself by reversing the jack on removal of the effort.

Example.—Find the force that must be applied to the end of the lever of a screw-jack in order to lift $\frac{1}{2}$ ton, if the length of the lever is 15 in., the pitch of the screw $\frac{1}{4}$ in. and the efficiency of the machine 40 per cent.

If the handle is given one complete turn, work done by the follower = 1120 x $\frac{1}{12}$ x $\frac{1}{4}$ ft.-lb. = 23·33 ft.-lb.

Since efficiency is 40 per cent., the amount of work put into the machine by the effort must be 23.33 x $\frac{100}{40}$ ft.-lb.

In one revolution of the handle, the effort moves 2π x 1.25 ft. = 7.85 ft.

Then, if x be the effort required,

$$x \times 7 \cdot 85 = 23 \cdot 33 \times \frac{100}{40}. \quad \therefore x = 7 \cdot 43 \text{ lb.}$$

Mechanical advantage of the jack = $\frac{1120}{7 \cdot 43}$ = 150 approximately.

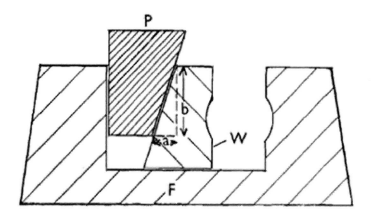

FIG. 301.—THE WEDGE

P, driver; **W,** follower; **F,** frame.

The Wedge.

The wedge (Fig. 301) is one of the oldest machines; it is fundamentally a modification of the inclined plane. While the wedge is being driven in a large distance, the follower moves sideways only a short distance. The velocity ratio (Fig.

301) is $\dfrac{b}{a}$.

Pulleys.

The pulley may be classified as a continuous lever. The fixed pulley (Fig. 302 *(a)*) simply changes the direction of action of the force, while the movable pulley illustrated doubles the force. The velocity ratio of a pulley tackle is always given by the **number of ropes supporting the lower block.** In rope tackle, the separate pulleys of each block are of the same size and are mounted side by side, each pulley being mounted loose on the axle. There is usually little advantage in increasing the velocity ratio above 7, for the mechanical advantage does not increase as rapidly as the velocity ratio, owing to the increased friction. For large mechanical advantages, the differential pulley (Fig. 302 *(c)*) may be used.

The Differential (Weston) Pulley.

In this tackle, the upper block consists of two pulleys of slightly different diameters, rigidly fixed

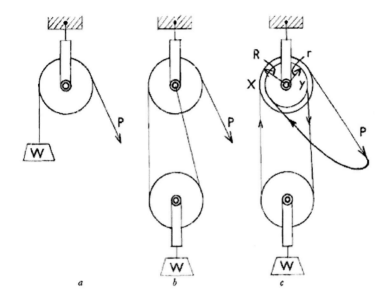

FIG. 302.—PULLEYS

a, fixed pulley; *b,* movable pulley; *c,* differential pulley. **P,** force; **W,** resistance.

together. An endless chain passes round the tackle in the manner illustrated. To determine the velocity ratio, consider one revolution of the upper block. If R be the radius of the larger pulley and *r* the radius of the smaller, then the driver moves a distance 2 πR. While this amount of chain is being pulled over the large pulley at *x,* a length of chain 2 πr is unwound from the small pulley at *y.* The change in the length of chain between *x* and *y,* supporting the lower

857

block, is therefore $2 \pi R - 2 \pi r$. The follower therefore moves upwards a distance $\frac{1}{2}(2 \pi R - 2 \pi r)$

$= \pi (R - r)$. The velocity ratio is therefore $\dfrac{2 \pi R}{\pi(R - r)} = \dfrac{2R}{R - r}$.

Example. —To find the pull required to raise I ton with a differential pulley in which the diameter of the large pulley is 6 in., that of the smaller pulley $5\frac{1}{2}$ in and the efficiency of the tackle 75 per cent.

Work done by the effort, P, in one revolution

$$= P \times 2\pi \times \frac{3}{12} \text{ ft.-lb.} = \frac{P\pi}{2} \text{ f..-lb.}$$

The follower in this time moves

$$\pi (R - r) = \pi \frac{(3 - 2\frac{3}{4})}{12} = \frac{\pi}{48} \text{ ft.}$$

Work done on resistance $= \frac{\pi}{48} \times 2240$ ft.-Ib., and since the efficiency is 75 per cent., this is 75 per cent, of the work done by the effort.

$$\therefore \frac{100}{75} \times \frac{\pi}{48} \times 2240 = \frac{P\pi}{2}$$
$$\text{and } P = 124 \cdot 4 \text{ lb.}$$

The Hydraulic Press or Jack.

An important fundamental principle of hydrostatics, which permits an understanding of the action of such machines as the hydraulic jack and the power lift on a tractor, may be stated as follows:

If a fluid at rest has any pressure applied to any part of its surface, that pressure is transmitted equally to all parts of the fluid.

For example, if the vessel shown in Fig. 303 is filled with oil and a given pressure is exerted on the small cylinder, the same pressure will

FIG. 303.—ILLUSTRATING A FUNDAMENTAL PRIN-CIPLE OF HYDRAULIC LIFTING SYSTEMS

be exerted *all over* the large cylinder. Thus, if the small piston is $\frac{1}{4}$ in. in diameter and the large is 2 in., then the area of the large piston is 64 times that of the small, and a pressure of 1 lb. exerted on the small piston will support a weight of 64 lb. on the large piston or ram.

The pressure in the system (neglecting friction) is obtained by dividing the total force on either piston by the area. Thus, considering the larger piston, it is $\dfrac{64}{\pi \times 1^2}=20\cdot4$ lb. per sq. in.

The hydraulic jack is a simple practical example of the application of this principle. The pump is hand-operated, and consists of a tiny plunger which is caused to reciprocate in a cylinder (Fig. 304). Fluid is sucked from a reservoir into this cylinder through a non-return valve, and is forced *via* another non-return valve into the ram cylinder, which is comparatively large. The result is that the application of a small force to the jack lever can cause the slow raising of a very heavy load. Control of the jack is simple. The action of the non-return valves is entirely automatic, as with the usual type of piston pump (Chapter Seventeen), and the only control that the operator has to manipulate is a valve which releases the oil from the ram cylinder when it is desired to lower the jack. This valve returns the oil to the reservoir from which the pump draws its supply.

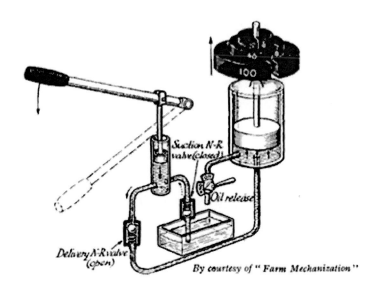

Suction N-R valve (closed)

Oil release

Delivery N-R valve (open)

By courtesy of "Farm Mechanization"

FIG. 304.—DIAGRAMMATIC ILLUSTRATION SHOW-ING OPERATION OF HYDRAULIC JACK

The action of a tractor's hydraulic lift system is similar in essentials, though the pump is in this case power-driven and the control more complex.

REFERENCES

(41) Andrews, E. S. *Mechanisms.* Univ. Tutorial Press.

(42) Duncan. *Applied Mechanics for Beginners.* Macmillan.

* The horse-power actually required would be much greater, owing to inefficiency of the pump.

APPENDIX TWO

THE TRANSMISSION OF POWER

One of the oldest and most common methods of transmitting power from one rotating shaft to another is by means of belting. The power is transmitted through the agency of the friction between the belt and the pulleys. With an open belt (Fig. 305) the follower revolves in the same direction as the driver. With a crossed belt the driver and follower rotate in opposite directions and the **are of contact** is increased. The power that can be transmitted by belting depends upon the strength of the belt and the friction between the belt and pulleys. The friction is influenced by the following factors: (1) the width of the belt; (a) the tension of the belt; (3) the diameter of the pulleys; (4) the material and condition of the belt; and (5) the are of contact between the belt and pulleys. In order to secure efficient transmission of power, a belt of suitable width and working on pulleys of adequate size, with a sufficient are of contact must be employed. The tension should be adequate but not excessive. The velocity ratio of two shafts is given by:

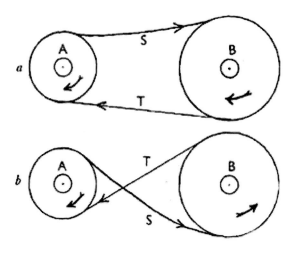

FIG. 305.—BELT DRIVE

a, open; *b,* crossed. **A,** driver; **B,** follower; **S,** slack side; **T,** tight side.

$$\frac{\text{Speed of follower}}{\text{Speed of driver}} = \frac{\text{Radius of driving pulley}}{\text{Radius of follower}},$$

i.e. the revolutions of the shafts are inversely proportional to the radii of the pulleys mounted on them.

When power is being transmitted, the tension on one side of the belt is greater than that on the other. If these tensions and the speed of the belt are known, the horse-power transmitted may be calculated. Thus, if T lb. is the tension on the tight side of the belt, *t* lb. that on the slack side, and V ft. per minute the belt speed, the horse-power transmitted is $\frac{(T-t)\ V}{33000}$.

The tension on the tight side is usually about $2\tfrac{1}{2}$ times that on the slack side. There is always some *creep* of the belt over the surface of the pulley due to the elasticity of the belt; excessive creep may cause a serious loss of power. Continual slipping of the belt over the pulleys must be avoided, for it causes rapid wear of the belt, as well as loss of power. The advantages of belting are that it is economical, adaptable and safe. In the driving of farm machinery, the fact that a belt will slip under unusual demands for power provides a useful safeguard against damage both to driven machines and to the engine driving them.

Belt pulleys for high speeds are usually *crowned* —i.e. the diameter of the pulley face is greater at the centre than at the edges (Fig. 306 *(a)*).

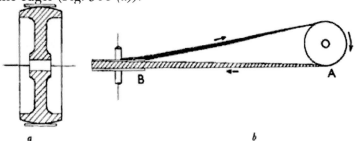

a   *b*

FIG. 306.—A, CROWNED PULLEY; B, DRIVE FOR AXES AT RIGHT ANGLES

A, driving pulley; **B**, driven pulley.

This construction helps to prevent the belt working off the pulley, for when the. pulley is rotating at high speed, the centrifugal force imparted to the belt causes it to move to that part of the pulley which is most distant from the shaft, viz. the centre.

It is sometimes necessary—e.g. with a horizontal buhr-stone mill—to arrange a drive between two shafts whose axes are at right angles to one another. This can only be achieved if the middle part of the belt, as it leaves one pulley, is in the central plane of the other pulley (Fig. 306 *(b))*. Shafts whose axes are inclined to one another, as frequently occurs in driving a straw elevator from a threshing machine (Fig. 307 (A)), may be connected by the use of guide pulleys. Jockey or idler pulleys may be used to increase the are of contact when a large pulley drives a small one close to it, as may occur in driving a hammer mill by means of a slow-speed fixed engine.

Pulleys are usually made of cast-iron for slow speeds, but steel is generally used for high-speed work. The crown may be faced with leather or fibre to increase the coefficient of friction. Where the pulley diameter is very small, e.g. on hammer mills, compressed paper pulleys, which have a high coefficient of friction, may be employed.

For short belt drives by electric motors the Rockwood system of mounting the motor is ideal. The motor is mounted on a hinge, and is so arranged that the greater the power requirement, the more the motor is pulled down to its work. The result is a remarkably steady and effective drive, with greatly reduced slipping.

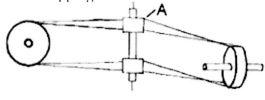

FIG. 307.—IDLER PULLEYS

A, guide pulleys for inclined axis; **B,** idler pulley used to increase are of contact.

Types of Belting.

The most reliable belting material for general purposes is leather which has been made of selected hides, tanned and cemented together with a waterproof cement. The hair side should be run next to the pulleys, on account of its higher coefficient of friction.

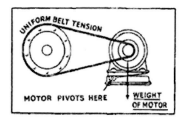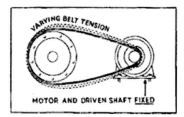

FIG. 308.—*ILLUSTRATION OF (LEFT) HINGED MO-TOR DRIVE (ROCKWOOD) AND (RIGHT) DRIVE WITH FIXED CENTRES*

A leather belt is made up of a large number of fibres that slide over each other as the belt rounds the pulleys. It is important that leather belts should be cared for by the regular application of a suitable proprietary or home-made dressing which lubricates the leather internally. A useful home-made dressing is a mixture of two parts of beef tallow and one part of cod-liver oil. The use of resin should generally be avoided.

Canvas belts consist of several plies of cotton, canvas or duck, sewn together. They are cheap, strong, and generally satisfactory for intermittent indoor service. For use out of doors they must be treated with a non-drying oil and then painted. They are unsuitable for many *agricultural purposes,* owing to variations in length due to the effect of moisture. *Balata* belts are made of canvas which has been treated with a gum called balata. This gum waterproofs the canvas and at the same time gives it a good friction surface. Such belts are cheap

867

and can be used almost anywhere about the farm. Rubber-impregnated canvas belts are well suited to wet conditions. They have an excellent friction surface, and can often be used in difficult situations where other types fail.

Belt Fastenings.

The best method of determining the length of belt required is to pass a tape round the pulleys when they are in position. The length may, however, be estimated by adding twice the distance between the centres of the pulleys to the product of 3.14 times the mean diameter of the two pulleys. If any means of adjusting the tightness of the belt is available, as by adjustable or jockey pulleys, the belt may be made endless by splicing, cementing or stitching. Leather belts may be laced with raw-hide thongs, but the operation is one requiring care and skill. For most farm belts, except those used for the engine drive to a thresher, chaff cutter or other machine where endless belts are advisable, the simplest and quickest method of fastening is to use a flexible steel fastening such as the *Alligator* fastener (Fig. 309).

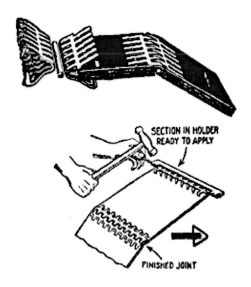

SECTION IN HOLDER
READY TO APPLY

FINISHED JOINT

FIG. 309.—METAL BELT FASTENER

For situations where shafts are very close together, **V-belt** drives are often suitable. The belt is made of rubber and canvas, and often runs on split pulleys which can be adjusted to take up slack or to vary the speed. When the sides of the pulley are adjusted closer together the belt runs farther from the centre, and the effective diameter of the pulley is increased. On machines such as combine harvesters, variations of fan and drum speeds may be provided for by such an adjustment.

Rope Transmission

is sometimes used for small powers and high speeds, as, for example, when a $\frac{1}{2}$-h.p. portable electric motor is used for driving a small centrifugal pump. The rope runs in V-shaped grooves on the pulleys. Care must, of course, be taken to ensure that the splice is strong and smooth.

Chain Drives.

Chain drive is well adapted to the transmission of power at slow speeds, in situations where a positive drive is required and circumstances do not permit the use of gear wheels. A wide use is made of various forms of chain drive in agricultural machinery.

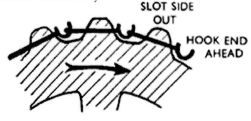

SLOT SIDE
OUT

HOOK END
AHEAD

FIG. 310.—ILLUSTRATING HOW HOOK-LINK CHAIN SHOULD BE PLACED ON SPROCKET

The **hook-link** type of chain (Fig. 310) is a simple form that is in common use. The links are made of malleable iron or steel. They should run with the hook ends of the links ahead in the direction of travel, and with the open parts outwards. The sprockets are usually made of cast-iron. The chain may tend to climb the teeth of the sprocket unless the pitch of the teeth on the driving sprocket is slightly greater than that of the chain, while the pitch of the driven sprocket is slightly less, so that the chain is in contact with the teeth as it leaves the sprocket in both cases.

Pintle chain is sometimes used in preference to the hook-link type where greater strength is required. Neither hook-link nor pintle chain is suitable for speeds exceeding about 400 ft. per minute. On many modern machines, especially binders, the old types of chain have been displaced by **roller** chains made of hardened steel parts. These are

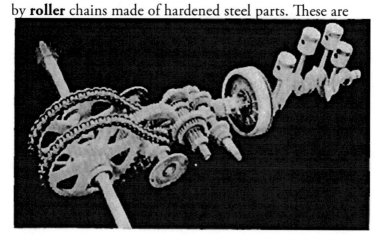

FIG. 311.—ROLLER CHAINS USED IN THE FINAL
DRIVE OF A TRACTOR. (CASE.)

efficient and durable, and may be operated at speeds up to 600 ft. per minute. They run on accurately-machined steel sprockets.

Silent chain drives are sometimes used where shafts are close together and ordinary belting cannot be employed. The chain has, on the inner side, teeth which engage with similar teeth on specially cut sprockets. The disadvantages of this type of drive are the expense and weight of the chain, and unsatisfactory performance at high speeds.

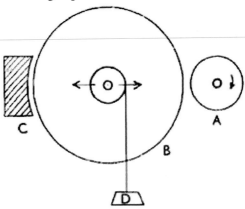

FIG. 312.—FRICTION DRIVE

A, friction cylinder; **B,** driven cylinder **C,** brake; **D,** load.

Care of Chain Drives.

Chain drives require adequate lubrication to ensure efficient performance and long life. When working continuously in grit, external oiling may be worse than useless; the only satisfactory method of lubricating exposed roller chain drives is to remove them periodically, and soak them first in paraffin and then in lubricating oil to which a little graphite has been added. The most satisfactory performance is secured where it is possible to enclose the drive completely and run the chain in an oil bath. Chains should not be allowed to become so slack that they tend to jump the sprocket teeth; but excessive tension is equally detrimental, causing unnecessary wear of chains, sprockets and bearings, as well as loss of power.

Friction Drives.

Friction gearing (Fig. 312) is used on farms for driving hoists in the hay field or the granary. The source of power is connected to a cylinder covered with soft friction material, such as compressed paper, rubber or leather. The drive is effected by causing this cylinder to press against a second one made of cast-iron. The power that can be transmitted depends on the materials of the two cylinders, the bearing pressure between them, and the peripheral speed of the driver. The great value of a friction drive for hoisting apparatus lies in the ease of control. The load may be easily raised, lowered or braked in any desired position, by the slight movement of a lever, which moves the driven shaft sideways.

Toothed Gears.

In order that gears may transmit power smoothly, the teeth must be designed with mathematical accuracy so that they may engage without shocks, and may transmit power efficiently, with a minimum of wear and noise. The gears on such machines as potato diggers, mowers, etc., where the strains are not excessive, are often of cast-iron. For heavy work, such as on the modern tractor binder, machined steel gear wheels are generally used. Special alloy steels, hardened by careful heat treatment, are used in tractor gear-boxes and other situations where the strains may be very severe.

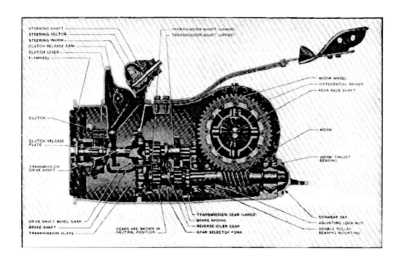

FIG. 313.—*TRANSMISSION SYSTEM OF TRAC-*
TOR WITH MULTIPLE-PLATE CLUTGH, 3-SPEED
GEAR BOX AND WORM-AND-WHEEL FINAL DRIVE.
(FORDSON, EARLY TYPE.)

tractor gear-boxes and other situations where the strains
may be very severe.

Spur gears are those which engage as cylinders rolling together.
The smaller wheel is often called the **pinion**.

FIG. 314.—BEVEL GEARS

Bevel gears (Fig. 314) engage as sections of cones. They may be used when it is necessary to connect two shafts whose axes are at an angle to one another and intersect. Such gears, connecting shafts at right angles to one another, are found in the binder, potato spinner, mower and many other farm machines. The **face** gear is a multiple bevel gear in which the velocity ratio may be varied. It is used on many modern drills for controlling seed rates. **Worm and wheel** gearing may be used for connecting shafts which are at right angles to one another and do not intersect. This gear gives a wide velocity ratio, and is used in the final drive of some makes of tractor as well as on many other common farm machines. The drive cannot usually be reversed, i.e. the wheel cannot drive the worm. But on some machines, such as the cream separator, a great increase of speed may be obtained by the use of a wheel driving a worm of wide pitch.

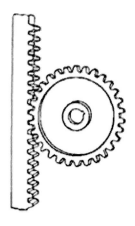

FIG. 315.—*RACK AND PINION*

The **Rack and pinion** represents an extreme case of spur gearing, where one of the wheels is infinitely large. If the pinion is mounted in stationary bearings, its rotation causes a rectilinear movement of the rack. This device is found in such diverse situations as the wheel adjustments of binders, the hopper adjustment of some cup-feed corn drills, and the adjustment for the fuel pumps of Diesel engines. **Helical** gears may be used to connect parallel shafts, shafts at right angles or shafts inclined at any angle and not intersecting. The teeth are cut so as to form helical curves around the cylinder of which the gear is a segment. The gears have great strength and operate quietly; they are being increasingly used in change-speed gear-boxes.

Epicyclic

gears are used on some modern drills and in differential gears. In its simplest form an epicyclic gear consists of one fixed wheel, the **sun** wheel, and a **planet** wheel which revolves around the sun wheel and at the same time rotates on its own axis. Fig. 316 illustrates a practical form of epicyclic train used as a two-speed gear on some combined corn and manure drills. When the outer casing is allowed to revolve, the whole gear rotates solid, and the speed of the follower is

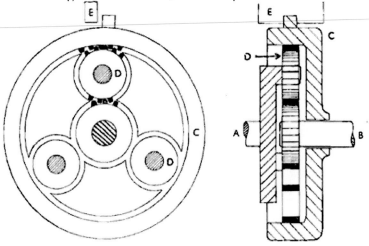

FIG. 316.—EPICYCLIC GEAR

A, driver; **B,** follower; **C,** outer casing; **D,** planet wheels; **E,** stop.

the same as that of the driver. When the outer casing is held stationary and the driver rotates, the follower is speeded up.

Velocity Ratio in Toothed Gears.

If a pair of spur gear wheels of diameters D and *d* are geared together, the relative speeds, N and *n*, are given by the equation $D \times N = d \times n$, i.e. the speeds are inversely proportional to the diameters. Since the *pitch* must be the same for both wheels, the number of teeth on each wheel is proportional to the diameter, and the general rule may be stated: "The speeds of two wheels geared together are inversely proportional to the number of teeth on each." When a train of gear wheels is used, the velocity ratio of the system may be obtained by dividing the product of the numbers of teeth on all the drivers by the product of the numbers of teeth on all the followers. The velocity ratio in epicyclic trains is not simple, and will not be discussed here.

Shafting.

On farms where several power machines are operated in a group, it is usual to drive them from a single engine by means of a line of shafting. Cold rolled steel is the most common shafting material. The shaft should be strong enough to transmit the required torque, and stiff enough to resist the bending effects of the weight of the pulleys and the pull on the belts. For ordinary farm operations, 2-in. diameter shafting, supported by bearings about 8 ft. apart, is satisfactory. In general terms, the strength of shafting is proportional to (diameter),3 but the only safe guide to the proportions required is experience, since shocks and whipping stresses cannot easily be calculated.

Couplings.

Shafting is usually manufactured in lengths not exceeding 16 to 20 ft. When longer lengths are required the ends must be connected by means of couplings. The types most commonly used are flanged and split compression couplings. In the flanged type, a flange is screwed or keyed to the ends of the two lengths of shafting, and the two flanges are bolted together. The advantage of compression couplings is that they may be fitted without making key-ways or threads on the shaft; such split compression couplings and pulleys are increasing in favour, owing to the ease with which they may be fitted or removed.

It is frequently necessary to provide a driving connection

between two shafts which are not in the same straight line, as in transmitting power from a tractor power take-off to field machines. In such a case it is necessary to provide some flexibility to permit variations in the angle between the driving and driven shafts. A device called a **Universal** joint (or Hooke's joint) is used for this purpose. It consists essentially of two pivots fixed together at right angles; the driving shaft pivots on one of these and the driven shaft pivots on the other. By a combination of the two movements, the driving and driven shafts may be run at any angle (within limits) to one another. In practice, a double universal joint is generally employed, and this gives a good drive for angles up to about 22 degrees.

FIG. 317.—UNIVEERSAL JOINT

Bearings.

The construction of the bearings used to support rotating shafting is determined by the nature of the work done. Bearings may be divided into two major classes, viz. radial bearings, which prevent a shaft from moving sideways, and thrust bearings, which are used to withstand pressure in the direction of the axis of the shafting. Bearings of many farm implements, where the speed of rotation is low, consist of a plain hole made of chilled cast-iron. For moderate speeds, brass and gun-metal are suitable materials, while for high speeds special antifriction alloys are used in some circumstances, and ball or roller bearings in others. These high-speed bearings are dealt with in Appendix Three.

Where a line of shafting is used at infrequent short intervals to drive barn machines, a simple and cheap type of **plummer block** may be used to support it. A simple plummer block may consist of a plain brass or bronze bush with a wick-feed oiling device. Where shafting is used enough to warrant the extra expense, self-aligning ball-bearings are

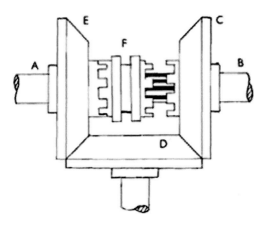

FIG. 318.—*DOG-CLUTCH IN A BEVEL GEAR RE-*
VERSING TRAIN, AS USED ON A CHAFF CUTTER

AB, main shaft, with bevel wheels, **C, E,** mounted loose on
it; **D,** bevel wheel on countershaft; **F,** dog, feather-keyed to
main shaft.

desirable, since they are very efficient and require
practically no attention. The type of support used to carry
the bearings must depend on the type of building and the
position in which the shaft is placed. Standards which
support the shafting just above the floor, brackets which fix
it to a wall, or hangers which suspend it from the ceiling
may be employed, according to the type and location of the
machines, and construction of the building.

Clutches.

A clutch is a mechanism that is used to engage or disengage a machine and its source of power. Clutches may be broadly divided into two main types, viz. (1) positive and (2) friction clutches.

Positive (dog) clutches are used on such machines as binders, mowers, potato diggers and many other farm machines, to put the mechanism in and out of gear. Fig. 318 illustrates a double dog-clutch such as is used to provide forward, neutral and reverse gears for a chaff cutter. The shaft, AB, has mounted loose on it two bevel wheels, C and E. The front faces of the bevel wheels are cut away to form claws. The sleeve, F, which is feather-keyed to the main shaft, has similar claws or dogs which may be put into gear with either wheel C or E, or may remain out of gear in the central position. The advantages of dog-clutches are their simplicity and the fact that they give a positive drive. A disadvantage is that the drive cannot be taken up gradually, and the machine must either be stopped or running slowly when the clutch is engaged.

Friction clutches are widely used because of their capacity to take up loads gradually and to transmit power efficiently. The simplest type of friction clutch is the conical one (Fig. 319). The shaft, A, has keyed to it a drum, B, of which the conical surface, C, forms the female part of the clutch. The male portion is feather-keyed to its shaft, D. It is normally kept pressed against the female part by the action of a spring, and is disengaged by means of a lever. Conical clutches are sometimes used on lines of shafting where machines must be started up slowly on account of the inertia of the parts. Large conical clutches are also employed in the transmission systems of a few types of tractor which have horizontal engines.

The single dry plate type of clutch (Fig. 320) is commonly employed on tractors, motor lorries and cars, being situated between the engine flywheel and the gear-box. The plate, A, is rigidly keyed to the shaft, B, which is connected to the gear-box. There is no positive connexion between the shafts, B and C. Attached to A are two friction discs, and when the pedal is released these are pressed into contact with the metal plate by the spring, D. The shaft, B, is then driven by the friction between the discs and the metal plate. The friction discs are generally lined with a special asbestos fabric. The clutch is run dry, and is very satisfactory, provided that it is kept adjusted and that no oil reaches the friction surfaces.

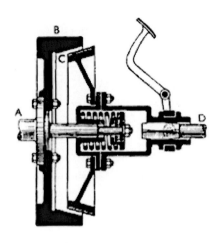

FIG. 319.—CONICAL CLUTCH

A, driving shaft; **B,** flywheel; **C,** conical surface; **D,** driven shaft.

The multiple-plate clutch is similar in essentials to the single-plate type, but there are two sets of metal friction plates running in oil. If there are ten driving plates and nine driven ones alternating between them, there are eighteen friction surfaces in contact; these are pressed together by means of a number of springs.

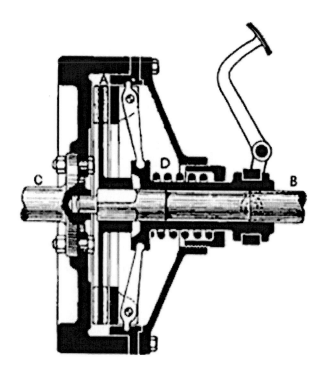

FIG. 320.—SINGLE-PLATE CLUTCH

A, fraction plate; **B,** driven shaft; **C,** driving shaft; **D,** spring.

Expansion and compression clutches are sometimes used for connecting line shafting to machines which must be started up gradually. In the expansion type, a split ring attached to the driven shaft is caused to expand inside a drum fixed to the driving shaft. Compression clutches are engaged by causing a band on one shaft to contract around a drum on the other.

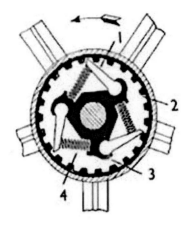

*FIG. 321.—PAWL AND RATCHET GEAR AS EM-
PLOYED ON THE MOWER*

1, pawl; **2,** ratchet; **3,** pawl casting; **4,** spring.

Pawl and Ratchet Drive.

The pawl and ratchet is a device for imparting intermittent motion in one direction. The ratchet is a toothed wheel or segment of a wheel, and the pawl is the part that engages with the ratchet teeth (Fig. 321). When motion occurs in one direction, the pawl engages with the ratchet, but when motion occurs in the opposite direction the pawl slips over the ratchet teeth. This device is used to connect the wheels of such machines as drills and mowers to the live axles which they drive; a similar mechanism is

888

also employed to drive conveyors intermittently at very low speeds, as in the farmyard manure distributor and some conveyor types of grass drier.

Cams and Eccentrics.

Cams and eccentrics are used for converting rotary motion into reciprocating or oscillating motion. A common type, often used on engines for the operation of the valves, is illustrated in Fig. 322. A rotating shaft drives the cam, and as the tappet follows the contour of the cam a reciprocating motion is imparted to it.

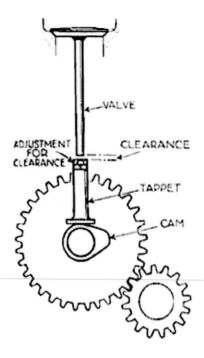

FIG. 322.—GAM EM-PLOYED IN VALVE GEAROF AN INTERNAL COM-BUSTION ENGINE

The eccentric (Fig. 323) is essentially a crank in which the crank-pin has been enlarged sufficiently to include the shaft. The shaft, A, is keyed to the eccentric, B, which rotates inside the strap, C. The rotary motion of the shaft is converted into a reciprocating motion, and the device is used for the operation of pumps, steam-engine valves, etc.

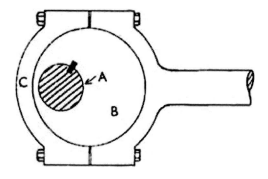

FIG. 323.—THE ECCENTRIC

A, rotating shaft; **B,** eccentric; **C,** strap.

Brakes.

A brake is a mechanism for absorbing the energy of motion of a body by converting it into heat. Brakes are used on vehicles in order to bring them to rest; they are also employed on tracklaying tractors to assist in the steering. It should be observed that brakes must absorb the energy that is apparently destroyed, and provision must be made for dissipation of the heat developed.

Band brakes are used on many farm vehicles. In this type a metal band, which may be fitted with a friction lining, is caused to grip a drum mounted on the wheel. On tractors, motor lorries and other motor vehicles commonly used on the road, internal expanding brakes are generally fitted. In these a band is caused, by the action of a cam, to press outwards against the internal face of a drum attached to the wheel. The expanding band is lined with a friction material which may consist of a mixture of wire and asbestos.

On modern tractors the brakes are usually situated on the intermediate high-speed differential shafts, rather than on the final drive shafts.

APPENDIX THREE

FRICTION, LUBRICATION AND BEARINGS

The chief cause of waste of energy in machines is the resistive force called **friction.** When two bodies are pressed together, resistance must be overcome before they can be made to slide over one another. If a body X press on a body Y with a normal pressure P (Fig. 324) no force, apart from that required to overcome friction, is required to cause sliding motion of X on Y, since the force P has no component at right angles to itself. In practice, a definite force tends to resist the sliding motion, and this resistive force is called the force of friction. Suppose a gradually increasing force, F, be applied to X; the resistive force, f, also increases up to a point when sliding commences. This point of limiting friction at which sliding commences is called the **limit of static friction.** When motion occurs, there is still a resistive frictional force, f', but this is not quite as great as the limiting static frictional force. It is called the force of kinetic friction, and its value depends to some extent on the speed of sliding.

893

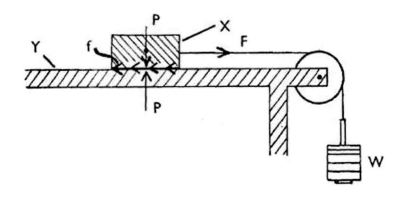

FIG. 324.—SLIDING FRICTION

Coefficient of Friction.

The quantity $\frac{f}{P}$ is called the coefficient of $\frac{f'}{P}$ static friction and the quantity — is the coefficient of kinetic friction.

It has been found by experiment that for two given materials with dry surfaces the coefficient of friction is practically constant at all pressures, but it varies with the kind and condition of the material, especially with the smoothness. It may be measured by finding the relation between the forces, *f* and P (Fig. 324) or by other methods.

Coefficients of static friction

Wood on wood	0.25 to 0.50
Steel on cast-iron (dry)	0.3 to 0.4
Smooth surfaces, well lubricated	0.05 to 0.008

The friction between two dry surfaces is, within limits, proportional to the pressure between them. It is almost independent of the velocity of sliding.

Rolling Friction.

The exact nature of rolling friction is not fully understood. In the case of pure rolling, there should be no frictional forces, since no sliding occurs, and frictional forces can only be caused by sliding. In practice, the roller must sink into the bed over which it moves, and may be considered, in a sense, as always going slightly uphill (Fig. 325), owing to the formation of a depression in the surface. The following general rules apply in the case of rolling friction:—

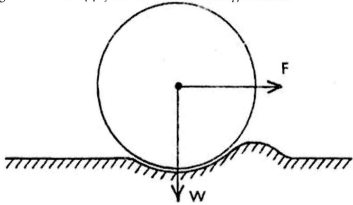

FIG. 325.—ILLUSTRATING A CAUSE OF ROLLING
FRICTION

1. The harder the surfaces, the less the rolling resistance.

2. The rolling resistance is proportional to the load.

3. The resistance is, within limits, inversely proportional to the diameter of the roller.

In most conditions, rolling friction is small compared with sliding friction, and advantage is taken of this in the use of ball and roller bearings, as well as in vehicle wheels.

Liquid friction is the resistance to flow of a liquid, or to sliding over one another of surfaces wetted with a liquid. It differs from dry friction in that it varies appreciably with the velocity, and is directly proportional to the area in contact. In many circumstances where surfaces are wet, the friction resulting does not conform either to the laws of dry or of liquid friction. In the design and operation of machinery, friction may be reduced by the use of materials with low coefficients of friction and by keeping them smooth and well lubricated.

Lubrication.

The object of lubricating moving parts is mainly to reduce friction, and so to minimize the energy absorbed and wasted by it. In addition to the straightforward purpose of reducing friction, lubricants are frequently required to fulfil other functions. For example, it has been estimated that the big-end bearing of an internal-combustion engine requires approximately 90 times as much lubricant to keep it cool as it requires for the actual reduction of friction. Another function of the lubricant in internal-combustion engines is the sealing of pistons, so as to make the combustion chamber gas-tight.

Lubrication is effected by imprisoning a film of oil between the two surfaces, so that friction between the metal surfaces is replaced by friction between the lubricant and the metals, and by friction within the lubricant. There must be no metal-to-metal contact, for if inadequate lubrication allows this to occur, the friction is greatly increased and a high temperature may rapidly be developed. The result may then be the running of bearings, the seizure of pistons in cylinders, or severe wear of the sliding parts. Adequate lubrication requires a proper grade of lubricant and a satisfactory method of applying it.

The Requirements of a Lubricant.

A lubricant should possess sufficient *body* to prevent the lubricated surfaces coming into contact under the maximum pressure to which they are subject. Requirements vary widely in this respect: for example, grease is required for a wagon wheel, while a thin oil is most suitable for a magneto bearing.

The oil used should always be as light as the conditions of its use permit, for in the case of machines such as the cream separator, a high fluid friction within the lubricant is very undesirable. When applied, the oil should produce a low coefficient of friction. For an internal-combustion engine it should be capable of dissipating heat and should not break down chemically when subjected to high temperatures. It should also have a high temperature of vaporization and should be reasonably fluid when cold, and free from grit.

The bulk of the lubricating oil in general use is produced from the heavier fractions obtained in the distillation of crude petroleum. After removal of the petrol and paraffin, the heavier fractions are washed with chemicals, filtered, and purified in various other ways. Oils from other sources are occasionally used as lubricants. Of these, lard and tallow are of animal origin, while castor oil is vegetable. Greases are produced by mixing with a heavy oil the soaps produced when various fatty acids are treated with lime or soda.

In the past it has not been possible to determine the suitability of a lubricant for a given purpose by recognized physical tests. Oils with such different lubricating properties as a good motor oil and cod-liver oil were difficult to distinguish. With increasing technical knowledge, disabilities in this respect are being rapidly overcome, and it is now possible to determine the properties of an oil within narrow limits by the employment of a sufficient number of different tests. Factors concerned are (1) specific gravity; (2) flash point; (3) viscosity at various temperatures; (4) solubility in various liquids; (5) emulsibility; (6) susceptibility to oxidation; (7) carbon residue; and (8) "oiliness". By these tests, the origin and properties of an oil may be estimated with a fair degree of accuracy.

The Effect of Temperature on Viscosity.

All liquids become thinner when heated, and in the case of lubricating oils the reduction of viscosity is marked. (The viscosity or internal friction of a liquid may be measured by observing the rate of flow through standard capillary tubes; the values obtained may be expressed as a number of seconds on the standard Redwood scale.) It is essential that the oil film should always remain thick enough to keep the metallic surfaces of bearings sufficiently apart, but on the other hand, too high a viscosity is undesirable. Other things being equal, the best oil, especially for an internal-combustion engine, is one whose viscosity changes least with increase of the working temperature. At 60° F., a normal starting temperature, the best oils have viscosities of 60 to 80 times those at 250° F., the average temperature of the cylinder walls; inferior oils are very much worse. Thus, since piston-cylinder lubrication generally depends on splash, the engine when starting from cold is at a disadvantage with respect to lubrication, though this is partly offset by the greater clearance between piston and cylinder. It is a mistake to buy oils that look thick, in the hope that they will have a good viscosity at high temperatures. This may be anything but the case; for some oils which are so thick at low temperatures that they cause starting troubles, are also very thin at high temperatures. Lubricants are often judged by their colours, but in general, colour is unimportant. Paleness often indicates a high degree of refinement, but is no guide to quality.

Modern Views on Lubrication.

In recent years, careful research by physicists has shown that the old idea that friction between two solids is simply due to the comparative roughness of their surfaces, must be modified. It has been shown that increased smoothness may sometimes lead to greater friction. It seems possible that the friction between two clean dry surfaces may be partly caused by the mutual attraction of the molecules comprising each surface.

It has been pointed out that the aim is to form a complete film of lubricant between the bearing surfaces; so long as this film exists, the condition is called fluid lubrication. With a given oil, the clearance between the lubricated surfaces becomes smaller with increasing loads or diminishing speeds, and when these reach a certain value the oil film becomes too thin to keep the surfaces properly apart, and a condition called **semi-fluid** or **boundary** lubrication exists. One of the chief types of motion in which the lubricant may be in the boundary state is reciprocating motion. For instance, an engine piston which slows down, stops, and reverses direction at the end of each stroke, makes the maintenance of a fluid film almost impossible. In conditions of boundary lubrication the value of an oil depends largely on a property called "oiliness".

Tractor Lubrication Problems.

In paraffin engines dilution of the lubricating oil by the fuel is a frequent cause of inefficient lubrication, and it leads to excessive cylinder wear. The cause is mainly the introduction of incompletely vaporized fuel into a cold combustion chamber, as is explained in Chapter Two. The addition of 6 per cent, of petrol to ordinary motor oil reduces the viscosity by half, but with petrol the dilution rarely exceeds this figure because an equilibrium occurs, and the addition of more petrol is offset by evaporation. With paraffin, however, there is no such equilibrium point, and serious trouble may arise if the crank-case is not regularly drained. Trouble with crank-case dilution rarely occurs in heavy-oil engines, but many of these, unfortunately, suffer from other lubricating difficulties. The operating conditions in such engines tend to cause the oil to break down and form a hard deposit, which may settle in the piston-ring grooves and cause them to stick. Various chemicals may be added to the oil to inhibit the formation of such hard products (inhibited oils), or to keep the solid matter in a fine suspension (detergent or H.D. oils).

In some tractor transmission systems, the pressures between gear teeth are extremely high, and oils which are particularly designed to maintain an oil film on the teeth in spite of this (extreme pressure or E.P. oils) must be employed.

The mixing of oils which contain chemical additives may cause the additives to come out of solution with harmful results, and it is a good rule never to mix lubricants.

The Application of Lubricants.

Failure to ensure adequate application of the lubricant may cause damage in a very short time. The types of lubricating devices vary widely in different circumstances. Examples of devices used are (1) wick-feed oil cups for plain plummer blocks; (2) ring oiling on electric motor and threshing machine shafts; (3) grease cups on pulleys, where the centrifugal force renders some types of lubricator unsuitable; (4) a pressure system such as is used for the main bearings of many internal-combustion engines; and (5) the high-pressure grease gun method now used on many farm implements and machines. On most farm machines adequate lubrication depends as much on the exclusion of all dirt from the bearings as on the application and retention of the lubricant. This is sometimes achieved by the complete enclosure of the moving parts, but where such a method is impossible, the use of felt washers may be effective.

A good grease gun has become an important item of farm equipment. Fig. 326 illustrates a modern gun which is sealed by a screw cap at the top after having been packed with grease. When the gun is pressed against the lubricator fitting. the nozzle moves inwards, and because of the small diameter of the part which moves into the grease reservoir, a pressure of around 2,000 pounds per square inch can be created in the reservoir by a moderate push on the gun. Grease is expelled during the stroke, and a spring returns the nuzzle to enable die process to be repeated.

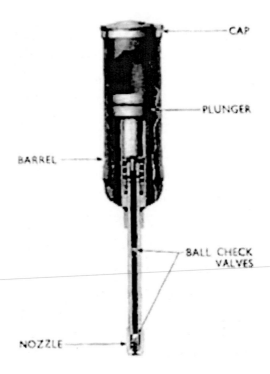

FIG. 326.—INTERNAL VIEW OF HIGH-PRESSERE GREASE GUN.

High-speed Bearings.

Bearings used for high speeds are usually either ball or roller bearings, or are made of special antifriction alloys.

For the big-end bearings of internal combustion engines, renewable bearings of white metal, babbitt metal or other anti-friction alloy of low melting point are employed. If lubrication fails, the metal of the overheated bearing melts before irreparable damage is done to the crankshaft.

These bearings illustrate the progress that is being made in improving the ease of maintenance of machinery. The procedure for renewing such bearings used to be complicated and difficult. The connecting rod had to be removed, and the molten metal run into a gun-metal *step* in the two halves of the bearing, after which an elaborate fitting process was necessary. The modern method is simply to remove the old "shell" bearings, and replace them by the new, no fitting of any sort apart from ensuring that they are correctly assembled being needed.

Ball bearings were originally used only for light loads and high speeds, but such great improvements have been made in the methods of manufacture that fittings suitable for the severest type of service available. The balls and the races in which they run are usually made of high-carbon or chromium alloy steels. There are many stages of grinding, hardening and polishing in the manufacture before the balls and races are finally burnished in a special dry atmosphere, and then immediately plunged into grease. In the best grades of bearings the tolerance allowed is only 0. 00005 in. in the diameter. A cage or spacer of soft material is frequently used to prevent the balls coming into contact with one another.

The simplest type of ball bearing has a single row of balls, with two-point contact between the balls and the two races. Some types have three-point and some four-point contact. The bearing may be made self-aligning by making the outer race a segment of a sphere which has its centre coincident with that of the shaft. Ball bearings may be designed to care for either radial or thrust loads, or for a combination of the two.

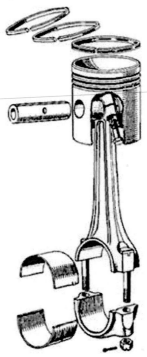

FIG. 327.—BIG-END BEARING OF INTERNAL-COM-BUSTION ENGINE SHOWING RENEWABLE BEARING SHELLS

Roller bearings are used and constructed in the same general way as ball bearings, but instead of point contact there is line contact between the rollers and the races. For bearings designed to carry only radial loads, cylindrical rollers may be used, but where there is a combination of radial and thrust loads the tapered roller type may be used with advantage. With cylindrical rollers, serious end-thrusts may result if the rollers do not run exactly parallel to the shaft.

APPENDIX FOUR

THE MEASUREMENT OF POWER

Power has been defined as the rate of work. For its measurement it is therefore necessary to determine the amount of work expended in a given period of time. In order to do this, it is necessary to determine the magnitude of the force acting, and the distance through which it acts in a known time. Thus, in measuring the power developed by a tractor in ploughing, the force required to pull the plough and the speed of the outfit must be determined. For example, if the outfit moves at $2\frac{1}{2}$ m.p.h. and the force required to pull the plough is 1,500 lb.,

$$\text{Work done per minute} = \text{(Force)} \times \text{(Distance travelled per minute)}$$
$$= 1,500 \text{ lb.} \times 220 \text{ ft.}$$
$$= 330,000 \text{ ft.-lb.}$$

and since 1 h.p. is equal to 33,000 ft.-lb. per minute, the horse-power used in pulling the plough is $\frac{330,000}{33,000} = 10$ h.p.

Dynamometers.

Dynamometers are devices for measuring power. **Traction** dynamometers are used for measuring the power expended in haulage operations, when any implement or

908

vehicle is being pulled by an independent source of power. Though the dynamometer may measure all the quantities required for the calculation of the power exerted, viz. force, distance and time, the term dynamometer is generally applied to devices which may give only a record or an indication of the force exerted. The power at any given moment may be calculated from simultaneous values of the force or **drawbar pull** and the speed. In practice, it is usual to employ average drawbar pulls and speeds in calculating the power. The average speed is obtained by noting, by means of a stop-watch, the time taken for the outfit to travel a known distance, and the average drawbar pull is obtained by use of a dynamometer.

The simplest type of dynamometer is the spring balance. This suffers from two serious defects, viz. the great difficulty of reading or recording the pulls, owing to rapid oscillations of the pointer, and the danger of damage to the balance, owing to the violent fluctuations in pull sometimes encountered, especially in starting. The spring balance has no damping effect to minimize the inevitable fluctuations in pull. It may be used for the measurement of drawbar pulls in certain special circumstances, but the best type of dynamometer for agricultural work is a hydraulic type. In the most common hydraulic type, the pressure set up in the fluid contained in a hydraulic link is transmitted to a small recording cylinder, in which a plunger acts against the force of a spring. The movement of the plunger may be read directly, or may be recorded on a moving chart. The recording type of dynamometer is sometimes employed for accurate research work, but the direct indicating type is more suitable for rapid tests and demonstrations.

The employment of the hydraulic link for dynamometers has many features to commend it. Among these is the fact that the sensitivity may be varied at will by variation of the viscosity of the fluid used. Thus, by the employment of various grades of oil from a thin machine oil to a thick gear oil, almost any desired degree of damping may be achieved. Another desirable feature is that the instrument may be adapted to use over a wide range of pulls, by using springs of various strengths in the indicating mechanism. Moreover, if the dynamometer is well made, it is robust, and accurate to within 5 per cent. The tractive horse-power of tractors is generally called the drawbar horse-power to distinguish it from the power delivered at a belt pulley or through the power take-off.

FIG. 328.—INDICATING TYPE HYDRAULIC DRAW-BAR DYNAMOMETER. (N.I.A.E.)

Electrically recorded strain gauges have many uses in modern agricultural engineering research. These can measure the forces exerted in particular components of machines without replacing the components. Examples of their use include measurement of the forces set up in the three links of a tractor-mounted plough, and determination of the shock loads occurring in various parts of a tractor's hydraulic lift system when the tractor runs over bumpy land with a heavy implement in the raised position.

The Measurement of Brake Horse-power.

The brake horse-power of an engine or tractor is the power the engine can deliver for the performance of work at the crankshaft. It may be measured by means of an absorption dynamometer, the simplest examples of which arc the **rope** brake and **prony** brake. In these, the energy developed by the engine is absorbed in the form of heat, and where large powers are measured, water cooling is necessary. The prony brake is one of the simplest devices for the measurement of brake horse-power (Fig. 329). A pulley is connected to the crankshaft, and has brake blocks fitted to it. These blocks can be set to grip the pulley as tightly as required by adjusting thumb-screws. A counterpoise balances the weight of the brake arm, which is connected through a spring balance to a rigid support.

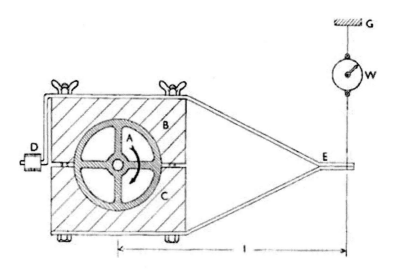

FIG. 329.—PRONY BRAKE FOR MEASUREMENT OF BRAKE HORSE-POWER

A, pulley; **B, C,** brake blocks; **D,** counterpoise; **E,** brake arm; **W,** spring balance; **G,** support.

In measuring the brake horse-power developed at various speeds and loads, the engine is first started with no load, and the speed of the pulley determined by means of a tachometer. The load is then increased in stages by tightening down the thumb-screws. At each stage, the speed and the force registered by the spring balance are recorded. Beyond a certain limit, it will be found that additional loads reduce the speed of the engine to such an extent that it is in danger of stalling. When this point has been reached, no further load should be applied, for the brake horse-power thereafter falls with increasing load

913

until stalling occurs.

If the length of the brake arm be *l* ft., the radius of the flywheel *r* ft., the load registered by the spring balance *w* 1b. and the total frictional forces round the rim of the pulley *f* 1b., taking moments about the centre of the pulley:

f x *r* = *w* x *l.*

The work done against the frictional forces in one revolution of the pulley is *f* x 2πr ft.-1b. If the speed of the pulley is *n* revolutions per minute, the work done per minute is *f* x 2 πr x *n* ft.-1b., and this is equal to 2 π *nwl* ft.-1b. (substituting *wl* for *fr*).

$$\therefore \text{b.h.p.} = \frac{2\,\pi nwl}{33,000}.$$

By plotting the values of the brake horse-power for various speeds or loads, a curve indicating the maximum power is obtained.

When large powers are to be measured, mechanical absorption by rope or prony brakes presents some difficulty, and it may be more convenient to employ a water brake or an electrical method. In any event, the horse-power developed is calculated from purely mechanical measurements involving the speed of rotation and the **torque reaction.** For example, when the power is absorbed electrically, the engine may drive a generator with a separately excited field and a controllable bank of resistances; the out-put of the generator may then be varied to absorb any reasonable power. The generator frame is mounted on bearings and tends to rotate with the armature, but its rotation is prevented by means of a spring balance attached to the bed plate. The torque reaction is then given, as in the case of the prony brake, by the product of the reading on the spring balance and the distance from the centre of the armature.

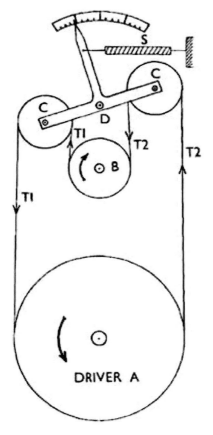

FIG. 330.—TRANSMISSION DYNAMOMETER

Belt Horse-Power.

Belt horse-power is the power which a tractor or other engine delivers at the belt pulley. It is measured in much the same way as brake horse-power, and will be exactly the same where the pulley is mounted directly on the engine crankshaft. Where, however, gearing is interposed, the loss in transmission can be appreciable. Belt horse-power is usually most conveniently measured by making the power unit drive an electric or hydraulic dynamometer by means of belting. This introduces the further complication of losses in the belt drive, and is one of the reasons why it is necessary to arrive at standard definitions and standard testing conditions for tractors.

Transmission Dynamometers.

In circumstances where it is necessary to measure the power delivered to a shaft without absorbing it, the Froude or Thorneycroft or some similar kind of transmission dynamometer must be employed. In Fig. 330, A represents a driving pulley and B the driven pulley. A continuous belt passes over the idler pulleys, C, C, which are mounted on an arm, pivoting at D. Neglecting frictional losses due to the guide pulleys, the tensions of the belt will not be altered by passing around them, and the forces T1, T1 pull on the left end of the lever, while the forces T2, T2 pull at the other end. The force required to balance these pulls is exerted by the spring, S, and is measured on the scale. By calibrating the scale, the difference in tension equal to any deflection may be calculated, and if the speed of the belt is measured, the horse-power transmitted may be calculated as follows:

$$\text{h.p.} = \frac{\text{Difference in tension (lb.)} \times \text{Speed of belt (ft. per. min.)}}{33,000}$$

Indicated Horse-power.

It is possible to measure the work done by the working substance on the piston of steam or internal-combustion engines by means of an engine **indicator.** This is an apparatus which measures and records the pressure in the engine cylinder at every stage of the cycle. Fig. 331 illustrates the type of diagram that might be obtained with a 4-stroke internal-combustion engine. Starting at A, the pressure rises (along AB) as the gases arc compressed. There is a rapid rise in pressure up to B as ignition occurs. On the working stroke, the pressure falls until the end of the stroke, C, and continues to fall slightly on the exhaust stroke as the gases escape from the cylinder. On the induction stroke, from D to A, the pressure is below atmospheric. The average force exerted on the piston during the cycle may be calculated from the diagram. If this force is P 1b. per sq. in., the length of the stroke is L ft., the area of the piston A sq. in., and the number of working strokes per minute N, the Indicated

Horse-power (i.h.p.) $= \dfrac{P.L.A.N.}{33,000}$.

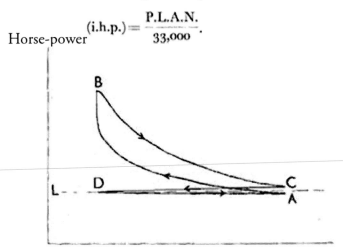

FIG. 331.—INDICATOR DIAGRAM OF A 4-STROKE ENGINE

DA, induction; **AB,** compression; **BC,** working stroke; **CD,** exhaust; **L,** atmospheric line.

The indicated horse-power gives a measure of the total work done on the piston by the fuel. The brake horse-power gives a measure of the useful power exerted at the crankshaft for the performance of work.

The ratio of these quantities, viz. $\dfrac{\text{b.h.p.}}{\text{i.h.p.}}$ gives the **mechanical efficiency** of the engine. It is usually from 80 to 90 per cent, in internal-combustion engines.

APPENDIX FIVE

MATERIALS USED IN THE CONSTRUCTION OF FARM MACHINERY

It is becoming ever more necessary for farmers to know something of the properties of the materials used in the construction of their equipment. It is impossible to understand the selection and operation of machinery without a knowledge of the type and magnitude of the forces to which a machine may be subjected, and of how the various materials may be expected to behave under these forces. In the design of machines, an estimate of the sizes of members required may be made if the properties of the material and the forces acting are known.

When a part of a machine is called upon to transmit a force, it resists deformation by internal forces called **stresses**. There are three simple types of stresses, viz.:

1. **Tensile** stresses, in which the forces tend to stretch the body upon which they act, e.g. a rope supporting a weight.

2. **Compressive** stresses are produced by forces which tend to shorten or compress the body, e.g. a column supporting a

roof.

3. **Shearing** stresses are set up by forces tending to cause one part of a body to slide over an adjacent part, e.g. the pin that connects a tractor to an implement drawbar.

In addition to the above three simple types of stresses, some parts of machines may be subjected to bending and other complex stresses which may consist of a combination of the simpler stresses. For example, the connecting-rod of a mower or of an internal-combustion engine is acted upon by forces causing very complex stresses.

The Strength of Materials.

Materials differ widely in their capacities to resist forces. The capacities of various materials may be tested by experiments with **testing** machines, in which known forces can be applied. The maximum force per square inch of cross section that a material can withstand is called the **ultimate strength** of the material. For instance, if a steel rod 0.5 sq. in. in cross-sectional area requires a pull of 20 tons to break it, the ultimate tensile strength of the material is 40 tons per square inch.

Elastic Strength and Yield Point.

When a body is subjected to a force that tends to change its shape, the deformation (strain) produced in the body is, within certain limits, proportional to the force applied (Hooke's law). For example, if a steel test-piece be subjected to tensile stresses in a testing machine, the elongation produced is at first directly proportional to the force applied; but if the load is gradually increased, a point called the **yield** point is presently reached, where Hooke's law no longer applies and the extensions cease to be proportional to the load, but increase at a more rapid rate. Up to this point, the test-piece will, on removal of the load, return to its original length by virtue of its property of elasticity. If loading is carried beyond the yield point (or **elastic limit**) the bar becomes permanently extended and is said to be "overstrained".

The elastic strength of a material is defined as the load per square inch of cross-sectional area at the elastic limit. In practice, the maximum stress permitted on any member of a machine should be well below that corresponding to the elastic limit, so that the member always behaves as if it were perfectly elastic. The **factor of safety** of any member is given by the ratio $\dfrac{\text{breaking load}}{\text{working load}}$. The magnitude of the factor of safety required in machine design depends upon many things, including the reliability of the material and the nature of the loading and shocks that may be experienced. High factors of safely must be used for materials such as cast-iron and wood, or in situations where the load is suddenly applied or is variable.

TABLE XI.—THE ELASTIC AND ULTIMATE STRENGTHS OF SOME COMMON MATERIALS (LB. PER SQ. IN.) (APPROXIMATE)

Material	Ultimate Strength			Elastic Strength	
	Tension	Compression	Shear	Tension	Compression
Cast-iron	20,000	95,000	20,000	10,000	25,000
Malleable iron	35,000	42,000	20,000	—	—
Wrought-iron	55,000	50,000	40,000	30,000	28,000
Steel (0·15% C.)	65,000	—	48,000	42,000	40,000
Steel (0·8% C.)	105,000	—	80,000	57,000	63,000
Chrome nickel steel	200,000	—	—	170,000	—
Brass castings	20,000	12,000	20,000	—	—
Bronze ; gun-metal	35,000	30,000	35,000	—	—
Aluminium castings	15,000	12,000	12,000	6,500	3,500
Timber	12,000	6,000	—	—	—

Hardness.

Hardness is an important property which may give some indication of the resistance of materials to abrasion, cutting or indentation. It also has some connection with the ultimate and elastic strength of the material. Hardness tests may be carried out with many types of apparatus, and the result is given as the hardness number of the material for the particular test. For instance, the Brinell hardness number is obtained by measuring the size of the indentation produced in the material when a standard pressure is applied to a standard steel ball situated on a flat surface of the material. Hardness numbers usually give a good indication of the resistance of materials to abrasion, but substances with high hardness numbers are unfortunately not invariably markedly resistant to abrasion.

The Ferrous Metals.

The ferrous metals include cast-iron, malleable iron, wrought-iron and all grades of steel. They constitute the most important group of materials used in agricultural machinery. The starting-point in the production of all the ferrous metals is the pig-iron produced when iron ore is smelted in a blast furnace.

Cast-iron is extensively used in the construction of many farm machines, owing to its cheapness and to the ease with which it can be cast into intricate shapes. In the production of castings, pig-iron is heated with coke and a variable quantity of scrap iron and other materials in a special furnace called a cupola. An air blast passes through the cupola, and the iron melts and runs down to the bottom of the furnace, where it is drawn off in a molten state. It is then run into moulds of the desired shape, which have previously been formed in a special moulding sand by use of the appropriate patterns. The quality of the castings produced depends upon the grade of pig-iron used and the amount and quality of scrap-iron and other material included in the melt. On an average, cast-iron contains from 3 to $3\frac{1}{2}$ per cent, of carbon, 1 to $3\frac{1}{2}$ per cent, of silicon and varying quantities of phosphorus, sulphur and manganese. The properties of castings vary greatly but, generally speaking, cast-iron is granular, brittle, and liable to break suddenly if large stresses are set up. It has, however, a tensile strength sufficient for many purposes, and possesses a great resistance to compression.

The properties of cast-iron are influenced by the rate at which cooling of the castings occurs, as well as by the chemical composition of the melt. When heated above 690° C., iron combines with carbon, producing iron carbide, $Fe_{24}C$. On cooling, the iron carbide splits up into two or more of the following:—

1. **Pearlite**, a cutectoid of Fe_3C in iron. The presence of pearlite gives toughness to castings.

2. **Cementite** (Fe_3C), which makes the castings hard but brittle.

3. **Ferrite** (pure iron), which produces soft spots in the

castings.

4. **Graphite** (pure carbon), which in large amount leads to a low tensile strength, but high resistance to compression.

The proportions in which the above four constituents are present depend largely upon the proportions of silicon and manganese in the melt. If a high proportion of silicon is present, this combines with the carbon and prevents the iron doing so, with the result that only small amounts of pearlite and cementite are formed. Manganese tends to neutralize the effects of silicon, a high proportion of manganese tending to produce a high proportion of combined carbon. Sulphur, in the absence of plenty of manganese, makes the castings very brittle, owing to the formation of ferrous sulphide.

The normal type of casting is grey when fractured, owing to the presence of some free graphite. Such castings are fairly soft and easily machined. When the carbon is present mainly in the combined form, the fracture is white and the castings are hard and brittle. A new field has been opened up by the addition of such metals as aluminium, nickel, chromium, molybdenum, etc., to castings, with subsequent heat treatments to produce great hardness and strength.

Chilled Cast-iron.

If cast-iron is cooled very quickly, most of the carbon remains in the combined state and the casting produced is white, hard and brittle. It is frequently desirable to produce castings that have one or more faces very hard, in order that these parts may better resist abrasion, while the bulk of the casting is of grey iron, for toughness. This may be effected by the use of an iron **chill** instead of moulding sand on those faces of the mould where hardness is required. Where the molten metal comes into contact with the chill, it is cooled rapidly; this part of the casting becomes white and glass-hard, but is prevented from shattering easily by the tougher grey interior or backing. The chemical composition of the melt must be carefully controlled if a chill of uniform depth is to be produced. The process of chilling is of special importance in the manufacture of plough shares.

Malleable Iron.

Ordinary grey cast-iron has little resistance to shocks. In situations where castings need to possess some of the strength and toughness of forgings, grey cast-iron is unsuitable, but malleable iron may be employed. It is made by annealing or decarbonizing white cast-iron. The first step in the process is the production of hard white castings. These are then made malleable by one of two main processes, viz. the Black Heart or the White Heart (Reaumer) process. In the Black Heart process, the castings are packed in burnt clay, bone ash or sand, and heated for 3 to 4 weeks in an annealing furnace. The cementite of the original castings is thereby decomposed into ferrite and free **amorphous** carbon. In the White Heart process, the castings are packed in iron ore and heated for about 10 days; this is an oxidizing process, the carbon of the castings being oxidized by the ferric oxide in the ore. In both processes for the production of malleable castings the final effect is a casting that is tough, ductile and suitable for machine parts that are required to withstand shocks. Such castings are used in farm machines for a wide range of couplings and fastenings where hammering, bending or tensile stresses must be withstood. When malleable castings fail, it may be due either to unsoundness in the original casting or to incomplete annealing.

Wrought-iron.

Wrought-iron is manufactured from pig-iron or cast-iron by the removal of carbon and silicon in a reverbatory or puddling furnace. The hearth of the furnace is lined with iron ore, and as the heating proceeds, the carbon is oxidized away and the silicon produces a molten slag, which may be squeezed out of the spongy ball of wrought-iron produced. After much squeezing, hammering, rolling, welding and reheating, bars of wrought-iron are obtained. The metal is fibrous, ductile and eminently malleable. It can be bent, sheared, punched, worked under the hammer when hot, and welded at 1600° F. It is never molten. It is used where an easy weld is necessary, as in chains, and in situations where considerable strength is required and the form is simple. During recent years, it has been largely superseded by mild steel, which is cheaper.

Steels.

Steels are produced by the removal from pig-iron of a large proportion of the carbon and as much as possible of the silicon, phosphorus and sulphur. Two main processes are employed, viz. the Bessemer and the Open Hearth processes. The furnace is lined with lime, and the phosphorus in the melt combines with the lime and produces basic slag, a valuable by-product, widely used as a fertilizer. Steels may be divided into a few main classes, viz. cast steel, mild steel, high-carbon and special steels.

Cast Steel contains about 0.25 per cent, carbon. It is harder and stronger than cast-iron and is ductile. Great care is needed in the production of steel castings, for molten cast steel is not as fluid as cast-iron and tends to form a honeycomb structure owing to bubbles. The castings are made under pressure and are subsequently annealed. They are employed in preference to cast-iron where great strength is required and the form is only moderately intricate.

Mild Steel (0.1 to 0.5 per cent, carbon) is very commonly employed for beams, rods, etc., where the shape is reasonably simple. It can be bent, sheared and forged, and can be machined by high-carbon steel or special tool steels. Its breaking tensile stress is about 30 tons per square inch.

High-carbon and Special Steels.

Steel which contains a high proportion of carbon may be hardened by heat treatments. Very hard steels are required for such purposes as cutting other steels or resisting continuous abrasion. In steels used for high-speed cutting tools, the percentage of carbon may be diminished and various other metals added in the manufacture. For instance, the addition of manganese or chromium produces steels that may be used for cutting tools or high-grade ball-bearings, and tungsten helps cutting tools to remain sharp when they become very hot. Nickel steels are elastic and ductile. The elastic strength of some of the special steels is very high.

The Heat Treatment of Steels.

The properties of steels may be modified by processes of heating and cooling called **heat treatment**. For example, if a steel containing a sufficient proportion of carbon be heated to 800° C. (a red colour) and quenched in water, brine or oil, the steel is hardened by the process. Any steel with a carbon content of more than 0.5 per cent, is appreciably hardened by such treatment, and the higher the carbon content up to about 1.2 per cent., the greater is the effect. But after hardening, the material is often very brittle, and it must be tempered if resistance to shock is required. Tempering consists of heating to about 500° C. (a blue colour) and quenching from this temperature. The material loses most of its brittleness and a little of its hardness in the tempering process. For a given steel, the more rapid the quench, the harder (and more brittle) the material becomes. Oil has a slower quenching effect than brine, and brine is slower than water.

FIG. 332.—SOFT-CENTRE STEEL

The hardening and tempering of parts of farm machines has become an important process in their manufacture. Steel plough breasts and shares, cultivator points, plough axles and horse rake teeth are given suitable heat treatments under carefully controlled conditions, in order to produce the necessary resistance to abrasion or bending. The effect of heat treatment on a typical plough beam with 0.68 per cent, carbon, 0.62 per cent, manganese, 0.03 per cent, sulphur and 0.03 per cent, phosphorus is shown in Table XII.

TABLE XII.—*THE EFFECTS OF HEAT TREATMENTS ON THE TENSILE PROPERTIES AND HARDNESS OF A STEEL BEAM**

Test	As received	Quenched from 1600° F.	Tempered at 1000° F.
Elastic Limit (lb. per sq. in.)	48,200	99,500	93,500
Ultimate strength (lb. per sq. in.)	98,600	152,600	139,500
Brinell hardness number	192	302	275

* From Bornstein, H., *Agric. Engineering*. St. Joseph, Michigan, U.S.A., August 1934.

Annealing.

In the manufacture of steel forgings or castings, enormous internal stresses may be set up, either by the actual forging process or by unequal contraction on cooling. These forces may be so great in steel or malleable castings that if the castings are allowed to cool before annealing, they may fly (break) in the cooling. Annealing consists of heating the articles to about 900° C. for half an hour, and then turning out the furnace and allowing the oven to cool very gradually. This process, which takes place before hardening, if that also is required, effectively disposes of the internal stresses in the materials.

Case Hardening.

Wrought-iron and mild steel are relatively soft, and cannot be appreciably hardened by ordinary heat treatments. They can, however, be given a hard surface by a process called case hardening. By this process it is possible to produce articles that are both resistant to shocks and extremely hard at the surface. The objects to be hardened are packed in iron boxes with a carbonaceous material, such as animal charcoal or a mixture of 3 parts of wood charcoal and 1 part of powdered bones, and are heated to about 900° C. for 5 to 10 hours, according to the depth of hard case required. The carbon combines with the iron of the materials, producing the hard cementite, $Fe3C$. When they are cool, the articles arc unpacked from the boxes and are then reheated alone to about 800° C. and quenched in cold water in order to harden the cases.

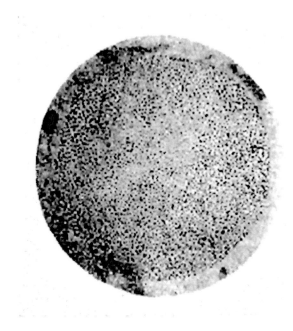

FIG. 333.—*CASE-HARDENED STEEL*

A very thin, hard case may be applied to small articles by dipping them in molten potassium cyanide. The knotter bills of binders are sometimes hardened in this way. The process is, of course, a dangerous one in unskilled hands, owing to the highly poisonous properties of potassium cyanide.

Non-ferrous Metals.

Copper is a soft, fibrous metal which is easily bent and sheared, and may be brazed with powdered borax as a flux. It is used in the form of wire, especially for electrical equipment, and in the form of small pipes, which may be bent cold. At low temperatures, it is safe at $1\frac{1}{2}$ tons per square inch, but it becomes "rotten" at high temperatures.

Zinc.

One of the most important properties of zinc is its resistance to atmospheric corrosion. It is used as a protective coating for iron in the form of plain sheets, corrugated sheets and wires. Zinc-coated iron is termed **galvanized** iron.

Lead is a soft, heavy metal that is used alone for pipes and sheets where resistance to corrosion is required. It is also extensively used in some of the alloys employed in high-speed bearings.

Tin is a very malleable metal, widely used as a protective coating for such iron-ware as comes into contact with human foods. Dairy utensils are frequently constructed of tinned iron, and tin is also an important constituent of certain alloys. It is, however, an expensive metal, and its use for dairy utensils is diminishing, in favour of such materials as stainless steel and aluminium.

Aluminium is a very light metal which is becoming widely used for dairy utensils, electric cables and alloys.

It is cheap, and its lightness (0.09 lb. per cubic inch) and moderate strength make it an attractive material for a variety of purposes.

Alloys.

Alloys are mixtures of two or more metals, typical examples being brasses, bronzes, babbitt metal and white metal.

Brasses are alloys of copper and zinc. There are two main types of brass, one having about 70 per cent, copper and 30 per cent, zinc, and the other having about 60 per cent, copper and 40 per cent. zinc. The first type is soft, and is used in the manufacture of tubes, wires, etc., while the other is much harder, and may be used for castings. Various metals added to brasses modify the properties. Thus, the addition of a little lead makes for easier machining, tin gives increased resistance to corrosion, and manganese imparts additional strength to the alloy.

Bronzes are alloys of copper and tin, the most important being gun-metal, which has an average composition of 88 per cent, copper; 10 per cent, tin; 2 per cent. zinc. It is used for bearings in which steel shafts revolve, and also for engine fittings. It is easy to cast, stronger than cast-iron, and does not corrode easily; but it is expensive on account of the tin contained in it. Phosphor bronze has an average composition of 89.5 per cent, copper; 10 per cent, tin; 0.5 per cent, phosphorus. It has all the properties of gun-metal, but is much stronger, and is used for pump rods, special bearings, etc.

White Metal.

The average composition of a good white metal is 90 per cent, tin; 6 per cent, antimony; 4 per cent, copper. It is an expensive metal which is sometimes used in high-speed bearings such as the "big-end" bearings of internal-combustion engines. One of its most important properties is its low melting-point, which causes it to "run" if lubrication fails. **Babbitt** metal is a white metal in which part of the expensive tin is replaced by lead. Its properties are similar to those of white metal.

Solder consists of a mixture of about 40 per cent, tin and 60 per cent, lead. In the process of soldering, a flux is generally used with the objects of cleaning the metal, reducing the surface tension (which causes the drop of solder to remain spherical) and preventing oxidization. Zinc chloride is an easy flux, but is not safe for electrical work owing to its hygroscopic properties. Rosin in alcoholic solution is fairly satisfactory, and some of the proprietary fluxes consist mainly of overheated rosin with a little zinc chloride.

Cement and Concrete.

Cement is a pulverized material composed principally of silica, alumina and lime; when mixed with water, it undergoes a chemical change, forming new compounds which develop the property of setting into a solid mass. Concrete is a mixture of cement with a quantity of inert materials which, when mixed with water, will harden to form an artificial stone. Mortar is a mixture of cement, sand and water. In making concrete, the inert material may be sand, pebbles or broken stone, and the term applied to these materials is the **aggregate**. The aggregate should be clean, for impurities of an organic nature may prevent the mortar or cement from hardening and attaining its maximum strength. For concrete work where strength is required, it is generally desirable that the aggregate should be well assorted as to size of particles. The best aggregate contains just a sufficient number of smaller particles to fill the interstices between the larger ones, and the better this filling of spaces is achieved, the lower the proportion of cement required. Where great strength is required and cost is a minor consideration, a rich mixture should be used, but this condition rarely occurs in the use of concrete for farm work.

Typical mixtures for various types of work are:—

1. Strong.—1 part cement; $1\frac{1}{2}$ parts sand; 3 parts coarse aggregate. This may be used for columns.

2. Standard.—1 cement; 2 sand; 4 coarse aggregate; used for reinforced floors, engine foundations and similar purposes.

3. Medium.—1 cement; $2\frac{1}{2}$ sand; 5 coarse aggregate; used for ordinary floors, foundations and walls.

4. Lean.—1 cement; 3 sand; 6 coarse aggregate; used in large masses for foundations and thick walls.

The importance of thorough mixing of concrete cannot be overemphasized. Every grain of the aggregate must be coated with a film of cement if the maximum strength is to be achieved. The concrete should be used in a medium to wet condition where possible, but if mixed with only a minimum quantity of water, care must be taken to ensure that it is well rammed. It is generally stronger if used rather wet, but takes longer to set hard. It should be put in place within about forty-five minutes of mixing, and should remain undisturbed for at least ten hours.

Concrete has great strength in compression and is also fairly strong in shear, but it is rarely used in tension. In reinforced structures, where tension must be considered, only the tensile strength of the reinforcement should be taken into account.

Timber.

Wood is well adapted to resist shocks and vibation, and is frequently used for machine parts receiving such treatment; e.g. the pitman of the mower or binder. It is strong for its weight, and has the great advantages of being easily fashioned by unskilled labour and also of being home-grown on many farms. A few of the hard and durable woods are used for special purposes in the construction of farm machinery, but mild steel is found on many modern machines where wood used to be employed. In spite of this, wood is still an important material in agriculture; for carts, wagons, fences, gates and buildings are often almost entirely constructed of timber.

The quality of wood is determined to a large extent by the selection of the trees and the care used in cutting, seasoning and preserving. It only remains a reliable material, however, so long as it is preserved from decay. Seasoned timber kept indoors, free from contact with the ground, is generally too dry for the attack of fungi, but where conditions of moisture and temperature favour the growth of bacteria and fungi, wood that is not effectively protected may be attacked and destroyed. Some timbers, notably oak and larch, are naturally durable, while alder, beech, birch, ash, elm and Scots pine are undurable unless protected. Paint is the preservative mostly applied to farm implements, but neither this nor tar is as good as impregnation with creosote for wood that is continually damp.

There are three main methods of treating timber with creosote, and the least satisfactory of these is the surface treatment achieved by brushing or dipping. The process widely employed by timber merchants is impregnation with hot creosote under pressure inside a sealed cylinder. This produces good results, but is hardly practicable on the farm. A method that can be recommended for the preservation of fencing material, etc. on the farm is the **open tank** method. The timber is placed in creosote in a suitable metal drum, and a fire kindled beneath. The temperature of the creosote should be raised to about 180-200° F. in two hours. During this time, air and moisture are driven off from the timber. Cooling is then allowed to take place, and after about twenty-four hours the timber is usually well impregnated. The cost of this method of preserving timber is trifling compared with the effects. Fortunately, sapwood and naturally undurable wood takes up the creosote very readily if the bark is peeled off.

Paints.

Paint is a material which, though liquid when applied, dries after application and forms a thin film that is comparatively hard. On ageing, it becomes gradually harder and more brittle, and eventually it fails. The length of life varies considerably, but there is no permanent paint. Factors influencing the ultimate breakdown are the composition of the paint, the surface to which it is applied and the weathering influences.

With both wood and metal equipment it is desirable to paint at once, while the surface is still new and clean, and before rust or cracks do irreparable damage. Wood, being a porous material, absorbs a good deal of paint, and needs three or four coats for full protection. It should be cleaned and wiped down smooth before painting with a primer (first coat) such as a mixture of white lead, orange lead and oil, the oil being mixed in the proportions 75 per cent, turpentine; 25 per cent, linseed oil, with a little drying oil added. The. use of plenty of turpentine in the primer gives a good penetration and a smooth surface which prevents subsequent blistering. Later coats of paint should contain plenty of white lead. The orange or red paint used for farm carts can be made entirely of white lead and orange lead, and greys or stone colours may also contain a high proportion of white lead, but blues and greens do not usually have as much "body".

Paint used for metal surfaces must possess certain distinctive characteristics, among which are rust prevention, distensibility, a high spreading rate, durability and case of application. The primer must be one adapted to metal surfaces, for paints that are suitable for wood or are good finishers may be of little use as primers for metal. Paints commonly used for metals include red lead, lead chromate, zinc oxide, metallic aluminium powder and metallic zinc powder. A mixture of 4 parts of metallic zinc powder with 1 part of zinc oxide and linseed oil is a good rust-inhibiting paint that is finding an increasing use in agriculture in the United States. It has excellent sticking properties and may be satisfactorily used for galvanized surfaces on roofs. The surface to be painted should be dry, warm and free from dust, and any rusted areas should be vigorously brushed before painting. The colour of the zinc paint is grey, and it may be used both as a primer and finisher.

Since the cost of labour frequently exceeds that of the paint, it is generally advisable to use a high-grade paint, whether for wood or metal. Repainting should be carried out before failure is far advanced, for over a period of years, good paint applied at sufficiently frequent intervals is the cheapest.

APPENDIX SIX

INTERNAL-COMBUSTION ENGINES

Heat engines are machines in which the potential energy contained in fuels may be employed to do useful work. They may be divided into two main classes as follows:—(1) engines in which the working substance receives its heat outside the working cylinder (e.g. steam-engines); and (2) those in which combustion of the fuel takes place inside the working cylinder (internal-combustion engines). In the latter, the working substance is a mixture of air with petrol, paraffin, oil, gas, alcohol or other fuel. The combustion of the fuel within the cylinder sets up a pressure which acts on the engine piston. The piston is forced down the cylinder, and by a suitable arrangement of mechanisms, part of the heat energy contained in the fuel may be converted into work.

The First Law of Thermodynamics states that "heat and work are mutually convertible"; but whereas the whole of a given quantity of work may easily be converted into heat, no engine will convert the whole of the heat energy supplied to it into work. The criterion of efficiency in a heat engine is the maximum amount of work obtainable from a given amount of fuel. The overall efficiency of an engine may, therefore, be obtained by expressing the heat equivalent of the useful work done by the engine as a percentage of the heat value of the fuel used. In general terms, internal-combustion engines are about twice as efficient as steam-engines in this respect. Efficiencies of 10 to 15 per cent, in steam-engines are good, and the best efficiency obtainable is about 25 per cent. Overall efficiencies of 22 to 25 per cent, are not unusual in petrol engines, and some Diesel engines reach values of 35 per cent. The highest efficiencies (about 42 per cent.) have been obtained by a combination of a Diesel engine and a steam-engine, in which the waste heat from the Diesel engine is employed to assist in the raising of steam.

The advantage of the internal-combustion engine with respect to efficiency, combined with an ever-increasing reliability, has enabled it to supplant the steam-engine for most agricultural power purposes. The steam-engine has its advantages for certain special operations, but it compares unfavourably with the internal-combustion engine in adaptability, lightness, convenience and general ease of operation.

Three main types of internal-combustion engines are commonly used on farms, viz.: (1) petrol; (2) paraffin; and (3) Diesel or semi-Diesel engines. There are few differences between engines which employ petrol alone as a fuel and those which start on petrol and then run on paraffin. But there are important differences between these two types and Diesel engines.

Working Principle of the 4-stroke Petrol Engine.

Various working cycles have been employed in internal-combustion engines, but although a few engines employ the 2-stroke cycle described later, the 4-stroke or **Otto** cycle is the one generally employed in petrol and paraffin engines. The complete working cycle occurs once in every four strokes of the piston or two revolutions of the crankshaft, as follows:—

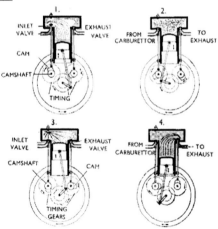

FIG. 334.—DIAGRAM ILLUSTRATING THE 4-STROKE CYCLE

947

1, induction stroke; 2, compression; 3, working stroke; 4, exhaust.

1. The first stroke of the cycle is the induction stroke (1 in Fig. 334). The piston moves down the cylinder with the inlet valve open, and a quantity of explosive mixture is drawn from the carburettor into the cylinder.

2. The compression stroke follows, and in this the charge is compressed by the upward movement of the piston, which occurs with both valves closed.

3. At the end of the compression stroke, the charge is exploded, generally by means of an electric spark. The pressure developed by the combustion forces the piston downwards on the working stroke.

4. On the fourth stroke, the exhaust valve is open as the piston moves upwards, and the exhaust gases are pushed out of the cylinder into the exhaust pipe. This stroke completes the working cycle.

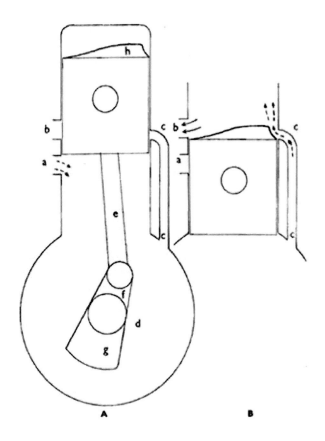

FIG. 335.—DIAGRAM ILLUSTRATING THE 2-STROKE CYCLE

a, inlet port; **b**, exhaust port; **c**, transfer port; **d**, crank-case; **e**, connecting rod; **f**, crankshaft; **g**, balance weight; **h**, deflector.

The 2-stroke Cycle.

In the 2-stroke cycle, induction, compression, ignition and exhaust are crowded into two strokes of the piston. This should theoretically permit the development of twice as much power as in the 4-stroke, for a given size of engine and speed, but in practice many limitations prevent the achievement of this. No valves (except a compression-release valve) are employed, induction and exhaust taking place through openings in the cylinder wall called **ports**. Fig. 335 (A) shows the position just after the firing of the mixture, and Fig. 335 (B) the position later in the downward stroke. The crank-case is gas-tight, and as the piston moves upwards and compresses the gases in the combustion chamber, a vacuum is created in the crank-case, so that when the lower edge of the piston uncovers the inlet port, gases from the carburettor are sucked into the crank-case. After combustion (Fig. 335 (A)), the piston is forced downwards, and just before it reaches the position shown in Fig. 335 (B) the exhaust port is uncovered, and exhaust gases rush out into the exhaust pipe. Immediately afterwards, the transfer port is also uncovered, and the mixture in the crank-case, which has been compressed by the downward stroke of the piston, passes into the combustion chamber. A deflector cast on the top of the piston causes the new gases to sweep upwards, so that they do not rush out of the open exhaust port. The entry of the new gases assists in the "scavenging" of the exhaust gases. Then the piston moves upwards again, closing the transfer and exhaust ports, compressing the new charge in the combustion chamber, and uncovering the inlet port so that a further charge of new gas is drawn into the crank-case.

One of the chief disadvantages of the simple 2-stroke petrol engine is that the scavenging of the exhaust gases is never complete, and that a proportion of the new gases escapes through the exhaust port in the scavenging process. Though most farm engines employ the 4-stroke cycle, 2-stroke engines are used on some small walking tractors. The 2-stroke, with a power impulse for every revolution of the crankshaft, gives smoother running than the 4-stroke. Moreover, it is more simple, in that there is no valve mechanism. The 2-stroke Diesel cycle is described on p. 613.

Some Constructional Features of Petrol Engines.

The cylinder is usually made of hard, fine-grained cast-iron. This wears well and takes on a high polish, but cylinders made of specially hardened materials may in future displace the softer cast-iron. Modern tractor engines are usually fitted with renewable cylinder **liners**. The closed end of the cylinder forms the combustion chamber. In the

4-stroke petrol engine two valves and a sparking plug fit into this, while in Diesel engines the fuel injection nozzle is present instead of the sparking plug. The **bore** of the cylinder is the internal diameter. The **stroke** is the distance the piston moves in travelling from top **dead centre** to the bottom dead centre.

The Cooling System.

The amount of cooling necessary depends upon the size and type of engine and the conditions in which it is used. Air-cooling is used on many small farm engines, fitted with a suitable fan and cowling, but some form of water-cooling is usually employed on the larger sizes. A common method is to employ the **thermo-syphon** system in conjunction with a large cooling tank, or a radiator and fan, A light casting, the water jacket, surrounds the cylinder, and this is connected by pipes to the radiator or tank. Owing to the fact that the specific gravity of hot water is lower than that of cold, the water that is heated by contact with the cylinders rises up the upper pipe to the cooler. As it is cooled it passes to the bottom of the cooler, and eventually back into the water jacket again through the lower pipe. Efficient cooling of a powerful tractor engine requires devices which will vary the cooling effect according to ambient temperature, and also to the amount of work the tractor is doing. Fig. 337 illustrates the cooling system of a modern tractor engine. A fan sucks air through the radiator, which consists of rows of gilled tubes. Circulation of water by thermo-syphon action alone would be inadequate in hot weather with the engine hard at work; so a centrifugal water pump is employed to speed it up. If the pump were allowed to circulate the water at maximum speed in cold weather with the engine running light, over-cooling would result. To prevent this, a thermostat is installed between the engine and the top of the radiator. This consists of a capsule, the length of which varies considerably with temperature, which operates a mushroom-shaped valve. When the water surrounding the capsule is cold, length of the capsule is short, and the valve impedes water circulation. When the water is hot, the valve is open and circulation is free. Another important temperature regulating device is the manually-controlled radiator shutter, which varies the amount of cooling air that passes through the radiator.

FIG. 336.—RENEWABLE CYLINDER LINER

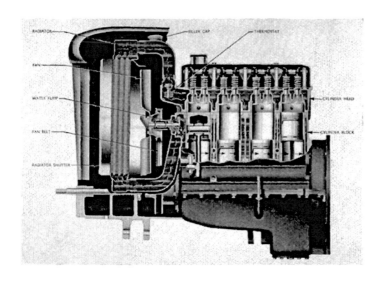

*FIG. 337.—COOLING SYSTEM OF A MODERN TRAC-
TOR ENGINE. (FORDSON)*

The radiator has a filler cap which seals the system, so that a slight pressure is developed in normal operating conditions. A pre-set relief valve ensures that no excessive pressure is built up. Nevertheless, it is dangerous to remove the filler cap on a pressure radiator until the water has been allowed to cool down.

A pressure cooling system raises the boiling point of the water in the radiator, and its advantages are that there is less loss of cooling water, while the higher boiling point permits effective cooling with a limited quantity of water in severe operating conditions.

The Piston and Connecting-rod.

The function of the piston is to transmit the work done on it by the exploded gases to the crankshaft, *via* the connecting-rod. It should be light, strong and gas-tight. Cast-iron has many advantages as a piston material and is still widely used for paraffin engines, but on modern high-speed engines aluminium alloys are being increasingly used. Gas-tightness of the combustion chamber is secured, without excessive friction, by the use of piston rings. These are plain rings of cast-iron, cast to a diameter slightly greater than that of the piston, and with a small segment cut away. They fit into grooves cut in the circumference of the piston, and when in position they press outwards against the cylinder wall because of their springiness and greater original diameter.

The connecting-rod must be reasonably long, in order that only a small amount of side-thrust is exerted on the piston. Its length is generally about four times that of the crank. It is subject to bending as well as to pure compressive stresses, and must be strong in order to resist these. It is usually constructed of I-section steel, which gives a high strength: weight ratio. The gudgeon pin connects the **little end** of the connecting-rod to the piston, and this bearing is usually made plain for the sake of lightness. Lightness of the piston and connecting-rod is necessary because of the inertia of the moving parts at high speeds.

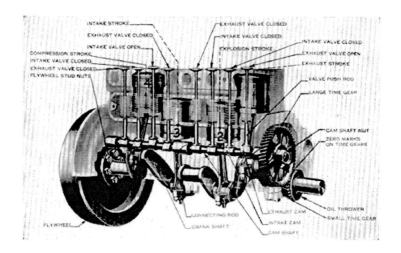

FIG. 338.—SECTION SHOWING CYLINDER AND
VALVE ASSEMBLY OF A 4-CYLINDER ENGINE

(Firing order, 1—2—4—3.)

The Crankshaft and Flywheel.

A power stroke occurs only once in two revolutions
of the crankshaft in a single-cylinder 4-stroke engine,
and the piston drives the crankshaft for only one stroke
out of four, while the crankshaft drives the piston for the
next three strokes. It is enabled to do this by the use of a
flywheel, consisting of a circular mass of metal attached to
the crankshaft. When the flywheel rotates it stores energy,
which is used to turn over the engine by assisting the piston

over the dead centres. The greater the number of cylinders and the more uniform the torque, the lighter the flywheel need be.

By employing four cylinders and arranging their sequence of operations in a suitable way, one power impulse takes place during every half-revolution of the crankshaft. The crankshaft is arranged so that the two end pistons descend as the intermediate ones rise, and vice versa (Fig. 338). The firing order usually employed is 1—3—4—2 or 1—2—4 3 where the cylinders are numbered consecutively from one end. Four cylinders give a reasonably smooth torque, but the torque of 6- and 8-cylinder engines is, of course, smoother.

The Valve Gear.

The orthodox 4-stroke engine employs mushroom-shaped poppet valves. The circular head serves to close a port through which the gases enter or leave the cylinder when the valve is lifted off its seating. The seating has a bevelled edge, and the valve-head is ground to give a gas-tight fit when the valve is closed. The valve is normally opened by means of a cam, and is returned to its seating by a stiff coil spring. Since each valve only opens once for every two revolutions of the crankshaft, the camshaft which operates the valves must be driven at half the speed of the crankshaft. This is achieved by a 2:1 gearing called the timing gears. Fig. 339 *(a)* shows a typical arrangement of the valve-operating mechanism in a side-valve engine. The camshaft, with the cams attached, causes the tappet rods to move upwards once every revolution. The valve spring keeps the tappet roller in contact with the cam, and rapidly returns the valve to its seating. The length of the tappet rod is adjustable by screwing in or out a screw at the upper end of the tappet rod. This is held in position by a lock-nut.

Overhead valves are generally operated by push-rods and rocker arms, the push-rods being actuated by cams in the usual way. On some high-speed engines the camshaft may be situated above the cylinders instead of in the crank-case, and it then operates the overhead valves through rocker arms only.

Timing of the Valves.

The valves are not arranged to open and close when the piston is precisely at the top and bottom dead centres, for if they were timed thus, the engine would run satisfactorily only at very low speeds. The sequence of the valve-opening periods is illustrated graphically by denoting the corresponding position of the crankshaft at which the various operations occur. For example, Fig. 340 represents the timing diagram in the case of a typical medium-speed petrol engine. The inlet valve usually opens at or just after top dead centre, and by the time the piston has travelled on the induction stroke to bottom centre, a column of mixture in the inlet ports is rushing into the cylinder with considerable velocity. Owing to the inertia of this column of gas it will continue to pass into the cylinder after the piston has passed the bottom of the stroke. The inlet valve is therefore kept open until about 30° after bottom dead centre, and this allows a large quantity of mixture to enter the cylinder.

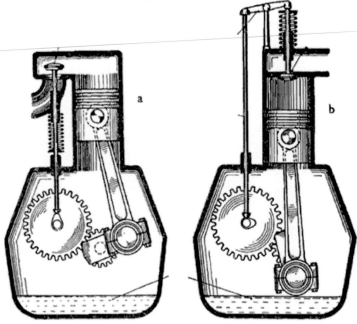

FIG. 339.—ENGINE WITH (A) SIDE VALVES, AND (B) OVERHEAD VALVES

The exhaust valve opens about 45° before bottom centre

and closes a few degrees after top centre. The early opening wastes very little power, and allows the exhaust gases to start escaping while the piston is still moving downwards slightly. The late closing enables the inertia of the exhaust gases down the exhaust passages to continue to suck out more gas, though the piston has started on the downward stroke.

Carburettors.

For economical and reliable operation, the fuel must always be mixed with air in the correct proportion for complete combustion. Mixtures that are too *rich* (i.e. contain too high a proportion of fuel) do not burn completely, and cause loss of power, explosions in the exhaust pipe, and sooting of the sparking plugs, valves and combustion chamber. Mixtures that are too *lean* burn slowly and may still be burning when the inlet valve opens. This may cause a flame to flash back through the inlet valve to the carburettor. Fig. 341 illustrates a simple carburettor. Fuel is fed from the tank to a chamber where a float and a needle valve ensure that a constant level of fuel is maintained. The height of the petrol in the float chamber is such that the level in the jet is just below the tip, when the engine is not running. The jet is a vertical pipe of small diameter, fitted at its upper end with a plug which has a fine hole through it, and this is situated in a constricted tube or venturi through which air is sucked into the cylinder.

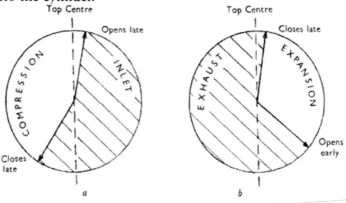

FIG. 340.—*CLOCK DIAGRAM OF VALVE TIMING A, INLET VALVE DIAGRAM; B, EXHAUST VALVE DIA-GRAM.*

When a gas or liquid flows past a constriction in a tube, the pressure at the constriction is lower than that in the wider part of the tube, and the difference in pressure varies with the rate of flow. In the carburettor, the air that is drawn into the cylinder on the induction stroke causes a reduction of pressure as it flows past the tip of the jet. Since the pressure on the surface of the petrol in the float chamber is atmospheric and that at the tip of the jet is less, petrol flows out of the fine jet as a spray, and is carried along with the air into the cylinder.

The dimensions of the carburettor venturi tube and the jet are proportioned so as to give about the correct ratio of air to petrol, viz. about 15 parts of air: 1 part of petrol, by weight. The carburettor must, however, be so constructed that richer or leaner mixtures may be produced when necessary, and so that the mixture does not vary with the speed of the engine. The simple carburettor described above does not give a constant mixture at varying speeds, mixtures tending to be too lean at low speeds and too rich at high. Various extra devices must be employed to provide an efficient and economical mixture at all speeds.

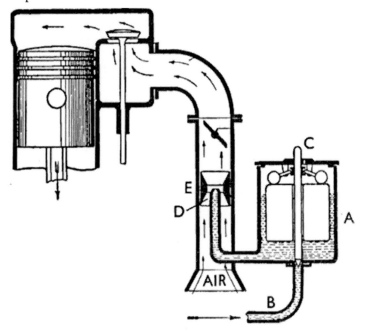

FIG. 341.—SIMPLE CARBURETTOR

A, float chamber; **B**, petrol supply pipe; **C**, needle valve; **D**, jet; **E**, choke.

The amount of mixture entering the cylinder is controlled by a butterfly throttle valve. The richness of the mixture may be controlled by regulating either the air supply or the petrol supply, or both. In simple carburettors, both the air and petrol supply may be controlled independently by hand levers, but for such engines as are used in tractors it is desirable to employ devices which automatically ensure a satisfactory mixture. Such devices include compensating jets, auxiliary air valves or air bleeds, and pilot or idling jets. The compensating jet (Fig. 342) draws its petrol supply from a special well which is fed by gravity at a constant rate from the float chamber. The flow of petrol into this well is entirely uninfluenced by the engine speed, and if the compensating jet were used alone, weaker mixtures would be produced at the higher speeds. The main jet, on the other hand, gives a richer mixture at higher speeds, and by a suitable combination of main and compensating jets, a mixture that is constant at all engine speeds may be produced.

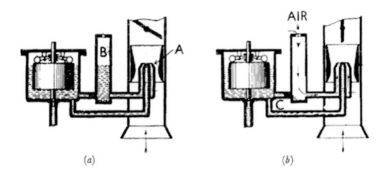

FIG. 342.—CARBURETTOR WITH COMPENSATING
JET

a, throttle nearly closed; *b,* throttle open. **A,** compensating jet;
B, well which is fed by gravity from float chamber.

Fig. 343 is a section of a typical modern tractor
carburettor, employing air bleeds and idling jet as well as a
main jet that works on the principle outlined above.

As the air is sucked through the venturi (1) it draws
fuel from the main discharge nozzle (3) and the main jet
(2), and air from the main air bleed (4). The amount of fuel
drawn through the main jet (2) corresponds to the engine
speed; but there is a definite limit to the amount that can
pass through, and beyond a certain point increase of suction
will not increase the amount of fuel passed by this jet. This jet
has an adjustment (7) by which the amount of fuel delivered
can be varied.

The air bleed system operates as follows. The faster the engine runs and the greater the suction in the venturi, the lower the level of fuel in the discharge nozzle. As engine speed increases the upper hole A is first uncovered, and air is admitted. The reduction of fuel level exposes more holes, and increases the amount of air bled into the discharge nozzle. Thus, the air bleed system keeps the mixture right by correcting the tendency to an increase in the ratio of fuel to air at high speeds.

The idling system operates when the butterfly valve is nearly closed, and provides a specially rich mixture. It is really a simple carburettor in itself, and delivers fuel to the engine at a point just above the butterfly valve. The system consists of the idling slot (8), the idling jet (5) and a connection to the main jet (a). On top of the idling jet chamber there is a hole extending to the air passage round the venturi (dotted in illustration). With the butterfly valve nearly closed, suction above it draws air from this passage and fuel from the idling jet. The idle air adjusting needle (6) controls the amount of air admitted.

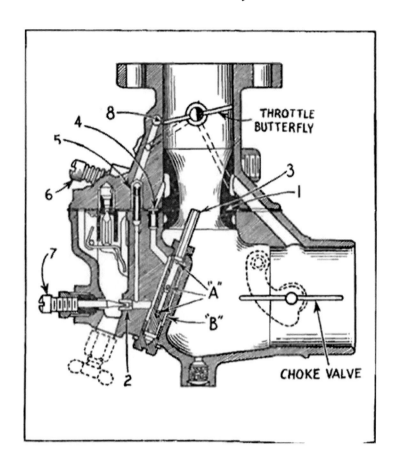

FIG. 343.—SECTION OF MODERN TRACTOR CAR-BURETTOR. (INTERNATIONAL.)

1, venturi; **2,** main jet; **3,** discharge nozzle; **4,** main air bleed; **5,** idling jet; **6,** idle air adjusting needle; **7,** main jet adjuster; **8,** idling slot; **"A"**, holes in discharge nozzle; **"B"**, accelerating well.

The idling system operates only from slow idle speed to fast idle speed. As soon as the throttle is operated for load conditions the effect of the system is reduced almost to nothing. Indeed, at wide open throttle the greatest suction exists in the venturi, and air flows from the space above the butterfly valve, down through the idling system to the passage in front of the main jet, and so acts as another air bleed.

Closing the choke valve shuts off the supply of air passing over the main discharge nozzle and increases the suction acting on the idling jet, producing a very rich mixture.

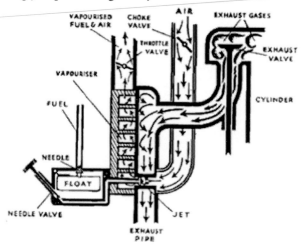

FIG. 344.—DIAGRAMMATIC SECTION OF PARAFFIN TRACTOR ENGINE VAPORIZER. (N.I.A.E.)

There is no fundamental difference between petrol and paraffin carburettors, but when paraffin is used as a fuel in internal-combustion engines, an efficient vaporizing device is required. The heat of the exhaust gases is normally employed to heat the mixture of air and paraffin in a special vaporizer before the mixture is introduced into the engine cylinder (Fig. 344). The petrol-paraffin engine commonly used on tractors starts on petrol and is only switched over to paraffin when the vaporizer has become sufficiently hot. Such engines should not be switched over to paraffin until the vaporizer is really hot, for the introduction of incompletely vaporized paraffin into the cylinders causes partial combustion, with the attendant carbon deposits and excessive crank-case dilution. Bi-fuel carburettors are discussed on p. 58.

Ignition of the Fuel.

In petrol and paraffin engines, the explosive mixture is usually ignited at the correct moment by a high-tension spark. On many farm engines, the spark is generated by a magneto, but on some tractors and motor lorries, where a dynamo and batteries are required, battery-and-coil ignition is adopted. The magneto is a self-contained electrical machine which is driven by the engine. The underlying principle in its construction, as well as in that of the dynamo, is the fact that if a wire is passed across a magnetic field, an electromotive force (E.M.F.) is induced in the wire. The most convenient method of producing such an electromotive force is to fasten the wire along the sides of a cylinder and rotate it between the poles of a magnet. In the magneto, the magnet is made of cobalt steel and is fitted with semi-circular soft iron pole-pieces. The wire is wound on an iron core which rotates between the hollowed-out pole-pieces on a spindle driven by gearing from the engine. The core with the wire wound on it is called the armature.

In the high-tension (h.t.) magneto, the armature carries two coils of wire which are insulated from one another, viz. (1) the primary coil, consisting of a few turns of thick wire; and (2) the secondary coil, consisting of many turns of fine wire. Rotation of the primary winding in the magnetic field induces in it a low-tension current of electricity. A device called the contact breaker is included in the low-tension circuit, and this regularly breaks the circuit, thereby setting up a high-tension E.M.F. in the secondary winding. If two coils of wire are close together and insulated from one another, the breaking or making of a current in one coil induces an E.M.F. in the other, and this E.M.F. is proportional to the ratio of the number of turns of wire in the two coils. The E.M.F. in the secondary coil is made so high that a current will flow even if there is a small air gap in the circuit. The coil is connected to the sparking plug, and when the contact breaker interrupts the current in the primary circuit, a spark jumps across the plug points inside the combustion chamber, thereby igniting the charge.

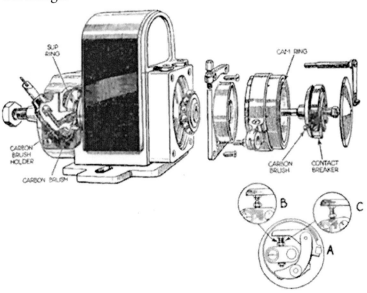

FIG. 345.—DETAILS OF A SIMPLE MAGNETO

A, correct bedding of contact breaker points; **B** and **C,** faulty contact.

The contact breaker has a gun-metal base which is screwed to the armature shaft and rotates with it. It carries a bell-crank lever which has at one end a small platinum point, and this is held in contact with a second platinum point by means of a flat spring. The fixed platinum point is connected to one end of the primary coil, and the bell-crank lever to the other. At the correct moment for ignition, the bell-crank lever rises over a cam on the stationary cam ring inside which the contact

971

breaker rotates, thus forcing the platinum points apart and interrupting the current in the primary circuit.

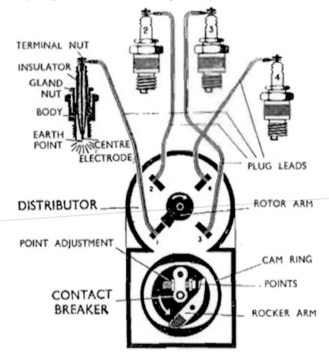

FIG. 346.—DIAGRAM SHOWING RELATION OF CONTACT BREAKER, DISTRIBUTOR AND SPARKING PLUGS. (N.I.A.E.)

The cam ring may be partially rotated in its casing in order to allow adjustment of the timing of the spark. In a single-cylinder 4-stroke engine, one spark is required for every two revolutions of the crankshaft, and the armature is driven at half the crankshaft speed. In some engines the magneto is driven at engine-crankshaft speed, with the result that an extra spark occurs near the end of the exhaust stroke. The advantage of this is that the magneto rotates twice as rapidly and so gives a much more intense spark when being started by hand. In the case of a 4-cylinder engine, the armature may be driven at the same speed as the crankshaft, and in this case the cam ring will have two cams situated opposite one another.

A condenser in the primary circuit is an essential part of the H.T. magneto, for when the current in the primary circuit is suddenly broken by the contact breaker, it tends to continue to flow, and to cause sparking at the platinum points. The condenser consists of alternate sheets of tinfoil and mica, one set of tinfoil sheets being connected to the primary winding and the other to *earth*. When contact is broken at the platinum points, the condenser acts as a buffer which forces the current back into the primary winding, causes a more sudden breaking of electrical contact, and permits the creation of higher secondary voltages.

One end of the secondary winding is attached to a brass collector ring and the other end to earth (i.e. to the iron of the engine or frame). A carbon **brush** presses on the collector ring and picks up the current, which is then conveyed *via* a safety spark gap direct to the sparking plug in the case of a single-cylinder engine, or to the distributor in the case of a multi-cylinder one. The distributor has a revolving arm that makes contact with a number of insulated metal segments as it rotates. Each of these segments is connected to its appropriate cylinder, so that a spark is delivered to each cylinder in the correct order.

Some adjustment of the timing of the spark is essential, for whereas the spark should occur after the piston has passed over the dead centre in starting, it should be advanced to occur before the piston reaches the top centre for the most economical running at high speeds. If it occurs too late, the piston has already commenced its downward movement when combustion takes place, with the result that the effective working stroke is lessened and the mean pressure is lower than it need be. Timing of the spark is effected by rotating the magneto cam ring, and on most farm engines the gearing to the magneto should be so arranged that the spark occurs at about top dead centre with the cam ring in its mid position, thus allowing adjustment in both directions.

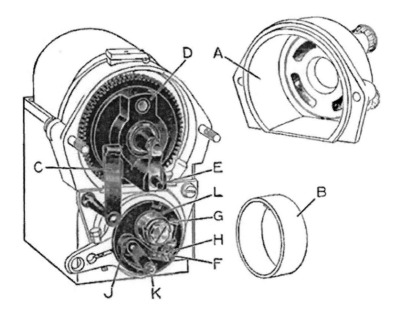

FIG. 347.—*ROTATING-MAGNET MAGNETO WITH
DISTRIBUTOR AND CONTACT BREAKER COVER
REMOVED. (LUCAS.)*

A, distributor; **B,** contact breaker cover; **C,** spring securing
contact breaker cover; **D,** distributor rotating arm; **E,** carbon
brush; **F,** contacts; **G,** locking sleeve; **H,** hexagon head; **J,**
contact breaker lever; **K,** nut securing locating spring; **L,**
cam.

Rotating-magnet magnetos are becoming increasingly popular
for farm engines. In this type the magnet system revolves
while, the less robust parts, such as the windings, condenser
and contact breaker remain stationary, and are therefore not

subjected to severe mechanical strains. The construction allows almost unlimited space for the windings, so that a heavy primary coil may be used and the secondary coil may be well insulated. These magnetos are robust, compact and cheap.

The modern magneto is very efficient and reliable, and with proper attention it will outlast the engine. But it has the disadvantage that it will only give a good spark at speeds above about 200 r.p.m., so that a comparatively weak spark is obtained on starting. The **impulse starter** is designed to obviate this difficulty. It contains a coil spring that is wound up during part of the revolution of the magneto drive, so that the armature is held back and is suddenly released, thus producing a rapid rotation and breaking of contact and, consequently, a more intense spark.

The battery-and-coil ignition system employs a 6- or 12-volt battery that is charged by a dynamo to furnish the low-tension current. A separate contact breaker, in conjunction with primary and secondary coils and a distributor, then furnish the high-tension current in a manner similar to that employed in the h.t. magneto. This system is generally used where the dynamo and battery are necessary for lighting purposes.

The reliability and performance of any engine depends on fitting a suitable type of sparking plug; and it is particularly important that the proper type of plug should be used in paraffin tractor engines. Plugs vary greatly in heat-conducting qualities. They should be self-cleaning, and a useful guide to performance may be obtained by examining the plug points. With an efficient plug there is generally a light brown discoloration on the insulator. A black or oily plug has remained too cool, while an ashy white colour indicates that the plug temperature has been too high. Information on the correct type of plug may be obtained from the instruction book or from the manufacturer of the engine concerned.

Diesel Engines.

The term Diesel is applied to a variety of types of engine which run without the necessity of an electric spark to cause ignition of the fuel. In 1890 Akroyd-Stuart took out patents involving a cycle in which pure air was compressed, and the fuel was injected later in order to avoid pre-ignition. These engines were of a low-pressure type, requiring a hot bulb or other external method to ignite the charge. In 1892 Dr. Diesel patented a cycle in which the fuel was caused to ignite spontaneously on its injection into the cylinder, owing to the high temperature set up by the compression of the air. Akroyd-Stuart and Diesel both contributed materially to the development of the modern heavy-oil engine. Many types of heavy-oil engine have been employed since their time, but the type to which the term Diesel is generally applied to-day is the solid (or airless) injection, compression-ignition (C.I.) engine, though many **semi-Diesel** and **hot-bulb** engines are also called Diesels.

The cycle of a typical modern 4-stroke Diesel engine is as follows:—

1. On the induction stroke pure air is drawn into the cylinder.

2. On the compression stroke this air is compressed at a high compression-ratio, so that the temperature is raised to a high value.

3. Towards the end of the compression stroke the fuel is injected into the combustion chamber in the form of a fine spray. The temperature of the air in the combustion chamber is sufficiently high to ignite the fuel as it enters, and combustion of the mixture of fuel and air produces the working stroke.

4. As the piston moves upwards on the final stroke the exhaust gases are expelled from the cylinder in the usual way.

The Diesel cycle thus differs in many important respects from the Otto cycle. The compression of pure air instead of a mixture of fuel and air makes it possible to employ high compression pressures without any danger of detonation. Thus, whereas the usual compression-ratios employed in petrol engines are $4\frac{1}{2}$: 1 to 6: 1, resulting in compression pressures of 70 to 80 lb. per sq. in., the compression-ratios employed in C.I. Diesels range up to 16: 1 and result in compression pressures of 500 to 700 lb. per sq. in. The temperature of the compressed air at the moment of fuel injection is generally about 750° C. (1382° F.).

The injection of the fuel takes place over a period equivalent to about 15 to 30 degrees of rotation of the crankshaft. Injection usually commences from 10 to 20 degrees before the piston reaches top centre, but both the point of commencement of injection and the duration are variable. The fuel is injected at pressures ranging from 750 to 3,000 lb. per sq. in. The gradual combustion of the fuel and expansion of the gases give an impulse to the piston that is more sustained than that produced in the petrol or paraffin engine, and the higher compression-ratio makes high efficiencies possible.

In the hot-bulb engine, lower compression pressures are used (e.g. 130 lb. per sq. in.) and the compression temperature is not high enough to ignite the charge without the help of a **hot spot** which is heated before starting by a blow-lamp, and thereafter remains red-hot so long as the engine is at work. The hot-bulb engine uses about 13 per cent, more fuel per b.h.p.-hour than a full Diesel engine.

Injection of the Fuel.

In the original Diesel engines the fuel was injected by the use of a high-pressure air blast. This was inconvenient and sometimes dangerous, and modern Diesels employ solid or airless injection, achieved by use of the fuel pump and injection nozzle. The fuel pump was invented by Alan Chorlton, and its commercial manufacture was brought about mainly by Robert Bosch. An important factor in the rapid progress achieved by Diesels in recent years has been the commercial production of reliable high-pressure metering fuel pumps and injection nozzles. Many types of pumps and nozzles are now obtainable.

The **fuel pump** comprises a hardened steel barrel and plunger, ground to a very high degree of accuracy. The plunger is operated by a camshaft, and it forces fuel, which enters the barrel through ports in the sides, through a spring-loaded delivery valve situated in the upper end of the barrel, and thence along steel tubing to the injection nozzle. The quantity of fuel delivered at each stroke of the plunger is usually varied by partially rotating the plunger by means of a rack and pillion device connected with the throttle and the governor (Fig. 348). A vertical channel leads from the top edge of the plunger to an annular groove, the upper edge of which forms a helix. When the plunger is at the bottom of its stroke the inlet ports are uncovered, and the barrel is filled with fuel. As the plunger rises on the delivery stroke the fuel is at first forced back through the inlet ports, until these are entirely covered, after which it is forced through the delivery valve. So long as the ports are kept closed by the plunger, injection of the fuel continues, but before the plunger reaches the top of its stroke the helical edge of the annular groove may partly uncover one of the ports, and as soon as this occurs the fuel above the plunger is free to flow down through the vertical channel and annular groove, back through the inlet ports. By rotation of the plunger the point at which delivery ceases may be varied between the point where no fuel is delivered and that where fuel is delivered throughout the whole of the stroke. The timing of the injection may also be varied, and this may be effected by rotation of the camshaft.

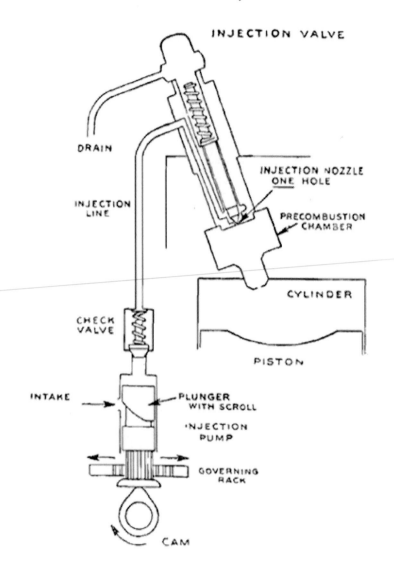

FIG. 348.—DIAGRAM OF A DIESEL FUEL-INJECTION SYSTEM. (CATERPILLAR.)

The fuel **injection nozzle** (Fig. 348) is an important part of a Diesel fuel system. In the usual type a needle valve is held on to a conical seating by a coil spring, until the pressure of the fuel in the delivery pipe is high enough to lift it. As soon as the pressure in the delivery pipe falls below the correct injection pressure, the spring forces the valve back on to its seating and injection ceases. The size, shape and length of the injection orifice influence the type of spray produced. The design of the combustion chamber and the method of injection influence the efficiency of combustion, and one of the objects in designing cylinder heads is to secure "turbulence" in order to mix the fuel and air as thoroughly as possible.

The design of the cylinder heads of Diesel engines is still in a state of flux. Among the successful types are the *comet* and *vortex* designs introduced by Ricardo. In both of these the fuel is projected across a mass of rapidly rotating air. In the comet type, the upward movement of the piston forces the air through a tangential passage to a partially separated combustion chamber, into which the fuel is injected. In the vortex type, an open type of combustion chamber is used, and rotational movement of the air is produced during the suction stroke by suitable formation of the inlet ports. Turbulence is frequently achieved in an open type of cylinder head by a suitable design of the top of the piston.

Many commercial Diesel engines employ the principle of the pre-combustion chamber. As the fuel is sprayed into the pre-combustion chamber it meets hot air and ignites; but only a small proportion burns immediately, due to the limited amount of oxygen in the pre-combustion chamber. As injection continues, however, the fuel is enveloped in flames and becomes a gas, a very high pressure is developed, and the mixture rushes into the main combustion chamber, where complete combustion takes place.

One of the essentials in designing and operating Diesel engines is to ensure that the fuel is kept perfectly clean. In addition to the usual coarse filters and sediment bowl in or near the tank, a modern Diesel tractor engine has a large-capacity full-flow fuel filter with a replaceable element which needs to be renewed regularly (Fig. 349). Air bleeds are provided at various points in the system, to enable all air to be expelled after servicing of the filter or injection nozzles. The fuel lift: pump is always capable of supplying more fuel than the engine needs. When the filter, the injection pump and the connecting pipes are full, the lift pump builds up a pressure in the connecting pipe which is sufficient to hold the lift-pump diaphragm down against its spring. This automatically disconnects the drive to the diaphragm. The fuel injection pump does not normally need any attention between overhauls, but injection nozzles need to be removed for testing and servicing at regular intervals. The mode of operation of the pneumatic governor on such an engine is described on p. 65.

The 2-stroke Diesel.

The 2-stroke cycle is especially attractive for adaptation to Diesel engines, for many of the disadvantages of the petrol-engine 2-stroke cycle are eliminated. The chief disadvantages of the 2-stroke cycle for petrol engines are incomplete scavenging of the exhaust gases and the loss of new gases through the exhaust ports. In the 2-stroke Diesel, scavenging can be made more effective because pure air is used, and a special compressor or rotary blower may be used to assist in this. Injection of the fuel does not occur until towards the end of the compression stroke, and there is no possibility of the escape of fuel such as may occur in the petrol engine. But the scavenging of the exhaust gases can never be as complete in 2-stroke as in 4-stroke engines, and the early opening of the exhaust ports leads to a further loss of power.

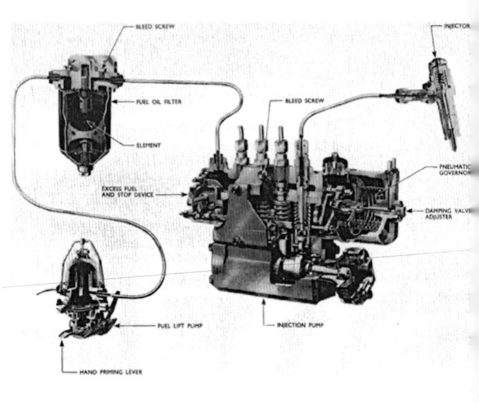

FIG. 349.—

Starting Diesel Engines.

The modern C.I. Diesel engine generally starts from cold without difficulty, and may be reckoned a more reliable starter than a petrol or paraffin engine; but with large powers, starting presents some difficulty, owing to the high compression-ratio against which the engine must be turned.

Large engines may be started by an electric motor, by a small petrol engine or by compressed air. Various devices may be used to facilitate the task. One widely adopted device is a compression-release which, allows momentum to be gained, and this is frequently used in conjunction with devices for increasing the heat, such as electrically heated plugs or heater cartridges. One method of increasing the heat at starting (a Lister patent) is to increase the compression-ratio by screwing in a special valve that shuts off part of the combustion chamber. This is, of course, used in conjunction with a compression-release valve.

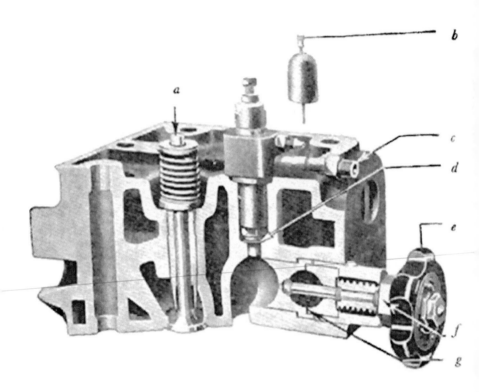

FIG. 350.—CYLINDER HEAD OF A DIESEL ENGINE, SHOWING COMPRESSION-RATIO CHANGE-OVER VALVE. (LISTER.)

a, exhaust valve; **b,** feeler pin and cover for injection nozzle: **c,** fuel pipe; **d,** injection nozzle; **e,** change-over valve handwheel; **f,** change-over valve; **g,** auxiliary chamber.

Many tractor engines of the indirect injection type employ a small hand pump on the dash panel to inject a spray of fuel into the induction manifold. This fuel is ignited by a brightly glowing electric heater plug situated in the mainfold, and substantially increases the temperature of the air charges entering the cylinder.

Some engines (e.g. International Diesels) employ a dual system which permits starting on petrol with electric ignition, and turning over to Diesel operation when the engine has warmed up. The main combustion chamber is built with a compression-ratio suitable for the Diesel cycle, and connected by a valve with the main chamber is a small one which, by the addition of its volume to that of the main chamber, produces a compression-ratio suitable for burning petrol. The small chamber is fitted with a sparking plug, and the usual magneto and petrol carburettor are provided. A branched induction manifold admits air either to the carburettor or direct to the main combustion chamber. In starting up, the engine runs on petrol, but later the valves connecting the main and ante combustion chambers are closed by the operator. The inlet manifold is thereby switched over to deliver air direct to the main cylinder instead of to the carburettor, the magneto drive is disconnected, the Diesel fuel pump is started, and the engine operates on the Diesel cycle.

Maintenance and Adjustments of Internal-

combustion Engines.

The reliability of engines may be maintained, and their annual running costs diminished, by regular attention to the essential parts. Future breakdowns may frequently be prevented by periodical checking of the lubrication, adjustment of the moving parts, etc. Cylinder heads and valves should occasionally be decarbonized, and the valves should be kept adjusted to the clearances recommended by the manufacturers. It is unnecessary to stress further the necessity of adequate attention to maintenance, or to indicate how running faults may be detected and remedied, for the information is available in the manufacturers' handbooks and in the many "Motor Manuals".

With Diesel engines the fuel injection system is the most frequent cause of trouble. Good clean Diesel fuel oil is not expensive, and it is advisable to use the best grade. Fuel pumps and injection nozzles should never be dismantled so long as they work efficiently, and when it does become necessary to take them to pieces the manufacturers' instructions should be carefully followed. A smoky exhaust in a Diesel engine is a sign of unsatisfactory working, caused either by overloading or by the fuel system not working correctly. If the engine runs for long in this condition, the piston, valves and injection nozzle become coated with carbon, and this will certainly cause inefficient running and may lead to more serious trouble. Diesel engines run best when the outlet temperature of the cooling water is about 160 to 180° F. Arrangements are normally made for controlling the flow of the cooling water by thermostat, and adjustments should be made to ensure that the engine runs at approximately the right temperature.

REFERENCES

(43) Duncan. *The Modern Diesel.* Iliffe.

(44) *Elementary Text-book of Automobile Engineering.* Ford Motor Co., Dagenham.

(45) Jones, F. R. *Farm Gas Engines and Tractors.* McGraw-Hill.

BIBLIOGRAPHY

Culpin, C. *Farm Mechanization: Costs and Methods.* Crosby Lockwood & Son, Ltd., London. 1951.

Nebraska Tractor Tests. The University of Nebraska, Lincoln, Nebraska, U.S.A. Bulletin 397. January, 1950 and supplementary data sheets on current models.

N.I.A.E. Tractor Tests. Reports of Tests on individual tractors and other farm machines are published periodically by the National Institute of Agricultural Engineering, Wrest Park, Silsoe, Beds.

Influence of Engine Loading on Tractor field Fuel Consumption. N.I.A.E., Silsoe, Beds. (1951).

Hine, H. J. *Tractors on the Farm.* Farmer and Stockbreeder, London. 1950.

Jones, F. R. *Farm Gas Engines and Tractors.* (An American Textbook.) McGraw-Hill. 1950.

Passmore, J. B. *The English Plough.* Oxford University Press. 1930.

White, E. A. *The Study of the Plough Bottom.* Trans. Amer. Soc. Agric. Engnrs., vol. xii, p. 42.

Farm Machinery

Nichols, M. L. *et al. The Dynamic Properties of the Soil.* Agric. Eng., July 1931, August 1931, August 1932 and November 1932.

Tractor Ploughing. H.M.S.O. (1952).

Bainer, R. *New Developments in Sugar Beet Production.* Agric. Eng., August 1943, vol. xxiv, no. 8.

Cooke, G. W. *Fertilizer Placement Experiments.* Agric. Eng. Record. Summer, 1949.

Ensilage. Ministry of Agriculture Bulletin, No. 37.

Moore, H. I. *Silos and Silage.* Farmer and Stockbreeder, London.

Threshing and Conditioning of Seed Crops. Ministry of Agriculture Bulletin, No. 130. H.M.S.O. 1951.

Threshing by British and American Machines. Cambridge Univ. Dept. Agric., Farmers' Bulletin, No. 10.

Harvesting by Combine and Binder. Cambridge Univ. Dept. Agric., Farmers' Bulletin, No. 9.

Hutt, A. C. *Combine Harvesting and Grain Drying.* London and Counties Coke Association, London.

Nicholson, H. H. *Field Drainage.* Cambridge University

Press.

The Small Automatic Hammer Mill. British Electrical Development Association, London.

Potatoes. Ministry of Agriculture and Fisheries Bulletin, No. 94.

Commercial Fruit Tree Spraying. Ministry of Agriculture and Fisheries Bulletin, No. 5.

Irrigation. Ministry of Agriculture and Fisheries Bulletin, No. 138.

Dixey, R. W. *Open Air Dairy Farming.* Oxford Univ. Agric. Econ. Res. Inst. 1942. 2s. 6d.

Modern Milk Production. Ministry of Agriculture Bulletin, No. 52.

Hosier, A. J. and F. H. *Hosier's Farming System.* Crosby Lockwood & Son, Ltd.

Roberts, E. J. *Grass Drying.* H.M.S.O., London. 1937.

Cameron Brown, C.A. *A Critical Study of the Application of Electricity to Agriculture and Horticulture.* Technical Report W.T. 2 of British Electrical and Allied Industries Research Association.

Electricity on the Farm. British Electrical Development Association, London.

Morris L. G. *The Steam Sterilizing of Soil.* N.I.A.E. Case Study, No. 14. (1951).

Courshee, R. J. *Fruit Grading Machinery.* N.I.A.E. Report. (1952).

Hine, H. J. *The Farm Workshop.* Crosby Lockwood & Son, Ltd.

Report of the Anglo-U.S. Productivity Council on Productivity of Labour in Farming.

Menzies-Kitchin, A. W. *Labour Use in Agriculture.* Camb. Univ. Farm. Econ. Branch, Bulletin No. 36.

Report of the British Agricultural Machinery Mission to Canada. H.M.S.O., London. 1949.

Davidson, J. B. *Agricultural Machinery.* Wiley (for American Equipment).

Smith, H. P. *Farm Machinery and Equipment.* McGraw-Hill. (For American Equipment).

Andrews, E. S. *Mechanisms.* University Tutorial Press.

Duncan, J. *Applied Mechanics for Beginners.* Macmillan &

Co., Ltd.

Duncan. *The Modern Diesel*. Iliffe & Sons, Ltd.

Lightning Source UK Ltd.
Milton Keynes UK
UKOW04f2052191215

265038UK00001B/27/P